ADVANCED FIBER ACCESS NETWORKS

ADVANCED FIBER ACCESS NETWORKS

CEDRIC F. LAM

SHUANG YIN

TAO ZHANG

ELSEVIER

ACADEMIC PRESS

An imprint of Elsevier

Academic Press is an imprint of Elsevier
125 London Wall, London EC2Y 5AS, United Kingdom
525 B Street, Suite 1650, San Diego, CA 92101, United States
50 Hampshire Street, 5th Floor, Cambridge, MA 02139, United States
The Boulevard, Langford Lane, Kidlington, Oxford OX5 1GB, United Kingdom

ISBN 978-0-323-85499-3

For information on all Academic Press publications
visit our website at https://www.elsevier.com/books-and-journals

Publisher: Mara E. Conner
Acquisitions Editor: Tim Pitts
Editorial Project Manager: Judith Clarisse Punzalan
Production Project Manager: Kamesh R
Cover Designer: Greg Harris

Typeset by STRAIVE, India

Contents

Foreword

When I started at Google as the head of Google Fiber back in 2010, Cedric Lam was one of the star engineers in the team and also one of the key drivers of innovation in the project. Over the years, the team was not content with simply using whatever vendors in the market offered, but in the tradition of Google, attempted to innovate in every area. Cedric, Shuang, and Tao led the way in optical network technology, and now through this book and efforts in the standards space, the rest of the world can benefit as well.

It is especially important to look at the innovations in the Fiber to the Home (FTTH) and access space now. The United States and other countries seek to play catch-up with aggressive national fiber deployments that have been proceeding briskly in Asia, especially China. Tens of billions of dollars have been allocated by the US Government to close the gap in connectivity in rural areas, with a focus on fiber deployment as well as by the investment community with its newfound love for fiber deployments in urban and suburban areas that serve homes, businesses, and wireless operators.

Rather than "building yesterday's networks tomorrow," the learnings and architectural approaches outlined in *Advanced Fiber Access Networks* should be the guide for new builds that will reduce cost, accelerate build rates, and enable redundancy and multiservice delivery over a common passive fiber network deployment.

While in many areas people have focused on 5G wireless as the access technology of the future, it is important to recognize that there is only one network, the wired network, with a little bit of wireless at the ends. In the limit, as cell sizes continue to shrink to drive higher and higher network capacity, the wireless network will look like an FTTH network with small Wi-Fi-like cells at the edge. As network deployments in Asia have shown, one cannot be a leader in 5G and especially 6G without first being a leader in fiber access to the home and business.

Informed by real-world experience and world-class optical engineering, this book is a great resource for those operators seeking to build the next generation of fiber access networks.

Milo Medin

Preface

After more than 30 years of developments from research to commercialization, fiber access networks are now mature technologies that have been widely deployed around the world. Optical access network technologies in the form of gigabit-capable passive optical networks (G-PONs) have been commoditized and are shipping in volumes of at least multiple tens of millions of units per year. Carriers around the world are busy rolling out fiber-to-the-home (FTTH) networks to prepare for new upcoming internet applications such as over-the-top (OTT) video streaming, augmented reality (AR)/virtual reality (VR), collaborative online real-time video gaming, and billions of new upcoming Internet of Things (IoT). China, the most populous nation in the world, boasts an FTTH penetration rate of more than 90%, representing 380 million households connected at the end of 2019, and they are busy moving into the next-generation PON technologies.

Most people think of fiber access networks as time-division-multiplexing (TDM)-based PONs. In fact, the word "PON" (almost always implied as some form of TDM-PON) is mostly used as a synonym of fiber access networks. However, TDM-PON is only one of the last-mile architectures most popularly adopted for the FTTH application because of its simplicity and economic benefits. For some access applications such as fronthaul in wireless networks, native TDM-PONs have their inherent challenges and may not be the suitable technology choice. Furthermore, in a real end-to-end broadband fiber access network, PON only represents the last mile connection from a carrier's central office (CO) to the end-user customer premise. Besides the PON last-mile connection, there are many other components coming into play for a well-designed broadband fiber access network. The readers will find out from this book that simply deploying the next-generation last-mile PON system may not actually address the right problems arising from the increasing customer bandwidth demands because of other network bottlenecks and limitations. There are also deployment scenarios (e.g., thin-fiber cables from COs or cable Multi-Service Operator network upgrades) that will require new fiber access network architectures, as we will see in this book.

Although we give a general overview of fiber access networks in this book, it is not intended as an entry-level text. This book is an advanced text assuming the audience already possesses the basic knowledge of fiber optic transmission and traditional PON knowledge such as the ITU-T G-PON and IEEE E-PON. (For an introductory text on PON, the audience is referred to Ref. [1].) In addition to covering the advanced next-generation PON technologies being worked on by standard bodies and the research community, we describe the overall end-to-end fiber broadband access network architecture and the various network components involved from the internet backbone to the end-user homes. We examine the scaling properties and how they affect the overall network performance and economics.

This book is roughly divided into three parts. Chapter 1 is the general introduction. Chapters 2 through 6 cover fiber access standards and physical optoelectronic technologies. Chapters 7 and 8 cover the network systems and the convergence of wireless and wireline access networks. The structure of the book was conceived when we were making the notes of an OFC (optical fiber communication conference) short course with the same title. Through this book, we would like to give the audience a balanced view of the physical fiber access network technologies and the overall access network structure and their evolutions in the broadband network industry.

We began writing this book when the 2019 Novel Coronavirus (2019 nCoV or COVID-19) pandemic was breaking out in the Hubei Province of China where Shuang's family came from. Wuhan, the capital of Hubei Province, was not only the center of this pandemic but the Optics Valley of China where most of the optical access components in the world were manufactured. To prevent the spreading of the virus, governments around the world implemented quarantines, travel restrictions, and social distancing everywhere. Civilians were asked to spend most of their time at home. During this special time, the internet, teleconferencing, and videos streaming delivered through fiber backbones and access networks became essential for people to stay connected, continue to work, purchase goods online, and entertain themselves. In China, during the national quarantine time, state-run cable networks unlocked many premium channels for people to enjoy at home and those were delivered through internet protocol television (IPTV) platforms.

Fiber optics is at the core of the internet, and fiber access networks are becoming more and more important to connect homes, businesses, and the booming 5G wireless infrastructure. While developing this manuscript during the pandemic, we hope the broadband technologies that we are working on will help humanity to (1) more efficiently tackle such emergencies, (2) improve our lives and productivities with better connections to the outside world and better access to the information and entertainment when we have to stay home, and (3) have the freedom to work from anywhere in the world.

Cedric, Shuang, and Tao
Silicon Valley, California
September 2020

Reference

[1] C.F. Lam, Passive Optical Networks: Principles and Practice, Elsevier, 2007.

Acknowledgments

This book is a result of over one decade of the direct experiences of research and development at Google Fiber [1], a program cofounded by one of the coauthors of this book. It is a summary of the dedicated hard work and creativity of many people directly and indirectly involved in Google Fiber. These include many of our coworkers, managers, our equipment vendors, friends, and families, so numerous that it is very difficult to list all of them one-by-one individually.

Nevertheless, we would like to particularly thank the following people for their significant influences and contributions in the works described in this book. Milo Medin, Vice President of Wireless Access Services at Google, who was also the first VP of Google Fiber; without his strong technical and executive support, the real production Super-PON network and technologies would not have come into existence. Milo was also instrumental for the wireless programs in Google and exposed us to the field of wireless communications while we were working on fiber access networks. Dr. Hong Liu, Google Fellow at Google Technical Infrastructure, who was a coinitiator of the predecessor of the Super-PON program named Kaleidoscope, which we built and successfully trialed on the Stanford University campus, an FTTH testbed carrying live service traffic. James Kelly, the first Product Manager on the Google Fiber program, who strongly promoted microtrenching as a low-cost deployment method even before the Google Fiber program was officially formed. James was also a strong promoter of our Kaleidoscope WDM-PON program. Tony Ong, the uber hardware engineering manager at Google Fiber, who led the hardware program and built many cutting-edge fiber access network systems and CPE equipment. Ke Dong, who led the software development for our WDM-PON and Super-PON systems and created the cloud OSS system used by Google Fiber. Xiangjun Zhao, a key contributor of numerous creative ideas enabling the realization of Kaleidoscope, Super-PON, and the Muxtender system described in this book. Yifan Gao, manager of the Google Fiber lab and the testing team, who led the testing, integration, and verification of many homegrown and third-party technologies in our network. Jack Wu, who headed the product definition of our first WDM-PON system that was deployed in Stanford for production. Claudio de Santi and Liang Du, who led the developments of Super-PON standard in the IEEE

802.3 Ethernet Standard Working Group and ITU-T Study Group 15. Claudio is the Chairman of the IEEE 802.3cs Super-PON standard task force. Sasha Petovic, Jhon Gaurin, and Jason Bone in the Google Fiber San Antonio team, who led the trial and deployment of the first real production Super-PON network in the world. Their faith and confidence in us were tremendous and will never be forgotten.

We would also like to thank the following Google executives: Walt Drummond (VP), Boon-lock Yeo (VP), and Ben Segura (Strategy Lead of Google Fiber). Together with Milo, these people recognized the value of the TWDM Super-PON project, both to Google Fiber itself and to the industry. They supported the continuation of the Super-PON program and many other innovation initiatives during a major force reduction in Google Fiber in 2017.

The first author would like to particularly thank the following people who worked in the Google Fiber architecture and technology team (in alphabetical order of last name): Adam Barratte, Pedram Dashti, Claudio de Santi, Liang Du, Joy Jiang, Scott Li, Muthu Nagarajan, Satrukaan Sivagnanasuntharam, Daoyi Wang, Shuang Yin, Tao Zhang, and Xiangjun Zhao.

Last but not least, we are very grateful to Google and its parent Alphabet. Google provided us with a nourishing ground to fertilize innovative ideas and necessary resources to test, refine, and realize those new solutions. We were surrounded by an environment of tremendous talents that coached and humbled us.

March 2021
Silicon Valley, California

Reference

[1] C.F. Lam, Google fiber deployments: lessons learned and future directions, in: Invited Paper presented at Optical Fiber Communications Conference 2021, 2021. San Francisco, CA.

CHAPTER 1

Introduction

1.1 Drivers and applications for broadband access

We live in an information society today where connectivity to data has become a utility like water and electricity to our modern human lives. Fig. 1.1 shows the annual growth of the Internet traffic between 30% and 37% from 2017 to 2022, as projected by the cisco VNI index [1]. Scalable, low-cost, and future-proof broad access infrastructure is key to the success in today's digital economy, promoting governments around the world to create strategies and policies to stimulate investments in broadband infrastructures and technologies, especially during economic downturns.

The importance of broadband was amplified during the recent COVID-19 pandemic. After the global outbreak of COVID-19 in the beginning of the year 2020, in order to curb the spreading of the highly contagious coronavirus, people are requested to stay at home by the lockdown orders issued by governments around the world. The Internet has become the main avenue for people to work remotely from the home, take online lessons, consult with their physicians, stay connected with their loved ones, and buy food and necessities. Zoom, the maker of a popular video conferencing tool with the same name, saw a huge boom in their business and their stock prices rose through the stratosphere against a free-falling stock market. According to Ref. [2], the Internet traffic on EU (European Union) telecom operators grew 30% to 60% in the first 3 weeks since the lockdown. Fig. 1.2 [3] shows the Internet data and traffic explosion after COVID-19. The prolonged COVID-19 pandemic will permanently change the way people work, play, and social, and the importance of the Internet will become even greater.

Optical fiber was originally developed for long-haul transmissions. The idea of using fiber optic for access was initially proposed in the 1980s, way before the Internet and broadband access became the norm of our society. After decades of developments, fiber access networks are now mature technologies deployed to hundreds of millions of users around the world. In 2019, China alone boasted more than 395 million fiber-to-the-home (FTTH) users. Besides directly connecting residential households with FTTH networks, new forms of fiber access networks are indispensable in

Advanced Fiber Access Networks
https://doi.org/10.1016/B978-0-323-85499-3.00008-4

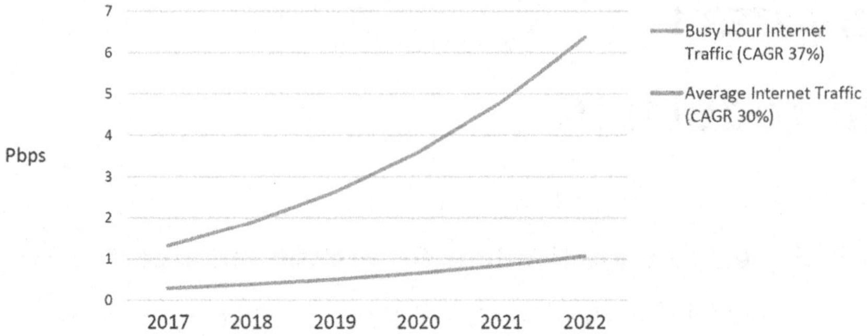

Fig. 1.1 Global busy hour vs average hour internet traffic Ref. [1].

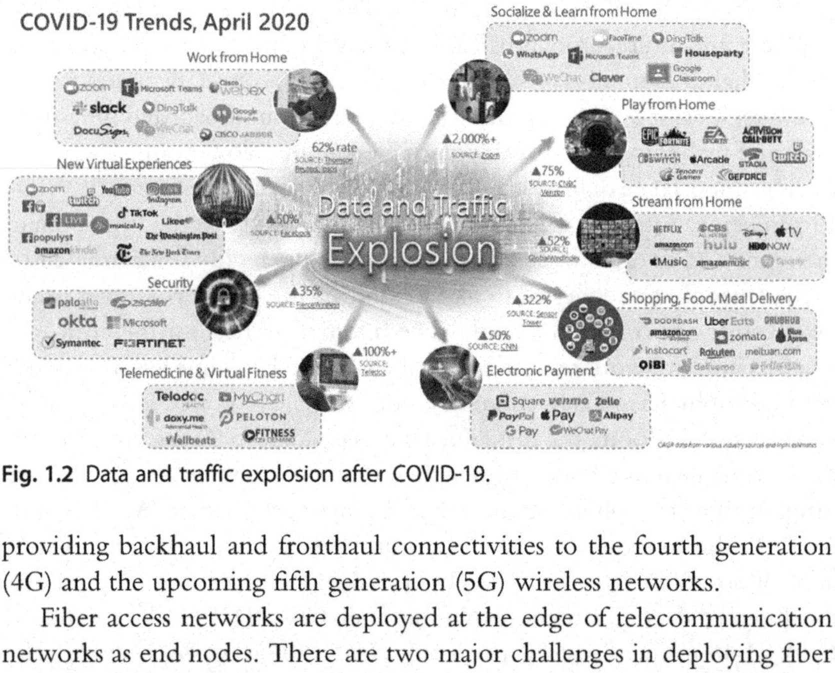

Fig. 1.2 Data and traffic explosion after COVID-19.

providing backhaul and fronthaul connectivities to the fourth generation (4G) and the upcoming fifth generation (5G) wireless networks.

Fiber access networks are deployed at the edge of telecommunication networks as end nodes. There are two major challenges in deploying fiber access networks. First, access networks are very cost sensitive, so the equipment cost has to be very low in order for it to be viable as a mass-deployed technology. The ways to achieve this are economy of scale and low-cost optoelectronics packaging techniques. The second major challenge in deploying fiber access networks is labor cost, and speed and ease of deployments [4]. Significant civil engineering cost is incurred especially in developed economies where (1) the labor cost is high, and (2) digging and trenching of infrastructure is not easy. Therefore, traditional incumbent carriers in developed nations would like to preserve their legacy copper infrastructure and delay the deployment of fiber in the last mile (from the central office) as much as possible. In developing economies or green field scenarios, there will be fewer architecture constraints, less legacy burdens, and more flexibility in technology and architecture choices. But those economies are also very capital cost sensitive and would like to leverage the low cost of existing, mature, and standard-based technologies. These challenges are guiding the design principles of fiber access technologies.

Broadband Internet access discussed earlier was the initial driving force for FTTH fiber access technologies. This was mainly propelled by the booming Internet applications, especially over-the-top (OTT) video streaming applications which offers any time and any place viewing experiences of on-demand contents. Higher resolution videos such as 4K will demand more bandwidths to end users, mainly in the downstream direction from carrier networks to the end users. New applications such as teleconferencing and high-resolution video surveillance will also command high uplink return bandwidth from end users to carrier networks and will affect the architecture of fiber access networks. For example, during COVID-19, many people worked from home. Broadband access networks with limited upstream bandwidths, such as DOCSIS[a] [5], would suffer from upstream bandwidth contention. People joining video conferences from DOCSIS cable modems often have poor image and/or sound quality due to the upstream access bandwidth contention. This was observed in a video conference among the first and second authors of this book and the publisher in England. The first author and the publisher were on FTTH access network, and the second author was on a DOCSIS cable network. During the teleconference, at the first author's terminal, the image of the publisher from London comes out crystal clear while the image from the second author who lived 20 miles away was very blurry. However, at the second author's terminal, the images from both the first author and the publisher were crystal clear because he had no downstream bandwidth problem. This is illustrated in Fig. 1.3.

One of the purposes of this book is to understand how to scale the overall end-to-end broadband fiber access network from an overall system

[a] Most legacy DOCSIS deployed occupies a very limited upstream bandwidth between 5 and 42 MHz on the coaxial cable RF spectrum [5], which is heavily shared among 500–1000 users. Newer systems can make use of a wider upstream spectrum and have small share group size. However, this requires significant network capital expenditure to upgrade the coaxial plant infrastructure.

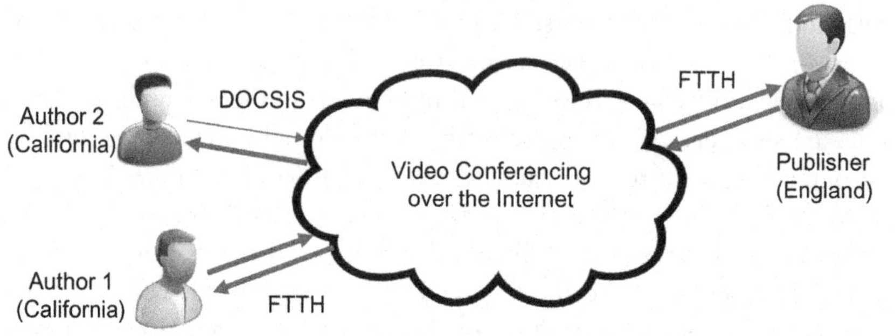

Fig. 1.3 The low upstream bandwidth from Author 2 causes his image to be blurred at Author 1 and Publisher while he perceives both Author 1 and Publisher as crystal clear.

perspective. FTTH was mainly provided by passive optical networks (PONs) [6]. In fact, PON is almost used as a synonym of FTTH although other FTTH implementations also exist. Most of the deployed residential FTTH networks are based on the IEEE 802.3ah EPON or ITU-TG.984 based G-PON technologies, with the latter being the most popular nowadays.

1.2 History and roadmap of broadband access development

Fig. 1.4 plots the peak bit rate of PON vs wireless (Wi-Fi and cellular) over the past four decades. Tables 1.1 through 1.3 summarize the major technology standards and peak data rate achieved by various generations of Wi-Fi, cellular, and PON systems. We can see from Fig. 1.4 that wireless access speed had a low starting point but is catching up very quickly (driven by the needs of both broadband applications and mobility). FTTH technologies, as represented by the PON speed development trends, are growing much more slowly, however. We are just at the point where fiber and wireless access speeds are converging. A detailed account of PON standards development can be found in [7].

Various standard bodies are involved in wireline and wireless access technology developments. Cellular network provides long-haul end-to-end wireless connectivity with international roaming functions. It is standardized by 3GPP (Third Generation Partner Project) under ITU-R (International Telecommunication Union—Radio). Wi-Fi, also called wireless local area

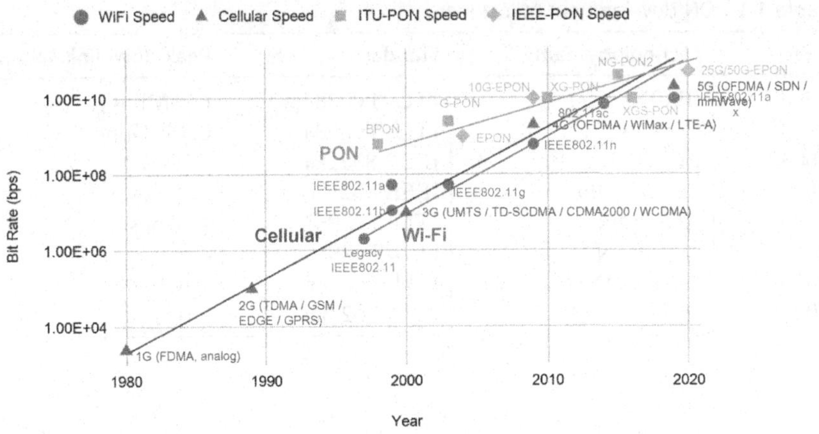

Fig. 1.4 Development trends of PON and wireless access technologies.

Table 1.1 Wi-Fi development milestones.

Year	Standard	Frequency band	Peak data rate
1997	Legacy IEEE 802.11	2.4 GHz	2 Mbps
1999	IEEE 802.11b	2.4 GHz	11 Mbps
	IEEE 802.11a	5 GHz	54 Mbps
2003	IEEE 802.11g	2.4 GHz	54 Mbps
2009	IEEE 802.11n	2.4 and 5 GHz (4 streams)	600 Mbps
2014	IEEE 802.11ac	5 GHz (8 streams)	6933 Mbps
2019	IEEE 802.11ax	2.4 and 5 GHz (8 streams)	9608 Mbps

Table 1.2 Cellular wireless development milestones.

Year (generation)	Technologies	Peak data rate
1980 (1G)	FDMA Analog	2.4 kbps
1989 (2G)	TDMA/GSM/EDGE/GPRS	10–100 kbps
2000 (3G)	UMTS/TD-SCDMA/CDMA2000/ WCDMA	100 kbps–10s Mbps
2009 (4G)	OFDMA/WiMax/LTE-A	150 Mbps-2 Gbps
2019 (5G)	OFDMA/SDN/mmWave	1–20 Gbps
2030 (6G)	Terahertz Transmission	Tbps?

network (WLAN), is the most prevailing indoor wireless access network technology with semi-mobility capability. It is a fast evolving technology developed by the IEEE (Institute of Electronic and Electrical Engineering) 802.11 standard group. In an FTTH network, a PON system is usually

Table 1.3 PON development milestones.

Year	Technology name	Standard	Peak downlink speed
1998	BPON	ITU-TG.983	622 Mbps
2003	G-PON	ITU-TG.984	2.488 Gbps
2004	EPON	IEEE 802,3ah	1 Gbps
2009	10G-EPON	IEEE 802.2av	10 Gbps
2010	XG-PON	ITU-TG.987	10 Gbps
2015	NG-PON2	ITU-TG.989	40 Gbps
2016	XGS-PON	ITU-TG.9807	10 Gbps
2020	25G/50G-EPON	IEEE 802.3ca	50 Gbps

terminated into a Wi-Fi access point (AP) at a customer premise, which provides the connectivity inside the residential home. Wireless signals are subject to high propagation losses, interferences and blockage by walls and obstacles. Although modern Wi-Fi systems boast multigigabit peak performance (Table 1.1) that is faster than the deployed G-PON or EPON FTTH line rates, as in a commercial FTTH network, Wi-Fi is usually the bottleneck that determines customers' experiences and perceptions. Carriers often get the most support calls related to the home Wi-Fi system rather than PON link issues. Nevertheless, Wi-Fi is not the covered subject of this book.

There are three major organizations working on PON standards. FSAN (Full Service Access Network) is a telecom carrier consortium that discusses requirements and ideas for new fiber access initiatives. The proposals will be submitted to ITU-T (International Telecommunication Union – Telecom) Study Group 15 (SG-15) for standardization. IEEE 802.3 LAN/MAN Ethernet Standard Group makes Ethernet standard-based PON (or EPON) standards, competing with ITU-T SG15.

The most widely deployed PON standards for FTTH or FTTP (fiber-to-the-premise) are the IEEE 802.3ah EPON and ITU-TG.984 G-PON. Although carriers have started deploying 10 Gbps based TDM-PON systems around the world, for residential FTTH uses, G-PON and EPON networks still offer enough bandwidth for at least another 3–5 years as we will see in a later chapter. Nevertheless, standard bodies have been busy working on PON technologies beyond 10 Gbps (i.e., 25/50 Gbps). IEEE 802.3 standard group just completed the IEEE802.3ca 25G/50G EPON in June 2020 [8] and ITU-T SG15 is busy working on a single-wavelength 50 Gbps high-speed PON standard [9]. The major driving forces for these higher speed PONs are enterprise services and wireless fronthaul and backhaul applications (Fig. 1.4). Furthermore, besides TDM, WDM (wavelength division multiplexing) has been introduced to further scale the speed and coverage of future PON networks [10]. These development trends will be discussed in the later chapters.

We can see from Fig. 1.4 that initially, the speed of wireline PON networks was far ahead of that of wireless networks by at least 2 orders of magnitude. However, there was a lack of killer applications to drive the fast development of PON access networks for residential users. The fast evolving wireless networks are now becoming a driving force for the developments of new fiber access technologies and standards, especially in the WDM domain. To support very high data rates, the 5G (fifth generation) wireless networks require cell densification and a pervasive fiber edge network to support the 5G base station infrastructure. PON and WDM will be helpful to solve the fiber congestion issues in 5G fronthaul networks.

We will see later in this book that in order to improve the cost efficiency and performance of wireless systems, the new wireless networks are deployed with distributed RF (radio frequency) front ends called RRHs (remote radio heads) or AAUs (active antenna units), which are connected to centralized baseband units (BBUs) by low-cost and high-speed fronthaul links. Such fronthaul links require very high data rate and play into the strengths of the next generation fiber access networks. However, wireless fronthaul requires precision timing control of the fronthauled signal which is very challenging for TDM-PONs that use traditional dynamic bandwidth allocations (DBA) algorithms for statistical multiplexing. Although techniques have been devised to accommodate TDM-PONs for wireless fronthaul [11,12] applications, WDM-PON has also arisen a simpler and more straightforward approach for wireless fronthaul, for example, in ITU-T G.698.4 [13].

Lastly, as PON speed and coverage increases, scaling PON systems by TDM alone is getting harder and harder due to the exponential increases in the transmission impairments of a single-wavelength high baud rate signal. Scaling in both the time domain and wavelength domain will help to ease some of the transmission difficulties at the expense of using the more expensive WDM optics. Electronic digital signal processing techniques [14] have also been discussed to overcome the physical transmission impairments especially for future optical access systems with line rates in excess of 25 Gbps. These techniques will be discussed in Chapter 6.

1.3 Fiber-to-the-home advantages and its future

One may look at Fig. 1.4 and start to wonder that if wireless communication is catching up so fast that the speed of wireless access starts to rival that of FTTH networks, then where the value of fiber access networks is, because in a wireless access network, there is no need to string fiber to customer premises, a process which is both time consuming and very costly especially in developed countries. To answer this question, we need to take a deeper look into the comparisons between wireless and FTTH systems.

1.3.1 FTTH vs wireless access network architectures

Fig. 1.5 shows a comparison between an FTTH network and a fixed wireless access (FWA) network. In an FTTH network, a completely passive optical distribution network (ODN) connects a customer premise to a central office which can be as far as 20 km apart with most FTTH network standards. The ODN connects an ONU (optical network unit, aka a fiber modem) to an OLT (optical line terminal) in the central office.

In an FWA network, a customer connects to the broadband wireless network using a CPE (customer premise equipment, aka a wireless modem), which connects to an RRH (remote radio head) at a cellular site wirelessly.

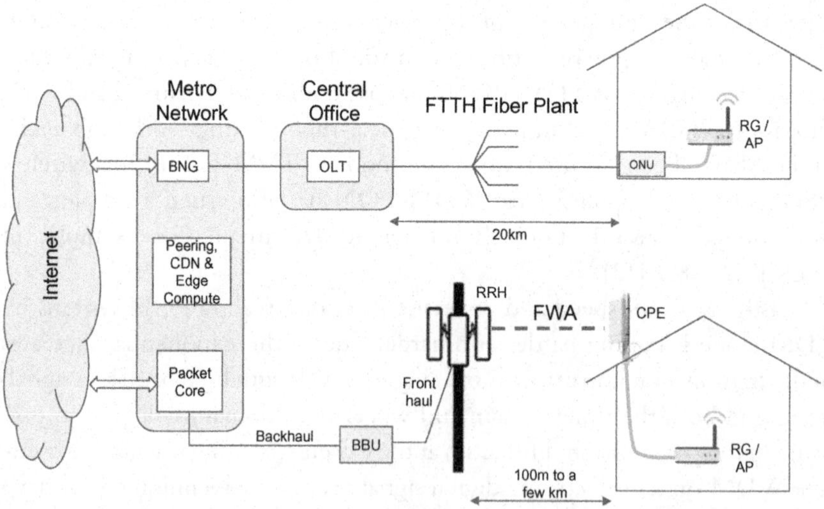

Fig. 1.5 FTTH vs FWA network. *AP*: access point, *BBU*: baseband unit, *BNG*: broadband network gateway, *CDN*: content distribution network, *CPE*: customer premise equipment, *OLT*: optical line terminal, *ONU*: optical network terminal, *RAN*: radio access network, *RG*: residential gateway, *RRH*: remote radio head.

Wireless signals suffer from free space signal losses, shadowing effects from buildings in urban environments, as well as scattering from trees and various obstacles in the propagation paths. Depending on the wavelengths of the wireless spectrum used for transmission (e.g., mmWave systems), they can also be affected by weather effects such as rain scattering. So the loss between the RRH and an FWA CPE is not only large but also varying due to environmental effects. This is also why we are comparing FTTH to FWA networks rather than mobile wireless access in this section. The quality of service of a mobile wireless network is even more unpredictable due to the mobility of the user terminal (called UE or user equipment). Nevertheless, mobile networks solve a completely different problem, that is, mobility, in addition to bringing bandwidth (which is what fixed broadband networks are designed for). So it is unfair to compare FTTH to mobile wireless systems.

In an FWA system, the distance between the RRH and CPE is usually limited to a few hundred meters in an urban environment. Experienced FTTH practitioners would have noticed that FWA is really replacing the drop fiber section from a pole or street handhole to a customer's premise (which is the last 100 m of the connection from the central office), with an active cellular tower and a wireless CPE. In order to get good system performance (a few hundred megabits per second) and larger coverage (e.g., from 400 m to 1 km by wireless standard), usually an outdoor CPE with directional antenna such as the one shown in Fig. 1.5 is used in an FWA network to ensure good signal-to-noise ratio (SNR) at the CPE. Professional installation of such outdoor CPE is often required, which is not necessarily cheap either.

In open rural environments, depending on the transmission wavelengths and required bit rates, wireless network coverage can be as wide as a few tens of kilometers. In fact, rural communication with sparse population is the strength of wireless networks because the capital cost of stringing fiber is very expensive.

To enhance the wireless coverage as well as capacity, other wireless architectures, such as wireless mesh where multihop connections through intermediate CPEs to neighboring customer premises, have also been proposed. Wireless mesh architecture is not only a hot active area of research but also many startup companies have been created to try to profit from it, because broadband access is so important to the modern digital economy and the capital cost of building pervasive and ubiquitous FTTH networks is so high, especially in OECD countries where labor cost is very high.

An example is the Terragraph project in the Facebook (now Meta) Connectivity Lab [15]. This subject is not the focus of this book and we will not have further discussions on this topic.

Coming back to the FWA network depicted in Fig. 1.5. The network segment from BBU to the CPE is called RAN (radio access network). The RRH is connected to a baseband unit (BBU) through fronthaul links. An RRH consists of RF modulators and antennas, the associated analog electronics that modulates the RF carrier signal and drives the antennas (i.e., power amplifiers (PAs) in the transmit direction and low noise amplifiers (LNAs) in the receive direction), and the DACs (digital-to-analog converters) and ADCs (analog-to-digital converters) that drive the PAs and LNAs. In the downstream direction, the BBU takes the user data payload, provides the digital signal processing (DSP) functions and, generates the modulating input to the RF carrier. In the upstream direction, the BBU takes the sampled received RF waveform and recovers the received data from the users using DSP.

Depending on the network architecture, the BBU may be colocated with the RRH in the same wireless antenna tower, or centrally located and shared among RRH from multiple towers such as in a C-RAN (centralized RAN) network [16]. We will see in Chapter 8 that in modern wireless networks, tremendous bandwidth is required for fronthaul connections. Modern wireless networks mostly use fiber for the fronthaul between RRHs and BBUs [17].

The connection from the BBU to the carrier backend network is called backhaul. The backhaul network feeds user data to the wireless RAN and it is almost invariably a fiber network. We can see that even if one wants to use wireless technologies for broadband access, a pervasive last mile fiber connectivity network is still needed. In fact, the rapid development in wireless network technologies is now in turn driving new applications and new approaches to fiber access networks, which we will see in Chapter 8.

1.3.2 Energy efficiency of FTTH networks

To illustrate the energy benefits of FTTH, we did a comparison of the energy required to transmit one bit of information on an FTTH network and an FWA network. For an apple-to-apple comparison, we include the network from the OLT to the ONU for the FTTH calculation whereas for the FWA calculation, we include the network from BBU to CPE.

Both systems used in the calculation are commercial systems of major brands. To protect the vendors, however, we omitted the vendor name and model numbers of the systems. The FTTH system we picked is a mature and more than 10-year-old OLT design with 256 G-PON ports (640 Gbps downstream and 320 Gbps upstream full duplex capacity), and a basic G-PON ONU with one WAN (Wide Area Network) port and one LAN (Local Area Network) port. The following table shows the energy to transmit one bit of information in the FTTH system.

The FWA system is a state-of-the-art RAN system with 64T64R massive MIMO (multiple-input multiple-out) antenna, operating in the 3.55–3.7 GHz CBRS (Citizens Broadband Radio Services [18]) band. The RRH can operate over 100 MHz occupied bandwidth (OBW) with 16 RF streams concurrently, each modulated with 256-QAM for a max RRH capacity of 9 Gbps. The RAN system is a time division duplex (TDD) system with both the upstream and downstream data sharing the same spectrum in the time domain. So the capacity above represents the maximum of both the upstream and downstream combined. The BBU processing capacity is also 9 Gbps. The CPE can handle a maximum of 1Gbps data throughput. Table 1.5 gives the energy required to transmit one bit of information in the FWA system.

We can see from Tables 1.4 and 1.5 that even using the best state-of-the-art microelectronic implementation, the energy required to transmit one bit of information wirelessly is still almost two orders of magnitude worse than that required for an FTTH system which is more than 10 year old. This is expected because of the significant losses in wireless signal transmission.

Table 1.4 Energy required to transmit one bit of data on a G-PON FTTH system.

FTTH	Value	Unit	Note
Power per OLT	2274	W	OLT has 256 G-PON ports
Downstream capacity of OLT	640	Gbps	
Upstream capacity of OLT	320	Gbps	
Power per ONU	4	W	1 WAN port and 1 LAN port
Downstream capacity of ONU	2.5	Gbps	
Upstream capacity of ONU	1.25	Gbps	
Energy per bit for OLT	2.37	nJ/bit	
Energy per bit for ONU	1.07	nJ/bit	
Total energy per bit (FTTH)	3.44	nJ/bit	

Table 1.5 Energy required to transmit one bit of data on a CBRS FWA system.

FWA	Value	Unit	Note
Power per RRH	700	W	100 MHz OBW, 64T64R massive MIMO and 16 streams
RRH capacity	9	Gbps	
Power of BBU	1200	W	
BBU capacity	9	Gbps	
Power per CPE	30	W	
CPE capacity	1	Gbps	
Energy per bit for RRH	77.78	nJ/bit	
Energy per bit for BBU	133.33	nJ/bit	
Energy per bit for CPE	30.00	nJ/bit	
Total energy per bit (FWA)	241.11	nJ/bit	

Bear in mind that the FTTH system has a transmission distance of up to 20 km and the bit rate is constant and guaranteed for ONUs separated up to 20 km apart. The maximum capacity of the FTTH system is also the achievable capacity. But for the FWA system, the achievable capacity/link speed is highly dependent on the separation between the RRH and the CPE as well as the buildings and the folios, etc. To achieve the maximum transmission rate, a separation of no more than a few hundred meters with very little blockage in between the RRH and CPE is necessary. Furthermore, there cannot be much interference from transmitters occupying the same RF spectrum nearby, which will degrade the signal-to-noise ratio. There are so many variables that will affect the performance of wireless systems that most of the time, a wireless system cannot achieve its designed maximum capacity.

We can see from this simple comparison that despite the fast catchup of wireless technology development, wireless systems, while offering their advantages, suffer from inherent energy inefficiency which will translate into ongoing operation costs (electricity bills). [For comparison and for the interests of the readers, we estimated the energy to transmit one bit on one of the transpacific Google undersea cables (about 11,000 km long) is only 0.67nJ, impressive! That's why optical fiber is such a nice transmission medium.]

1.3.3 Other consideration of Fiber vs wireless access networks

Spectrum availability

To operate a wireless network, the first thing that one needs is wireless spectrum. Wireless spectrum is a scarce public resource which is regulated by public agencies. There are free unregulated radio spectrum (e.g., ISM bands) which are usually used for very short distance indoor wireless communication purposes (e.g., Wi-Fi). There are very stringent radiation restrictions which constraints the useful transmission distances for these bands. Furthermore, these bands also suffer from much interference from many unregulated users sharing the same spectrum.

For serious operation of wireless systems, usually licensed spectrum is required. Acquiring such a spectrum usually involves complicated bureaucracy. In some countries, licensed spectrum comes with hefty price tags. For example, in the US, the FCC (Federal Communication Commission) auctions license spectrum to operators and billions of dollars are required in the bidding of such spectrum. Spectrum with good propagation properties also commands higher dollars. An example is the recent auction of the CBRS spectrum by FCC [19]. Carriers paid nearly $4.6B for 70 MHz of priority access licensed (PAL) spectrum in the CBRS band for a period of 10 years [20]. Such a license is not even exclusive nor perpetual. Furthermore, FCC has transmission power constraints for the CBRS band which limits the useful coverage for CBRS systems.

RF frequency, propagation, and capacity

Not all the RF frequencies propagate equally. In general, the higher the frequency, the higher the propagation loss and the shorter the transmission distance. The free space loss of RF signals is proportional to the frequency squared. So from a coverage perspective, lower frequency (hence longer wavelength) signals can cover large areas with fewer cellular towers. This is, however, a double-edge sword as it also means higher interference from neighboring cells and it is harder for spectrum reuse, which is critical to increase the capacity of wireless systems.

High-frequency signals propagate with higher losses. So it is easier to reuse the same spectrum over geographic areas to boost system capacity. Also, there are more bandwidths available in higher carrier frequencies. Assume 10% modulation bandwidth, at 1 GHz carrier frequency, the available spectrum is 100 MHz. However, at 10 GHz carrier frequency, there is 1 GHz (i.e., 10 ×) of the available spectrum to carry data.

According to the Shannon theorem [21], information capacity is proportional to carrier bandwidth, so there is more capacity at higher carrier frequencies. That's why modern wireless systems (e.g., 5G wireless networks) use heterogeneously small cells with high-frequency carriers such as millimeter waves (mmWave, i.e., 30–300 GHz) for capacity offload in high-density wireless hotspots and lower frequency macrocells for wide area coverage.

In general, wireless carriers with frequency less than 6GHz suffer from very little rain fade. Carriers higher than 6 GHz also suffer from rain fades and various atmospheric absorption effects (e.g., oxygen absorption at E-band or 60 GHz). Also, as the frequency increases and wavelength decreases, RF propagation is more and more like free space optics. Millimeter waves require line of sight operation. They are also easily blocked by obstacles, making them difficult to operate with.

None of the above is relevant to fiber networks. Optical fiber has about 53 THz of stable usable bandwidths (corresponding to wavelengths from 1260 to 1625 nm).

Upgrade

Demand for data capacity is always increasing in today's Internet world. When an FTTH system runs out of capacity, it is quite easy (and standard) to upgrade its capacity by overlaying a new backward compatible system on the existing fiber plant, using new transmission wavelengths. So the initial build cost of an FTTH system may be expensive, but once fiber is laid in place, it is future proof.

To upgrade a wireless system requires new radio spectrum or refarming the old spectrum with more efficient modulation methods. Since radio spectrum is shared, it is usually challenging for spectrum refarming. Getting new spectra means more spectral cost.

Besides adding more spectrum, wireless carriers can also densify the number of cells in the existing system to reduce the number of users per wireless cell. This means deploying new active cell sites and the associated RF planning and cellular engineering. It is usually very costly.

As we run out of low and mid-frequency spectrum, carriers are forced to explore the capacity of higher and higher frequency spectrum. Such systems are not only difficult to build (require densely populated base stations, as well as very high-speed analog and digital electronics) but also difficult to operate due to propagation constraints explained above.

Reliability

Fiber-to-the-home network uses a completely passive fiber plant between the OLT (usually located inside a CO) and ONUs at customer premises. Most FTTH systems actually have active electronics (i.e., OLTs and ONUs) located indoors (there are also some systems built with OLT deep in the field and located in street cabinets, for example, in portions of Chinese FTTH networks, in order to conserve fiber strands from the COs). Apart from occasional software bugs, most of the faults in FTTH systems are related to fiber cuts in the field.

Wireless systems, however, rely on many active elements (RRHs and BBUs) in the field. These systems have to withstand outdoor operating conditions such as extreme temperatures, condensation, ice and wind loading, making the maintenance of such systems challenging. Nevertheless, the untethered convenience, especially the mobility property makes wireless systems very desirable and indispensable to our modern society.

We can imagine in a converged wireline and wireless network, fiber is used not only for FTTH but also to provide the very much needed backhaul and fronthaul functions for an overlaid wireless network. The FTTH network is mostly used for residential fixed broadband access and the precious wireless spectrum reserved mostly for broadband mobile access most of the time. The same wireless infrastructure can also be used to provide FWA access during FTTH fiber cut (as long as the cut doesn't also bring down the wireless fronthaul and/or backhaul).

In Chapter 7, we will discuss FTTH network reliability and redundancy in detail.

Initial capital cost and time to market

The initial capital cost of deploying FTTH is usually very high. Most countries with successful FTTH deployments (e.g., China, South Korea, Japan, etc.) also have strong government subsidies and policy tilts. FTTH deployments incur significant upfront civil engineering efforts, which are both time consuming and costly. Despite all the fiber benefits, in developed countries, the labor costs of laying new fiber cables can be prohibitively high. Consequently, incumbent carriers try their best to prolong the life of their existing copper infrastructure and delay the deployments of FTTH. New carriers are often faced with not only high deployment costs, but also lack of the right of way, complex permitting processes, and competitions and headwinds from existing carriers.

Wireless access networks are much faster to deploy because it removes the need to connect the fixed wire to end users. It is used by many developing countries such as India to quickly get its population online, in spite of the unstable network performance. From a time to market perspective, wireless systems have their advantages and can quickly get subscribers and revenue. However, as the bandwidth demand and number of subscribers increase, wireless systems will quickly run out of capacity and need upgrades.

At a low take rate, the capital of FTTH deployments in terms of cost per subscriber can be prohibitively high. With the right frequency spectrum available, wireless networks could be much more cost effective. One strategy for a new carrier could be to use FWA to jump start the market, get the service revenue, and test the market. As the take rate increases and more and more subscribers are connected, then start to fill the subscribers with incremental FTTH deployments. We will describe such strategies in Chapter 8 when we discuss the converged fiber and wireless access networks.

Maintenance, operation cost, and complexity

As we had discussed earlier, once the network has achieved its stable state, the maintenance and operation of FTTH networks will in general be much cheaper than wireless networks because of the mostly passive footprint.

Wireless networks generally have much higher operation and maintenance costs. In addition to the electricity cost to run the active RAN network, there is recurring rent to the cellular tower or vertical asset owner where the RRHs are installed. The structure of an FTTH network is also much simpler than that of an FWA network. As a result, it is also much easier to troubleshoot a fixed line fiber access network than a wireless network which is also subject to weather and interference disturbances.

Demographics

In general, FTTH networks are well suited to urban and suburban areas where the density of subscribers is high enough to spread the costs of running fiber cables. In certain demographics such as Rome where there are many historic relics that makes laying fiber impractical, FWA may be the only viable way to provide access,

Also, in very rural areas, the cost of laying fiber and the associated infrastructure could be prohibitively high. Wireless or even satellite communication would be the only economically viable means of offering Internet access.

1.3.4 The future of FTTH

We have seen in this section that FTTH and wireless access are really complementing each other. Each has its own strengths and weaknesses. In most aspects, for residential broadband access networks where mobility is not a concern, fiber has unparalleled advantages in terms of network reach, quality, stability, energy efficiency, future proofness, as well as simplicity and ease of operation.

Both FTTH and wireless access will continue to advance and they will continue to complement each other. In fact, the fast development of wireless networks will also help to drive new developments in fiber access networks, both in architecture and new physical layers [15,16], which we will also discuss in this book. Table 1.6 summarizes the high-level comparison between fiber and wirceless access networks.

1.4 Status of FTTH developments around the world

Fig. 1.6 plots the global FTTH/B (FTTB: fiber-to-the-building) ranking in March 2019 [22]. What we can observe from this graph is that most of the world, especially the western world, still has very low FTTH penetration, far less than 50%. This indicates there are still ample business and innovation opportunities. Some major countries such as India, the world's second most populous country, have so low penetrations that they did not even make into the ranking in Fig. 1.6.

Table 1.6 Comparison between fiber and wireless access networks.

	Fiber (wireline)	Wireless
Spectrum	Not constrained	Constrained
Bit rate vs coverage	Guaranteed	Bit rate drops with distance and interference
Interference	Not applicable	Affect throughput and performance
Initial CapEx	High	Depends (spectrum cost could be very high)
OpEx	Low	High
Upgrade/future-proof	Easy	Difficult
Time to market	Slow	Fast
Mobility/convenience	None	Great

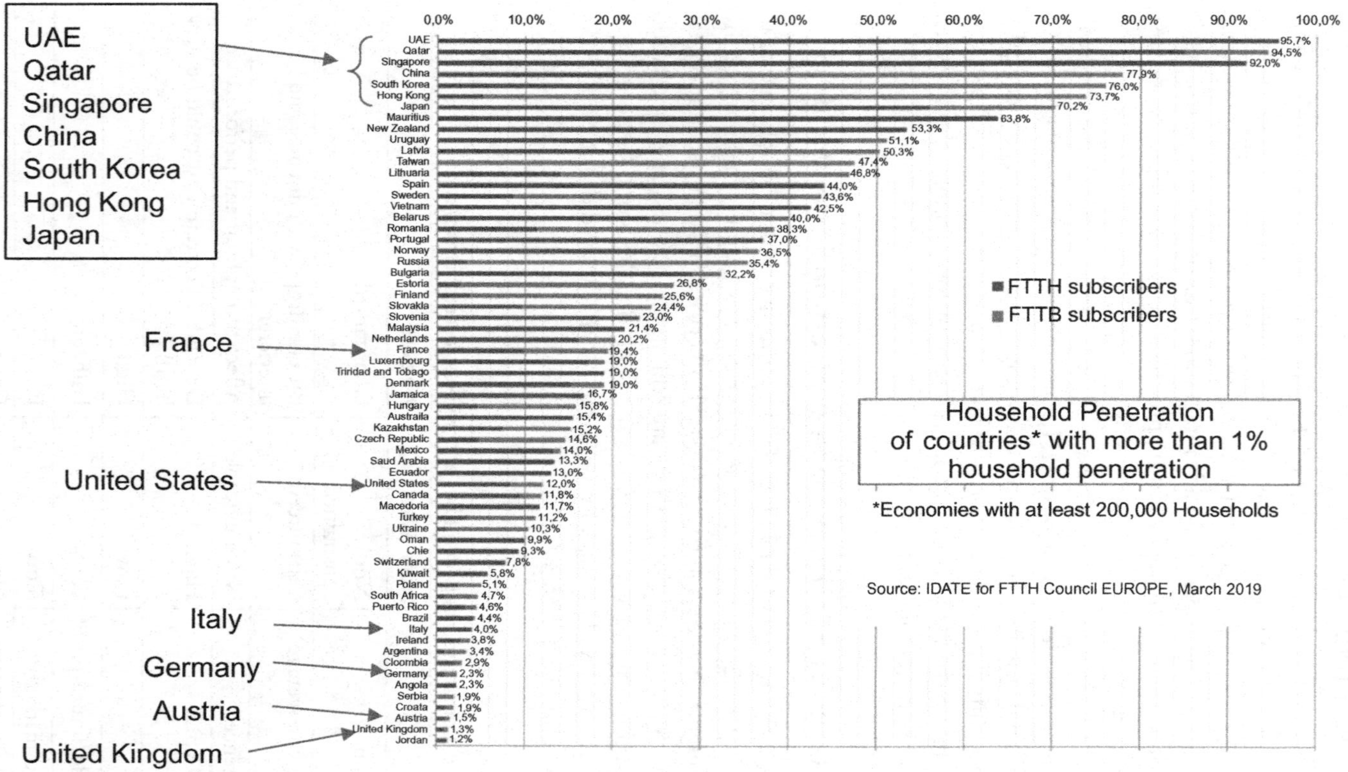

Fig. 1.6 Global FTTH/B ranking as of March 2019.

The most developed FTTH markets are all in Asia. China boasts most of the FTTH users in the world. The total number of FTTH subscribers in China (395 M in March 2019) accounts for more than half of the total FTTH users deployed in the world. The scale of the economy in China played a crucial rule in driving down the cost of FTTH technologies and making FTTH technologies so affordable in the rest of the world. Most of the Chinese FTTH systems are EPON (IEEE 802.3ah) and G-PON (ITU-TG.984) based. In fact, China is now onto the deployment of the next generation 10G PON systems which are split among the following systems: 10G-EPON (IEEE 802.3av [23]), asymmetric XG-PON1 (ITU-TG.987 [24]), and symmetric XGS-PON (ITU-TG.9807 [25]). Fig. 1.7 shows the Chinese FTTH growth from 2012 to 2019.

We can see from Fig. 1.7 that the Chinese FTTH market is already saturated with a market penetration of 92%. China is also leading in both 4G and 5G wireless networks. The fast broadband network and high Internet penetration stimulated the digital economy in China and created a nearly cashless society. In addition, nearly all the TV services in China are streamed through broadband networks as OTT programs from the Internet. Such an effect forms a positive feedback loop to the Chinese telecommunication technology development and widens the gap with other economies.

The Chinese example is a direct demonstration of how important pervasive broadband networks are to the modern economy. Fiber access networks play a very critical role in the digital infrastructure. Throughout this book, we not only want to introduce the latest fiber access technologies

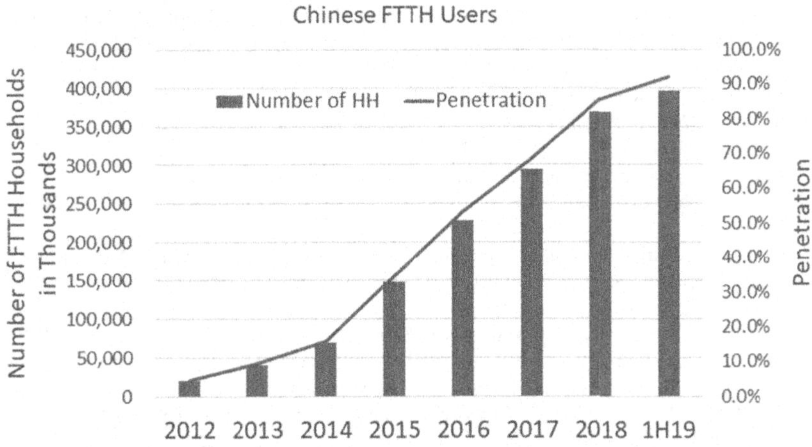

Fig. 1.7 Chinese FTTH growth from 2012 to 2019.

and their developments, but also want to shed light on how to orchestrate these latest developments to create the most efficient and scalable fiber access strategies and network infrastructures.

References

[1] M. Nowell, Cisco VNI forecast update, in: Presented at IEEE 802.3 NEA meeting on June 24, 2019. Available from: https://www.ieee802.org/3/ad_hoc/bwa2/public/calls/19_0624/nowell_bwa_01_190624.pdf.

[2] C. Labovitz, Pandemic Impact on Global Internet Traffic, NANOG, June 2020, Available from: https://storage.googleapis.com/site-media-prod/meetings/NANOG79/2208/20200601_Labovitz_Effects_Of_Covid-19_v1.pdf.

[3] Ford Tamer, Bandwidth in the Age of COVID-19, Inphi Blog Post (link no longer available).

[4] C. Lam, FTTH deployment—google fiber's perspective, in: Optical Fiber Communication Conference, OSA Technical Digest (online) (Optical Society of America, 2017), paper Tu2K.1. (Full Presentation Available by request), 2017.

[5] DOCSIS 3.0 Physical Layer Specification, Version C01, Available from: https://www.cablelabs.com/specifications/CM-SP-PHYv3.0.

[6] C.F. Lam, Passive Optical Networks: Principles and Practice, Academic Press, 2007.

[7] C.F. Lam, Fiber to the home: getting beyond 10 Gb/s, Opt. Photonics News 27 (3) (2016) 22–29. Available from: https://www.osapublishing.org/DirectPDFAccess/2885A151-DEFE-F5B8-D19A669E8CBFBE92_336783/opn-27-3-22.pdf?da=1&id=336783&seq=0&mobile=no.

[8] IEEE 802.3ca, 50G-EPON Task Force: Physical Layer Specifications and Management Parameters for 25 Gb/s and 50 Gb/s Passive Optical Networks, available from http://ieee802.org/3/ca/index.shtml.

[9] D. Zhang, D. Liu, X. Wu, D. Nesset, Progress of ITU-T higher speed passive optical network (50G-PON) standardization, J. Opt. Commun. Netw. 12 (2020) D99–D108. Available from: https://ieeexplore.ieee.org/stamp/stamp.jsp?arnumber=9123509.

[10] C. DeSanti, D. Liang, J. Guarin, J. Bone, C.F. Lam, Super-PON: an evolution for access networks [Invited], J. Opt. Commun. Netw. 12 (2020) D66–D77. Available from: https://www.osapublishing.org/DirectPDFAccess/C5D96D41-9F2E-D4A4-A07D9CBCD58AF2DB_432722/jocn-12-10-D66.pdf?da=1&id=432722&seq=0&mobile=no.

[11] D. Hisano, T. Kobayashi, O. Hiroshi, T. Shimada, H. Uzawa, J. Terada, A. Otaka, TDM-PON for accommodating TDD-based FrONUhaul and secondary services, J. Lightwave Technol. 35 (2017) 2788–2796.

[12] ITU-T G, Supplement 66(07/2019): 5G Wireless fronthaul Requirements in a Passive Optical Network Context, Available from: https://www.itu.int/rec/T-REC-G.Sup66/en.

[13] G.698.4, Multichannel Bi-Directional DWDM Applications with Port Agnostic Single-Channel Optical Interfaces, Available from: https://www.itu.int/rec/T-REC-G.698.4/en.

[14] A. Teixeira, D. Lavery, E. Ciaramella, L. Schmalen, N. Iiyama, R.M. Ferreira, S. Randel, DSP enabled optical detection techniques for PON, J. Lightwave Technol. 38 (2020) 684–695. Available from: https://www.osapublishing.org/DirectPDFAccess/2FB828F7-B136-EA35-97CD8FD8ECEAAB36_426644/jlt-38-3-684.pdf?da=1&id=426644&seq=0&mobile=no.

[15] Terragraph Project Homepage. https://connectivity.fb.com/terragraph/.

[16] I.A. Alimi, A.L. Teixeira, P.P. Monteiro, Toward an efficient C-RAN optical Fronthaul for the future networks: a tutorial on technologies, requirements, challenges, and solutions, IEEE Commun. Surv. Tutorials 20 (1) (2018) 708–769.
[17] X. Liu, N. Deng, Chapter 17—Emerging optical communication technologies for 5G, in: Optical Fiber Telecommunications VII, Elsevier, 2020, pp. 751–783. Available from: https://www.sciencedirect.com/science/book/9780128165027.
[18] FCC PART 96—Citizens Broadband Radio Service, Available from: https://www.ecfr.gov/cgi-bin/retrieveECFR?gp=&SID=0076fe7586178336d9db4c5146da8797&mc=true&n=pt47.5.96&r=PART&ty=HTML.
[19] FCC Auction 105: 3.5GHz, Available from: https://www.fcc.gov/auction/105.
[20] CBRS Spectrum auction maps: who won what, and where. LightReading Article, Available from: https://www.lightreading.com/5g/cbrs-spectrum-auction-maps-who-won-what-and-where/d/d-id/763837?_mc=RSS_LR_EDT.
[21] C.E. Shannon, A mathematical theory of communication, Bell Syst. Tech. J. 27 (1948) 379–423. 623–656. Available from: http://people.math.harvard.edu/~ctm/home/text/others/shannon/entropy/entropy.pdf.
[22] FOMSN New Article, Asian Countries Lead the FTTH-FTTB Global Ranking, Available from: https://www.fomsn.com/market-research/sobhana/asian-countries-lead-the-ftth-fttb-global-ranking/.
[23] 802.3av-2009, IEEE Standard for Information technology- - Local and Metropolitan Area Networks- - Specific Requirements- - Part 3: CSMA/CD Access Method and Physical Layer Specifications Amendment 1: Physical Layer Specifications and Management Parameters for 10 Gb/s Passive Optical Networks, Available from: https://standards.ieee.org/standard/802_3av-2009.html.
[24] ITU-T G.987, 10-Gigabit-Capable Passive Optical Network (XG-PON) Systems: Definitions, Abbreviations and Acronyms, Available from: https://www.itu.int/rec/T-REC-G.987/en.
[25] ITU-T G.9807, 10-Gigabit-Capable Symmetric Passive Optical Network (XGS-PON), Available from: https://www.itu.int/rec/T-REC-G.9807.1/en.

CHAPTER 2

Overview of fiber access architectures and mature PON standards

2.1 Introduction

Optical fiber was originally developed for long-haul transmissions. The idea of using fiber optic for access was initially proposed in the 1980s [1], way before the Internet and broadband access became the norm of our society. As we had seen from the last chapter, after decades of developments, fiber access networks are now mature technologies which have been deployed to hundreds of millions of users around the world. Besides directly connecting end customers with optical fibers in FTTH networks, new forms of fiber access networks are indispensable in providing backhaul and fronthaul connectivities to wireless networks such as the fifth generation (5G) wireless network constructions around the world.

Fiber access networks are deployed at the edge of telecommunication networks as end nodes. There are two major challenges in deploying fiber access networks. First, access networks are very cost sensitive, so the equipment cost has to be very low in order for FTTX to be viable as a mass-deployed technology. Two things are needed in order to achieve the required low equipment cost: (1) standard-based implementation with economy of scale and (2) low-cost optoelectronics packaging techniques. The second major challenge in deploying fiber access networks is labor cost and speed and ease of deployments. Significant civil engineering cost is incurred in deploying FTTX infrastructures around the world, especially in developed economies where (1) the labor cost is high, and (2) the digging and trenching involved in laying fiber cables is not easy. Therefore, traditional incumbent carriers in developed nations would like to preserve their legacy copper infrastructure and delay the deployment of fiber in the last mile (from the central office) as much as possible. In developing economies or greenfield scenarios, there will be fewer architecture constraints, less legacy burdens, and more flexibility in technology and architecture choices.

Advanced Fiber Access Networks
https://doi.org/10.1016/B978-0-323-85499-3.00007-2

But those economies are also very capital cost sensitive and would like to leverage the low cost of existing, mature, and standard-based technologies. These challenges are guiding the design principles of fiber access network infrastructures.

Broadband access was the initial driving force for FTTH fiber access technologies. As we have seen from the last chapter, this was mainly propelled by the booming Internet applications, especially over-the-top (OTT) video streaming applications which offer any time and any place viewing experiences of on-demand contents. Higher resolution videos such as 4 K will demand more bandwidths to end users. Also, the outbreak of the COVID-19 pandemic in 2020 has significantly increased bandwidth demands due to applications such as video conferencing, network gaming, and remote education.

2.2 FTTX architectures

FTTH was mainly provided by passive optical networks (PONs), especially time-division-multiplexed (TDM) PON systems. That however, does not mean TDM-PON is the only way to provide FTTH services. In this chapter, we give an overview of fiber access network architectures and mature PON standards existing in the market. Experienced readers who are familiar with FTTH technologies could skip this chapter.

2.2.1 Home-run fiber, TDM, WDM, and TWDM

In the broad sense, a passive optical network or PON means that the network between a carrier's central office (CO) (where the customer traffic is first terminated) and the end user has no active elements requiring electrical power.

In an FTTH system, the equipment in the CO terminating customer traffic is called OLT (optical line terminal) and the customer-end modem converting the optical signal from the CO to electrical signal is called ONU (optical network unit). Usually, there are three ways to achieve this (Fig. 2.1):

1. A point-to-point home-run fiber from the CO to every customer which makes the medium access control protocol and optical transmission the simplest. There is no contention for the dedicated transmission medium from the CO to each individual household and users are isolated by their own fiber, which guarantees last mile transmission security. The last mile transceivers only need to account for the transmission fiber loss with easy

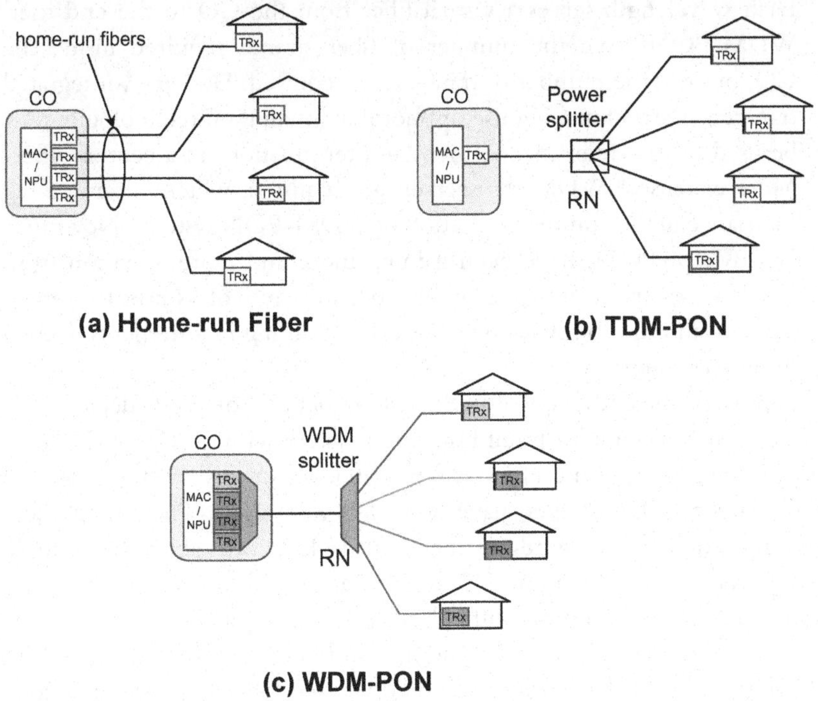

Fig. 2.1 Common passive optical access network architectures. *MAC*: medium access Control; *NPU*: network processor unit; *TRx*: transceiver.

dynamic range to handle. However, this architecture requires termination of a large number of fibers and transceivers inside the CO);

2. Using a power splitter in the field as a remote node (RN) to broadcast the signal from a common transceiver at the CO to the multiple end users. The bandwidth at the common transceiver is shared among the users using a TDM protocol. The TDM protocol employs a dynamic bandwidth allocation (DBA) algorithm to efficiently allocate the shared bandwidth of the common transceiver among the end users. TDM-PON takes advantage of statistical multiplexing of the traffic from different end users. It saves both the number of fiber strands and the number of transceivers required inside the CO, greatly simplifying CO spacing, power, and fiber termination requirements. This type of TDM-PON is the most common form of commercial deployments today, to which both the ITU-T G-PON [2] and IEEE 802.3 EPON [3] belong;
3. Using a wavelength router in the field as the remote node to distribute WDM wavelengths (in the form of virtual fibers) to end users.

Each wavelength serves a virtual fiber from the CO to the end user. WDM-PONs save the number of fiber strands required inside the CO, but not the number of transceivers required. However, integrated transceiver arrays making use of photonic integration technologies may be used to ease fiber management and terminations in a central office. More expensive WDM transceivers are required, which became a big hindrance in the commercialization of WDM-PON systems. Nevertheless, WDM-PONs have been gaining more and more attention from wireless carriers for fronthaul connections because of its protocol transparency and the ability to meet the precise timing requirements needed in wireless networks.

To appreciate the advantage of PON systems, let us look at the deployment comparison of a point-to-point home-run system with a TDM-PON network using 1:64 splitting ratio. A typical telco CO serves 10,000–30,000 users. Larger COs can even have 60,000–80,000 users. The largest cable used in Google Fiber's outdoor fiber plant is 432-core cable. An example of such cable is shown in Fig. 2.2. Bigger cables with higher number fiber strands are thicker and more difficult to handle and repair.

To support 30,000 users with home-run fibers, 70 such gigantic cables are required per CO. In reality, standard Google Fiber COs are only connected with six (6) 432-core cables to cover 40,000 users (including ample spare fibers and accounting for fiber termination inefficiencies). Furthermore, in a home-run network, every user needs a separate fiber termination

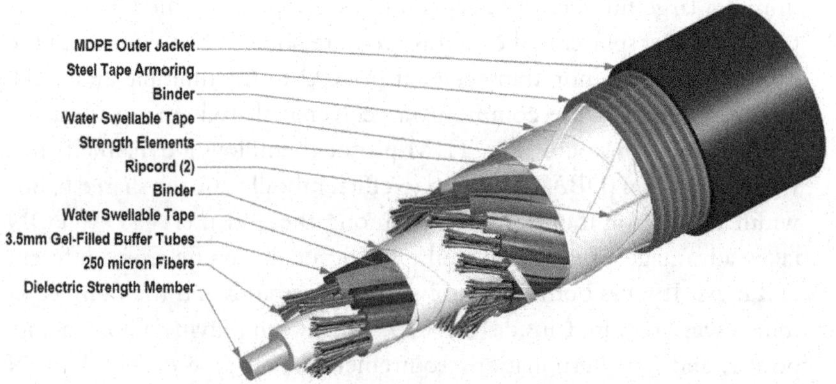

Fig. 2.2 A 432-core fiber cable. *(From Commscope Product Spec, 760181628 | O-432-LA-8W-F24NS outdoor cable, Commscope. [Online]. Available: https://www.commscope.com/product-type/cables/fiber-cables/outside-plant-cables/item760181628/ (Courtesy CommScope Inc.))*

Fig. 2.3 Large real estimate (more than 5 full 7-ft racks) is required to terminate point-to-point fibers in a home-run network.

inside the CO. Typical racks inside COs are 7-ft (213.36 cm) tall, which is equivalent to 42RU.[a] Assume that the field fibers are terminated by the compact LC[b] connectors, for example, using Corning's Edge fiber termination system (or something equivalent) with a highest density of 567 fiber termination in 4RU. To terminate 30,000 fiber alone would require more than five (5) 7-ft racks as shown in Fig. 2.3.

The space required by active OLTs is also significant for home-run systems. Space required by active equipment is often limited by the front panel density. As an example, a 10RU-tall point-to-point OLT chassis with 16 service cards, each having 32 active Ethernet ports, would be able to serve 512 home-run users. A 7-ft rack would be able to serve 2048 active users (Fig. 2.4). For a 30% take rate, a CO passing 30,000 households would require 4.5 active racks to accommodate the OLTs required to serve all the active users.

Assuming similar port density (limited by the front panel connector and active optical module form factors), if the same system is built using G-PON networks with 1:64 splitting ratio, we only need to terminate 469 (30,000/64) fibers (which can be accommodated by one 4-RU Edge housing described above), and one 10-RU active G-PON OLT chassis.

[a] Height of equipment in CO is measured in RU (rack unit). Each RU is 1.75-in. tall. Both 19″ wide and 23″ wide racks are used in telecom COs, with 23″ being more common. Datacom racks such as those found in datacenter networks are usually 19″ wide.

[b] LC is the name of a small form factor connector invented by Lucent Technologies. It actually stands for Lucent Connector.

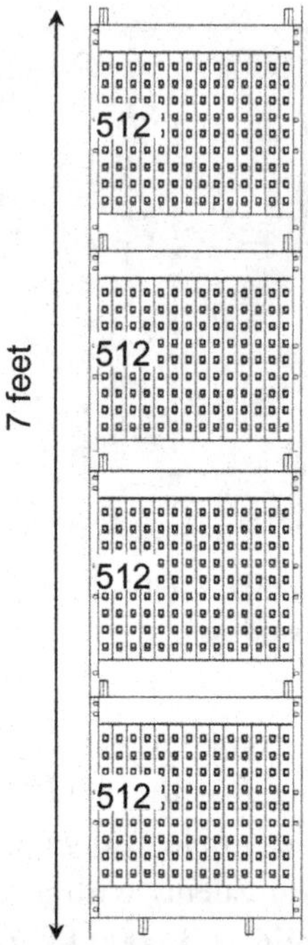

Fig. 2.4 A 7-ft rack can only accommodate 2048 active users. Assume 30% take rate, a CO serving 30,000 households would require 4.5 active racks to serve all the active users.

The whole CO is reduced from 10 racks to 1/3 rack (Table 2.1), a 30 × reduction of the required CO space. In addition, it leads to significant reduction of power required. The space that is saved would be valuable for system upgrade. During the transition period from one generation to the next, an operator is almost inevitably required to keep both the old system and the new system running in parallel.

To further consolidate fiber strands in the field, combinations of WDM and TDM PONs in the form of T-WDM PONs have also been implemented in the industry. Examples are the ITU-T G.989 NG-PON2 [4] and Google Fiber's super-PON technology [5] which, at the time of writing,

Table 2.1 CO space requirements using home-run point-to-point—gigabit Ethernet (GbE) system vs 1:64G-PON system, for a CO with 30,000 users and 30% take rate.

	Passive racks	Active racks	Total rack space
Home-Run GbE	5.2 rack	4.5 rack	9.7 racks
G-PON, 1:64	4-RU	10RU	14 RU (⅓ rack)

is being standardized by the IEEE in the 802.3cs Ethernet Standard task force. We will discuss these systems in detail in Chapter 4.

2.2.2 FTTX systems with distributed OLT in the field

In the last section, we described architectures with OLTs consolidated in COs. Such architecture has the advantages of active OLTs being centrally managed and operated in controlled environments. The outside plant (OSP) is completely passive. This makes network management, operation, and maintenance simpler and easier. However, as we have also seen in the last section, even with the aggregation offered by PONs, hundreds or even thousands of fiber strands are still needed at the central office. Such a large number of fiber strands at a CO are not always available, especially for legacy COs which were not planned for FTTH from day 1.

A distributed OLT architecture employs mini-OLTs deeper into the field and closer to the subscribers. These mini-OLTs are located inside a street cabinet. An example of a CO, the racks inside the CO, as well as a street cabinet is shown in Fig. 2.5.

As shown in Fig. 2.6, the mini-OLTs deep in the field are backhauled to the CO with a transport system, usually some kind of metro DWDM (dense wavelength division multiplexing) rings for both capacity and path diversity and fault tolerance. A mini-OLT usually has 8–32 PON ports, capable of aggregating to around 2000 users, therefore significantly reducing the number of outside fiber strands needed at a CO.

Although the distributed OLT architecture significantly reduces the number of fiber strands at a CO, it also has its own set of challenges. Street cabinets usually lack environmental controls. Equipment deployed in street cabinets needs to be hardened and is more expensive. The limited space and heat exchanging capability also makes network upgrade a challenge,

(a) a central office (b) racks inside a CO (c) a street cabinet with a mini OLT

Fig. 2.5 Photos of (A) a central office (the *green* (*gray* in print version) structure on the side is the backup generator), (B) the racks inside a CO, and (C) a street cabinet with a mini-OLT inside.

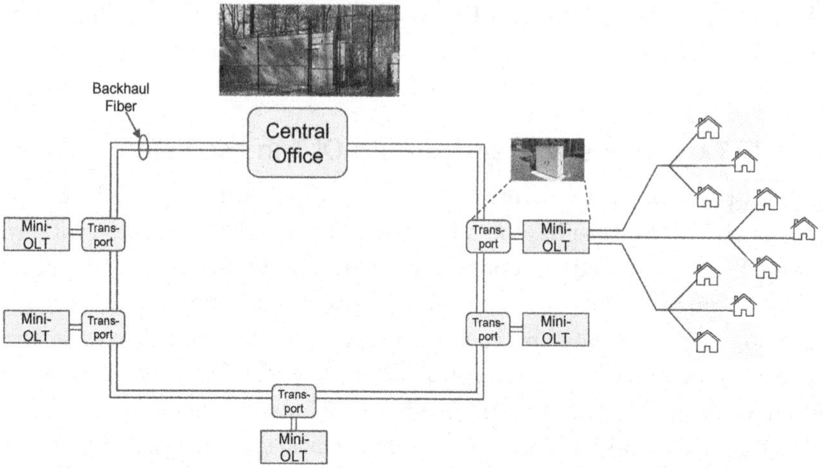

Fig. 2.6 Distributed OLT architecture with mini-OLTs deep in the field and located inside street cabinets.

let alone the need to visit multiple locations during network upgrades and maintenance. From a practical perspective, a CO is often secluded in a fenced or gated facility as shown in Fig. 2.5. It is thus less prone to vandalism and being damaged by runaway vehicles on the road.

An operator building networks with distributed OLTs needs to coordinate with electricity companies to install electrical power feed and metering to each cabinet location. Moreover, a CO offers high availability with backup generators and batteries (Fig. 2.5A). For example, starting from Feb 13, 2019, the FCC (Federal Communications Commission) extended the battery backup requirements from 8 to 24 h in order to provide telephony service with E911 emergency calls [6]. Backup generators and batteries are not easy with the distributed OLT architecture. Besides the limited space available in the cabinets, batteries require regular maintenance. In general, backup batteries need to be visited every 6 months for health evaluation. Lead-acid batteries need to be replaced every 3–5 years. Lithium iron phosphate batteries have a longer lifetime of 10 years but they do not like cold weathers. A battery self-diagnosis and maintenance system with management network should be installed in the cabinets offering high-availability services.

In summary, distributed OLT networks may be necessary and easier and quicker to construct in certain scenarios (especially if the operator is fiber strand limited), they are also associated with their own operational complexities and limitations, which is amplified by the large number of cabinets in the field. As discussed earlier, a properly designed PON network does not take much CO space (less than 1 rack) to construct and operate. In Chapter 4, we will discuss how to solve some of the fiber strand shortage issues with the TWDM super-PON architecture [7].

2.3 Mature PON standards

2.3.1 Burst-mode operation in TDM-PONs

The high-level architecture of a TDM-PON is depicted in Fig. 2.7 [8]. In a TDM-PON network, the downstream transmission is a point-to-multipoint operation where the OLT broadcasts the same signal to multiple ONUs. An ONU relies on the address field in the downstream frames to determine if the associated data are addressed to itself and discards those not intended for its consumption. The upstream direction, on the other hand, is a multipoint-to-point operation where the ONUs take turns to transmit to the shared OLT receiver. The OLT uses a dynamic bandwidth allocation (DBA) algorithm to schedule proper time slots for ONUs to transmit in the upstream direction. An ONU not transmitting must shut off its transmitter to avoid collision with signals from other ONUs. The OLT uses a burst-mode receiver with fast clock recovery to recover the signal from the ONUs. Each time a different ONU transmits, the OLT has to perform fast

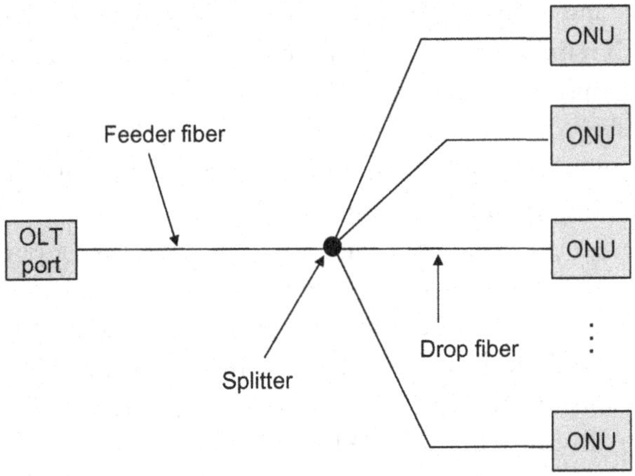

Fig. 2.7 1:32, 1:64 or 1:128. The standard distance from the OLT to ONUs is usually less than 20 km. Besides power budget, the transmission distance and splitting ratio are also limited by the TDM-PON protocol such as the size of the ranging window and available ONU address space.

clock recovery which adds overhead to upstream transmission. Therefore, guard times are reserved between upstream bursts from different ONUs.

2.3.2 IEEE 802.3, FSAN, and ITU-T

There are three major standard organizations responsible for making PON standards:

- IEEE 802.3 Ethernet Working Group, https://ieee802.org/3/
- FASN (Full Access Service Network), https://www.fsan.org/
- ITU-T SG15/Q2 (International Telecommunication Union—Telecom, Study Group 15, Question 2): Access Networks Protocols, https://www.itu.int/en/ITU-T/studygroups/2017-2020/15/Pages/default.aspx

IEEE 802.3 Ethernet Working Group makes the most popular data networking standard on this planet—the Ethernet. Before a project task force is authorized, the project objective along with the following five IEEE 802.3 CSD (Criteria for Standards Development) [9] needs to be established:

1. Broad market potential
2. Compatibility
3. Distinct Identity
4. Technical feasibility
5. Economic feasibility

IEEE 802.3 standards are consensus based and the meetings are well attended by equipment and technology vendors. The technical requirements defined in IEEE 802.3 standards are very comprehensive including detailed definition and description of managed objects and management interfaces to ensure unambiguous implementability. As a result, IEEE 802.3-based systems have excellent track records of interoperability, compatibility, and economic and implementation viability.

FSAN is a forum of major telecom service providers and equipment vendors to discuss the evolution of broadband access networks with a focus on PON. FSAN collects the requirements for next generation PON systems from operators and works hand-in-hand with ITU-T SG15/Q2 to create the relevant standards. IEEE 802.3 Working Group tends to be more equipment vendor-driven whereas FSAN and ITU-T SG15/Q2 are more operator-driven. The technical requirements generated by ITU-T are called recommendations as opposed to standards. In many cases, those are technical guideline documents with many implementation details left to the interpretation of the vendors who actually implement the products. It is up to the service providers to enforce compatibility and interoperability requirements, and perform the interop tests among vendors in service providers' system qualification labs.

Most telecom service providers regularly participate in ITU meetings and present their requirements and agenda there. They also have a preference over ITU-T PON standards to IEEE 802.3 PON standards, although most of the time, the technoeconomic feasibility and requirements are first studied, debated, and established inside IEEE.

2.3.3 Gigabit TDM-PON standards

TDM-PON is commonly specified with 20 km coverage between the CO and the end users, using 1:32, 1:64, or 1:128 splitters as the RN. In a PON system, the upstream and downstream signals are separated by wavelengths using a wavelength diplexer in the optical transceivers inside the OLT and the ONU.

The IEEE 802.3ah E-PON [3] is a PON standard that offers 1 Gbps aggregated bandwidth in both upstream and downstream directions, whereas the ITU-T G-PON [2] offers 2.5 Gbps and 1.25 Gbps in the downstream and upstream directions, respectively.

Both G-PON and E-PON have very similar physical layer characteristics. They both use 1310 nm wavelength for upstream transmission and 1490 nm for downstream transmission (Fig. 2.9).

The IEEE 802.3ah E-PON standard was the first PON standard created by the IEEE 802.3 Ethernet Standard Group. Before the IEEE 802.3ah E-PON and ITU-T G-PON, other lower speed PON standards had been created, for example, the ITU-T Recommendation G.983 Broadband Passive Optical Network (called B-PON) with 622 Mbps/155 Mbps downstream and upstream speeds. None of those systems were really deployed on large scales.

E-PON and G-PON are the most successful and most deployed PON standards in the world, especially G-PON, which is still being actively deployed. Compared to E-PON's symmetric 1 Gbps downstream and upstream PON capacity, G-PON has 2.5 Gbps downstream capacity, more than doubling that of E-PON's. As we will see in Chapter 7 that this 2.5 × downstream capacity allows G-PON to be used to offer gigabit access services using statistical multiplexing, which is a huge advantage over E-PON.

As a matter of fact, manufacturers have made optical transceivers for G-PON and E-PON identical. PON SOC (system on chip) manufacturer

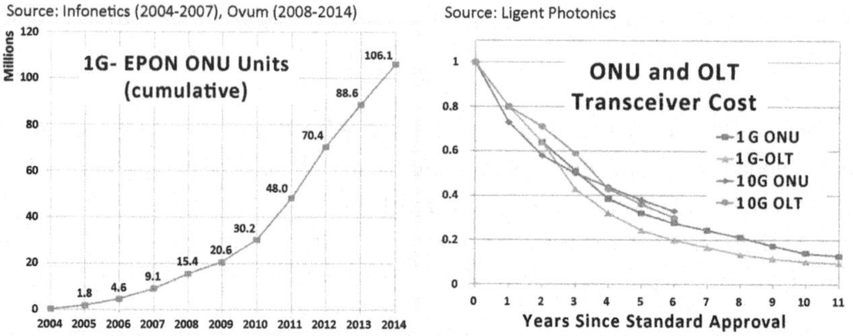

Fig. 2.8 Price erosion of PON optical transceivers with respect to deployment volume.

also made the PON MAC ASIC (application-specific integrated circuit) configurable among E-PON, G-PON, and point-to-point Ethernet modes. One can imagine that almost all the on-going PON deployments would be G-PON based after the G-PON technology matured (E-PON had a slight time-to-market advantage than G-PON because of the less stringent technical specifications).

Another major difference between E-PON and G-PON is in management protocols. E-PON uses the Ethernet OAM (operation, administration, and management) protocol for management and the Ethernet MPCP (multi-point control protocol) for PON layer management whereas G-PON uses the ITU-T G.984.4 OMCI (ONU management control interface) for the PON layer management between the OLT and ONUs.

Nowadays, incumbent operators around the world have pretty much standardized on G-PON deployments. The resultant economy of scale has significantly driven down the costs of G-PON equipment prices (Fig. 2.8).

E-PON, on the other hand, is preferred by the MSOs (multiple service operators, aka cable operators) in North America. Cable Labs (the research arm for cable operators) has specified DOCSIS (Data over Cable Service Interface Specifications) over the E-PON (DPoE or D-PON) architecture and operations [10,11], allowing MSOs to easily incorporate the IEEE E-PONs into their DOCSIS network.

2.3.4 10 Gbps TDM-PON standards

In 2009, the IEEE 802.3 Working Group ratified the standard for 10 Gbps capable E-PON or 10GE-PON as IEEE 802.3av [12]. This IEEE TDM-PON standard was proven to be the most practical 10 Gbps

PON technology, which enjoyed early commercial successes of volume deployments.

Symmetric IEEE 802.3av 10GE-PON adopts 1270 nm wavelength for 10 Gbps upstream transmission and 1577 nm wavelength for downstream transmission. These wavelengths are not overlapping with the G-PON and E-PON wavelengths so that one can use a wavelength multiplexer to overlay 10GE-PONs to either a legacy G-PON or E-PON network. Fig. 2.9 shows the optical spectrum allocation for existing TDM-PON standards.

Around the time that IEEE was developing the 10GE-PON standard, ITU-T has published the standards for XG-PON1 (with 10 Gbps downstream and 2.5 Gbps upstream capacities) and XG-PON2 (with symmetric 10 Gbps in both upstream and downstream directions) [13]. These standards were collectively called NG-PON (next generation PON) standards. While prototypes of these systems had been built, the very difficult-to-meet requirements for burst-mode timing, coupled with a lack of real demand, delayed commercialization of these standards. Yet, the ITU-T SG15/Q2, which is dominated by service providers, aggressively tried to create a "40 Gbps" TWDM-PON called NG-PON2 (or ITU-T G.989) [4], to leapfrog the IEEE 802.3av 10GE-PON. This effort, as will be discussed in the next section, was not very successful. In 2015, demands for 10 Gbps PONs systems started to emerge. Faced with the competition of the mature IEEE802.3av 10GE-PON technology (which was already in product manufacturing at the time), to quickly come up with its own implementable version 10 Gbps PON standard for product offerings, ITU-T neutered the NG-PON2 standard to create an interim "single-wavelength" symmetric 10 Gbps PON, called XGS-PON [14], which will be forward compatible with future full NG-PON2 standards when manufacturable technologies will be within reach. In fact, the physical layer of XGS-PON almost completely adopted the IEEE 802.3av 10GE-PON physical layer characteristics, except with the modification of burst-mode timing requirements. Early implementations of XGS-PON systems even relaxed the timing requirements and completely adopted 10GE-PON transceivers to jump start the business.

As explained earlier, telecom operators have a strong preference for ITU standards. Besides their own strong involvement and the vast amount of time and effort they vested in creating the ITU standards, the ability to continue to use the same network OAM approach and methodology and the ease of navigating through similar set of terminologies are strong reasons for

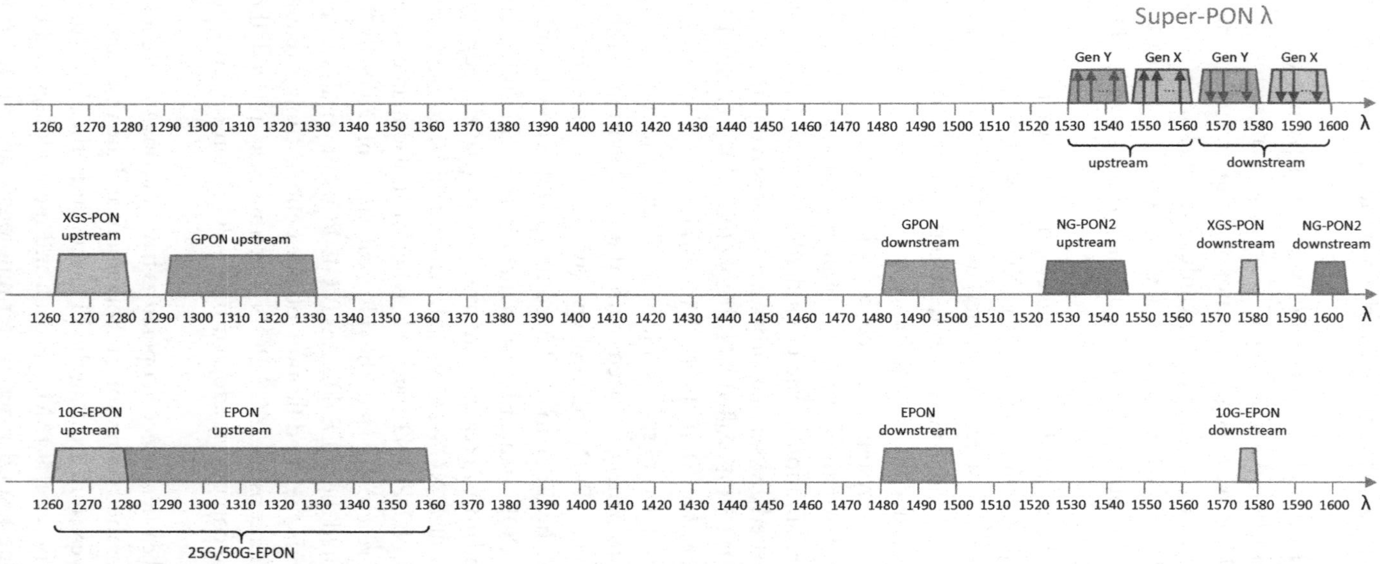

Fig. 2.9 Optical spectral allocation for common PON standards. The super-PON is being standardized in the IEEE 802.3cs task force at the time of writing.

traditional telecom operators to adopt an ITU-standard based 10G-PON system when they have a choice. Therefore, even though the XGS-PON standard was much behind the IEEE 802.3av, it is becoming the plan of record for incumbent telecom carriers.

It should be noted that although both the 10G-EPON and XGS-PON are boasting a nominal PON interface bitrate of 10 Gbps, the actual achievable data rate is around 8.5 Gbps in both upstream and downstream directions. We will see in Chapter 7 that this achievable bit rate is not good enough to offer 10 Gbps services.

When Google Fiber first launched its service in 2012, it used G-PON with 2.5 Gbps PON interface to cost effectively provide gigabit accesses with high successes. When this book is published, it will be 10 years after the launch of the first Google Fiber services. Many carriers will be using 10G-PON systems (whether it is 10G-EPON or XGS-PON) to offer multi-gigabit services (e.g., 2 Gbps or 5 Gbps) cost efficiently.

2.3.5 TWDM-PON standards

The ITU-T G.989 NG-PON2 combines wavelength division multiplexing and time division multiplexing (TWDM) together in the same architecture to offer a total capacity of 40 Gbps. It was the first time that dense wavelength division multiplexing (DWDM) had been adopted in a commercial access network standard. Transmitting a 40 Gbps TDM signal on a single wavelength was not easy at the time the NG-PON2 standard was created. So four (4 ×) 10 Gbps wavelengths between 1524 and 1544 nm are used to transmit the upstream signal and four wavelengths between 1596 and 1603 nm to transmit the downstream signal. The OLT uses an array transceiver with four transmitters and four receivers, all multiplexed together with an internal wavelength multiplexer/demultiplexer built into the OLT optical module. To be compatible with conventional PON networks, NG-PON2 inherited the optical power splitter in the field as the remote node. Therefore, all the four wavelengths are broadcast to every NG-PON2 ONU. ONUs in NG-PON2 are, however, only specified with a single 10 Gbps transceiver, with both a tunable upstream laser and a tunable receiver. Fig. 2.10 shows the architecture of NG-PON2.

The NG-PON2 design philosophy has several problems, however. For one, the new standard aims at supporting multistage splitting and high splitting ratios (perhaps as great as 1024 users per CO connection), in order to

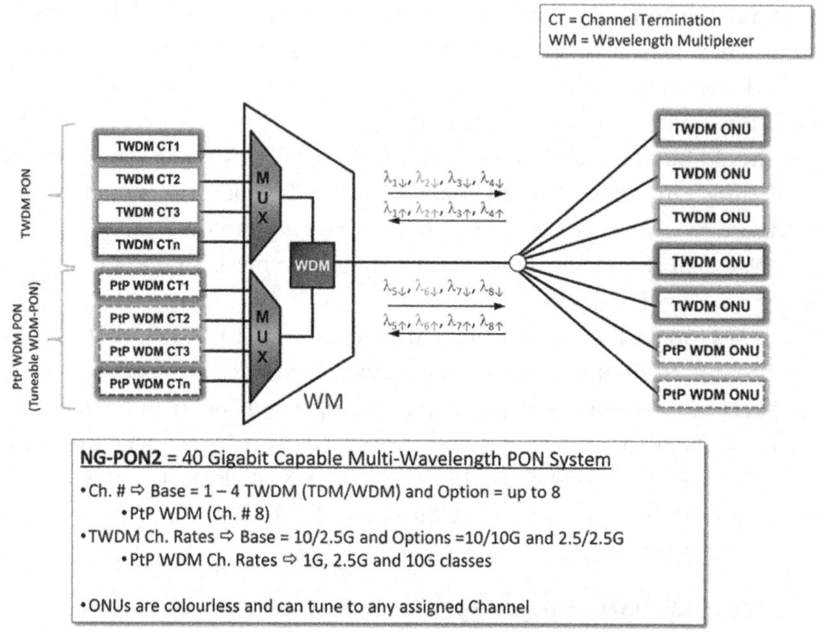

Fig. 2.10 Architecture of NG-PON2 [15].

efficiently use the vast bandwidth offered by NG-PON2 and to reduce system costs through better sharing of the expensive OLT optics and by decreasing the number of fibers required to connect to the CO. But not many of the embedded PON systems currently deployed have such large splitting ratios. Moreover, to support the very high splitting ratio, the OLT burst-mode receiver requires a high dynamic range (greater than 20 dB), a huge price to pay to make NG-PON2 backward compatible with embedded legacy PON systems while trying to achieve other new goals listed above.

A second problem is the potentially very high power budget needed for NG-PON2 implementation. Having to support very high splitting ratios will further strain the power budget requirements. The NG-PON2 standard specifies a loss budget of up to 35 dB between the OLT and ONU [16,17] (compared to the 28 dB Class B + power budget commonly used in G-PON deployments and 32 dB specified for class C++ of G-PON, the latter of which is already very difficult to produce with high yield, for a transmission of only 2.5 Gbps). This has not accounted for the extra losses from the WDM multiplexers inside the OLT and the tunable filter in the ONU optical module. As a result, lasers with very high transmitting power and receivers with high sensitivity will be needed, resulting in potentially significant increase in the system costs—even if the technology is achievable [18].

At the time of writing this chapter (January 2021, almost 8 years after the proposal of NG-PON2 standard in March 2013), optical component technologies still have not advanced to the point that NG-PON2 implementation is economically feasible for mass production. In the authors' opinion, the failure to recognize the technical feasibility in the creation of the NG-PON2 standard by members who are mostly from operators and lacking the input from actual technology vendors might be the reason leading to the dystocia of commercially deployable NG-PON2. The step from G-PON to 10-Gb/s TDM-PON was incremental, as the optical systems are quite similar, apart from the speeds of the optical transceivers. However, in moving from current 10-Gb/s TDM-PONs to NG-PON2, performance requirements of the optical components become significantly more stringent, the structures of the optical transceivers are much more complex, and packaging complexity becomes exponentially higher.

Small, low-loss, low-cost, and low-power tunable filters, required for ONUs in an NG-PON2 system, are not easier to manufacture than semiconductor tunable lasers (which are mostly monolithic structures). Innovations in photonic integration circuits are needed to solve these challenges, in the long run.

Burst-mode operation of the tunable DWDM laser in NG-PON2 ONU causes transient wavelength drift [19], as the sudden injection of current into the laser heats up the laser structure (unless an external modulator is used, which increases cost and optical losses). The drift increases with the laser bias current and laser output power and can be as large as 20 to 30 GHz, with time constants on the order of milliseconds. Such drift not only causes a penalty in OLT receiver sensitivity at the central office, but also crosstalk to other wavelength channels in a broadcast-and-select TDM-PON fiber plant. A pure WDM-PON without TDM overlay, on the other hand, does not require burst-mode operation and, thus, does not suffer from these challenges. We will discuss this in Chapter 4.

From the medium access control (MAC) protocol perspective, in addition to the usual DBA algorithm used in TDM-PONs, NG-PON2 adds a layer of optical wavelength complexity to manage. The millisecond-scale tuning speeds of the tunable laser and filters used in NG-PON2 will make it economically difficult to do fast wavelength switching on the packet level. Coordinating wavelengths with TDM time slots together complicates the DBA algorithms used in NG-PON2.

As we will also see in Chapter 7, in the foreseeable future, for residential FTTH applications, 10-Gb/s PON networks should provide adequate bandwidth to meet existing demands, so there is no immediate justification for dynamically adjustable wavelength allocations, especially with the additional cost and complexity. That fact may underlie ITU-T's decision in 2015 to create the XGS-PON (10-Gb/s symmetric PON with single wavelength) standard as an "initial stage" of NG-PON2.

Nevertheless, despite lack of commercial successes, NG-PON2 started the chapter of using parallel WDM wavelength channels to scale optical access networks. Google Fiber's TWDM super-PON architecture [20], which will be discussed in detail in Chapter 4, is another example of using WDM to scale FTTH networks. The super-PON TWDM system avoids some of the implementation challenges of NG-PON2 by sacrificing some flexibilities such as optical tunable receivers and broadcast ODN networks based on power splitting remote nodes.

References

[1] P.E. White, L.S. Smoot, S.E. Miller, I.P. Kaminow (Eds.), Optical fibers in loop distribution systems, in: Optical Fiber Telecommunications II, Academic Press, 1988, pp. 911–932.

[2] ITU-T, ITU-T Recommendation G.984.1: Gigabit-Capable Passive Optical Networks (GPON): General Characteristics, 2008, ITU-T, G.984.1. [Online]. Available: https://www.itu.int/rec/T-REC-G.984.1.

[3] IEEE, IEEE 802.3ah-2004, 2004, [Online]. Available: https://standards.ieee.org/standard/802_3ah-2004.html.

[4] ITU_T, ITU-T Recommendation G.989, 40-gigabit-capable passive optical network (NG PON2), G.989, 2015. Available: https://www.itu.int/rec/T-REC-G.989-201510-I/en.

[5] C. DeSanti, D. Liang, C. Lam, J. Jiang, Super-PON: Scale Fully Passive Optical Access Networks to Longer Reaches and to a Significantly Higher Number of Subscribers, Presented at the IEEE 802.3 NEA Meeting, Geneva, CH [Online], IEEE, 2018. Available: http://www.ieee802.org/3/ad_hoc/ngrates/public/18_01/desanti_nea_01a_0118.pdf.

[6] FCC Public Record DA 18-1205. https://docs.fcc.gov/public/attachments/DA-18-1205A1_Rcd.pdf.

[7] C.F. Lam, Scaling the next generation broadband access networks with super-PON, in: Presented at the OSA Advanced Photonics Congress (AP), 2019.

[8] C.F. Lam, Passive Optical Networks: Principles and Practice, Elsevier, 2007.

[9] IEEE 802.3 Criteria for Standards Development, [Online]. Available: https://www.ieee802.org/3/NGAUTO/public/jan17/NGAUTO_CSD_DRAFT_01_0117.pdf.

[10] Cable Labs, DoPE Architecture, Cable Labs, [Online]. Available: https://www.cablelabs.com/specifications/search?currentPage=2&sortby=false&order=&query=&category=DPOE&subcat=&doctype=&cONUent=false&archives=false.

[11] V. Blake, DOCSIS over PON, in: Presented at the OFC/NFOEC, 2008. San Diego, CA.

[12] IEEE, IEEE 802.3av-2009, 802.3av, 2009. [Online]. Available: https://standards.ieee.org/standard/802_3av-2009.html.
[13] D. Nesset, The PON Roadmap, 2016, https://doi.org/10.1364/ofc.2016.w4c.1.
[14] ITU-T, 10-Gigabit-Capable Symmetric Passive Optical Network (XGS-PON), ITU-T Recommendation G.9807.1," G.9807.1, June 2016.
[15] D. Nesset, NG-PON2 technology and standards, J. Lightwave Technol. 33 (5) (2015) 1136–1143.
[16] J.S. Wey, et al., Physical layer aspects of NG-PON2 standards—part 1: optical link design [invited], IEEE/OSA J. Opt. Commun. Netw. 8 (1) (2015) 33.
[17] ITU-T, 40-Gigabit-capable passive optical networks 2 (NG-PON2): Physical media dependent (PMD) layer specification, G.989.2, 2014, G.989.2, 2014.
[18] Y. Luo, et al., Physical layer aspects of NG-PON2 standards—part 2: system design and technology feasibility [invited], IEEE/OSA J. Opt. Commun. Netw. 8 (1) (2015) 43.
[19] D. Van Veen, W. Pohlmann, B. Farah, T. Pfeiffer, P. Vetter, Measurement and mitigation of wavelength drift due to self-heating of tunable burst-mode DML for TWDM-PON, in: Conference on Optical Fiber Communication, Technical Digest Series, 2014, pp. 1–3.
[20] L.B. Du, et al., Long-reach wavelength-routed TWDM PON: technology and deployment, J. Lightwave Technol. 37 (3) (2018) 688–697. Available: https://ieeexplore.ieee.org/stamp/stamp.jsp?tp=&arnumber=8395363.

CHAPTER 3

Advanced TDM-PON standards

3.1 Introduction

Besides the mature TDM-PON standards, for example, 10G-EPON, XGS-PON, etc., discussed in the previous chapter, optical access network is constantly evolving with the introduction of new standards, which is typically driven by the growth of subscriber traffic due to new applications. Among all the latest TDM-PON standards under discussions or close to publication, there are two which generate broad interests among Internet Service Providers (ISPs), system and component vendors, which in turn represent general consensus among those industry groups in the evolution path toward the next generation TDM-PON-based optical access networks. IEEE first initiated 802.3ca study group in 2016 to study toward 100G-EPON as the next generation system after 10G-EPON [1]. Around the same time period, FSAN/ITU-T SG15/Q2 started the G.sup.HSP (HSP refers to higher speed PON) project to look into technologies enabling beyond 10 Gbps per wavelength [2], for example, 25 and 50 Gbps. This chapter focuses on discussing these two standard efforts.

3.2 IEEE 802.3ca 25G/50G-EPON

The IEEE 802.3 working group initiated ad hoc discussions on next generation EPON (NG-EPON) in January 2014, which was expected to serve as the evolution path after 10G-EPON was introduced as IEEE 802.3av in 2009. This new standard was mainly driven by the anticipated increase of access network bandwidth demands for both residential and business applications, with the focus of guaranteed bandwidth for the latter case as shown in Table 3.1. Three main bandwidth drivers were thought to lead to the exponential growth of the total bandwidth demand [6]: (1) growing number of subscribers, (2) increasing number of connected devices per subscriber, and (3) proliferating bandwidth demand per device or application. In addition, cellular network back-haul and front-haul bandwidths also have been growing at an unabated pace during the evolution from 4G to LTE, and eventually 5G cellular networks [7]. Given the typical business access

https://doi.org/10.1016/B978-0-323-85499-3.00010-2

Table 3.1 Guaranteed access bandwidth requirement.

Subscriber type	Guaranteed access bandwidth range (2018–2025) (Gbps)
Small Business [3]	0.1–1.0
Medium/Large Business [4]	1.0–10.0
Cellular Backhaul [5]	1.0–5.0

Table 3.2 Required PON capacity for typical business access deployments.

Typical subscriber combinations on the same PON			
Small business	Medium/large business	Cellular backhaul	Required PON capacity
24	4	–	~35 Gbps
20	2	1	~40 Gbps
–	6	1	~51 Gbps

deployment scenarios shown in Table 3.2, usually without oversubscription unlike residential deployment, that is, including some business and cellular tower subscribers, guaranteed bandwidth requirement can easily add up to above 30 Gbps, which goes beyond 10G-EPON capacity. On the other hand, residential demand can also break the 10G-EPON capacity, especially in dense multidwelling units (MDUs) deployment with split ratio $\geq$1:128 [8]. However, due to the inevitably high system cost of the NG-EPON system at the time of early market penetration, it is commonly expected that such systems will be first adopted to serve business customers, before they are eventually deployed in residential markets as the cost being driven down by the increasing volume.

In parallel with the IEEE NG-EPON ad hoc discussion, FSAN and ITU-T SG15/Q2 were close to finalize the standardization work on NG-PON2 as the successor for their NG-PON standards, a.k.a., XG-PON1 and XG-PON2. Since the beginning of the NG-EPON discussion, both system and component vendors in IEEE 802.3 working group were fully aware of the technical difficulties in the NG-PON2 system, for example, transmitter laser wavelength drifting in burst-mode operation [9], tunable optical receiver integration [9,10], and the large power budget requirement [11], etc. In the meantime, bandwidth demands of intradata center and interdata center networks have been growing exponentially, which led to fast market adoption of multiple variations of 100GE-based transceiver standards for these applications, mostly in different flavors of 4×25G implementations, for example, 100GE CWDM-4 [12], 100GE LAN-WDM-4 [13], 100GE PSM-4 [14], etc. It is natural to leverage the high-volume components, for example, laser diode, laser/modulator driver, photodetector, etc., used by data center network at 25 Gbps or above per channel/wavelength to make higher serial rate above 10 Gbps a feasible option at the time of developing the NG-EPON standard. The NG-EPON call for interest (CFI) was completed in July 2015 [6], which moved the ad hoc discussion group into an official IEEE 802.3 NG-EPON study group. Project Authorization Request (PAR), Criteria for Standards Development (CSD), and project objectives were developed in subsequent meetings and approved by the 802.3 working group to officially form the IEEE P802.3ca 100G-EPON task force in November 2015. The initial target of the working group was to standardize 25/50/100G EPON as reflected in the project name of IEEE P802.3ca.

The planned timeline for standardization in IEEE of 25/50/100G EPON is depicted in Fig. 3.1 [15]. The standard was completed in June 2020, as initially planned. Originally, 802.3ca was intended to be a single standard for multiple new generations of PONs with the first new generation at 25 Gbps, the next new generation at 50 Gbps, and a third new generation at 100 Gbps [16]. All three new generations should be able to coexist over the same outside plant (OSP); and network equipment, such as ONTs, and should be backward compatible. Similar to previous PON system deployments, the initial deployment of symmetrical 25 Gbps system was envisioned for business applications, for example, cellular front- and back-haul and business subscribers. On the other hand, asymmetrical 25/10 Gbps system was considered for residential subscribers, which has been driven by headline speed competition among ISPs.

One of the initially agreed upon key project objectives of the IEEE P802.3ca task force [17] was to provide specifications for physical layers operation over a single SMF (standard single-mode fiber) strand and supporting symmetric and/or asymmetric MAC data rates of: (1) 25 Gbps in downstream and $\leq$25 Gbps in upstream, (2) 50 Gbps in downstream and $\leq$50 Gbps in upstream, and (3) 100 Gbps in downstream and $\leq$100 Gbps in upstream. However, significant challenges especially with regard to wavelength planning were observed during subsequent task force technical discussions, which were mainly due to the narrow low-dispersion window and 10G-EPON/XGS-PON coexistence requirements. Additionally, there were also time-to-market pressure for 25 Gbps PON from major US

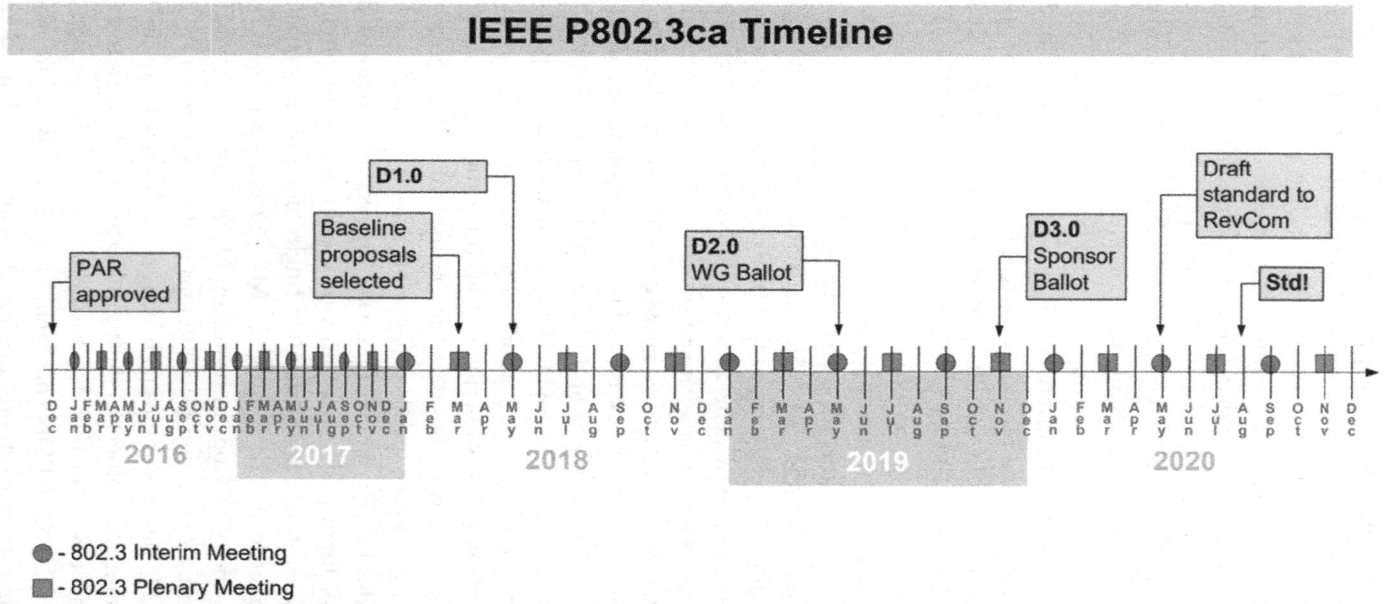

Fig. 3.1 25/50G-EPON standardization timeline [15].

operators. Consequently, "100 Gbps in downstream and $\leq$ 100 Gbps in upstream" was removed from the 802.3ca task force objectives in November 2017. Another significant objective change along the discussion was adding G-PON as one option for coexistence instead of only supporting 10G-EPON as described in the initial object [17]. This was mainly due to overwhelming popularity and very large existing footprint of G-PON systems, and some ISPs were considering skipping the generation of 10G-PON, for example, 10G-EPON, XGS-PON, etc., and transition directly into NG-EPON (25 Gbps) from existing IEEE EPON or ITU-T G-PON. The IEEE P802.3ca task force eventually settled with the following key objectives [18]:

- Support subscriber access networks using point-to-multipoint topologies on optical fiber.
- Provide physical layer specifications that:
  - Operate over a single SMF strand.
  - Support symmetric and/or asymmetric MAC data rates of:
    - 25 Gbps in downstream and 10 Gbps or 25 Gbps in upstream (25G-EPON).
    - 50 Gbps in downstream and 10 Gbps, 25 Gbps, or 50 Gbps in upstream (50G-EPON).
- Have a BER better than or equal to 10^{-12} at the MAC/PLS service interface (or the frame loss ratio equivalent).
- Support coexistence with select legacy PON technologies:
  - Optical power budgets to accommodate channel insertion losses equivalent to PR20 and PR30, as defined in Clause 75.
  - Wavelength allocation allowing concurrent operation with 10G-EPON, XG-PON1, and XGS-PON PHYs (1575–1580 nm downstream, 1260–1280 nm upstream).
  - Wavelength allocation allowing concurrent operation of 25G-EPON and G-PON reduced wavelength set (1480–1500 nm downstream, 1290–1330 nm upstream) PHYs.

We will review the following important aspects for IEEE P802.3ca, including system architecture, physical layer design, and channel bonding in the next few sections.

3.2.1 System architecture

The system architecture of IEEE 802.3ca is shown in Fig. 3.2. It adopts the 25 Gbps per channel/wavelength basis design and can include up to 2

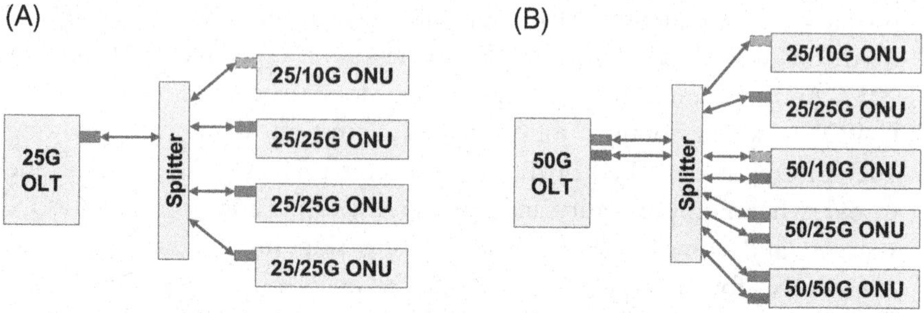

Fig. 3.2 25/50G-EPON system architecture, (A) 25G-EPON employing a single wavelength pair shown in *blue* color solid line (25G/25G) and *green* color dash line (25G/10G) and (B) 50G-EPON employing two wavelength pairs shown in *blue* color solid line and *pink* color dotted line.

channels to provide both symmetrical and asymmetrical operations for the following downstream/upstream data rate working modes, that is, 25/10 Gbps, 25/25 Gbps, 50/10 Gbps, 50/25 Gbps, and 50/50 Gbps. The capability of bonding multiple channels to provide higher capacity is one of the major improvements of IEEE 802.3ca comparing to the previous generation EPON standards. These working modes are identified by the Ethernet PMA/PMD (physical medium attachment/dependent) type selection register bits and extended ability register bits. There are in total 40 types of medium attachment units (MAUs) defined in the IEEE 802.3ca, including 5 aforementioned working modes, 2 coexistence options (G-PON or 10G-PON), 2 power budget class (medium and high), and 2 elements (OLT or ONU). For example, 25/10GBASE-PQX-U2 corresponds to 1 × 25.78125 GBd continuous mode reception (to ONU), 1 × 10.3125 GBd burst-mode transmission (from ONU), and medium-power class, while 50/25GBASE-PQG-D3 refers to 2 × 25.78125 GBd continuous mode transmission (from OLT), 1 × 25.78125 GBd burst-mode reception (to OLT), and high power class [19].

The IEEE 802.3ca standard specifies physical and data link layer of 25/50G-EPON following the IEEE Ethernet model, which are the lowest two layers in the open system interconnection (OSI) reference model as illustrated in Fig. 3.3 [19]. There are 5 major sublayers in this model, that is, from low to high, physical medium dependent (PMD), physical media attachment (PMA), physical coding sublayer (PCS), multichannel reconciliation sublayer (MCRS), and medium access control (MAC) layer. The combination of the first three sublayers, PMD, PMA, and PCS, is typically regarded as physical layer device (PHY). In the physical layer, the PMD

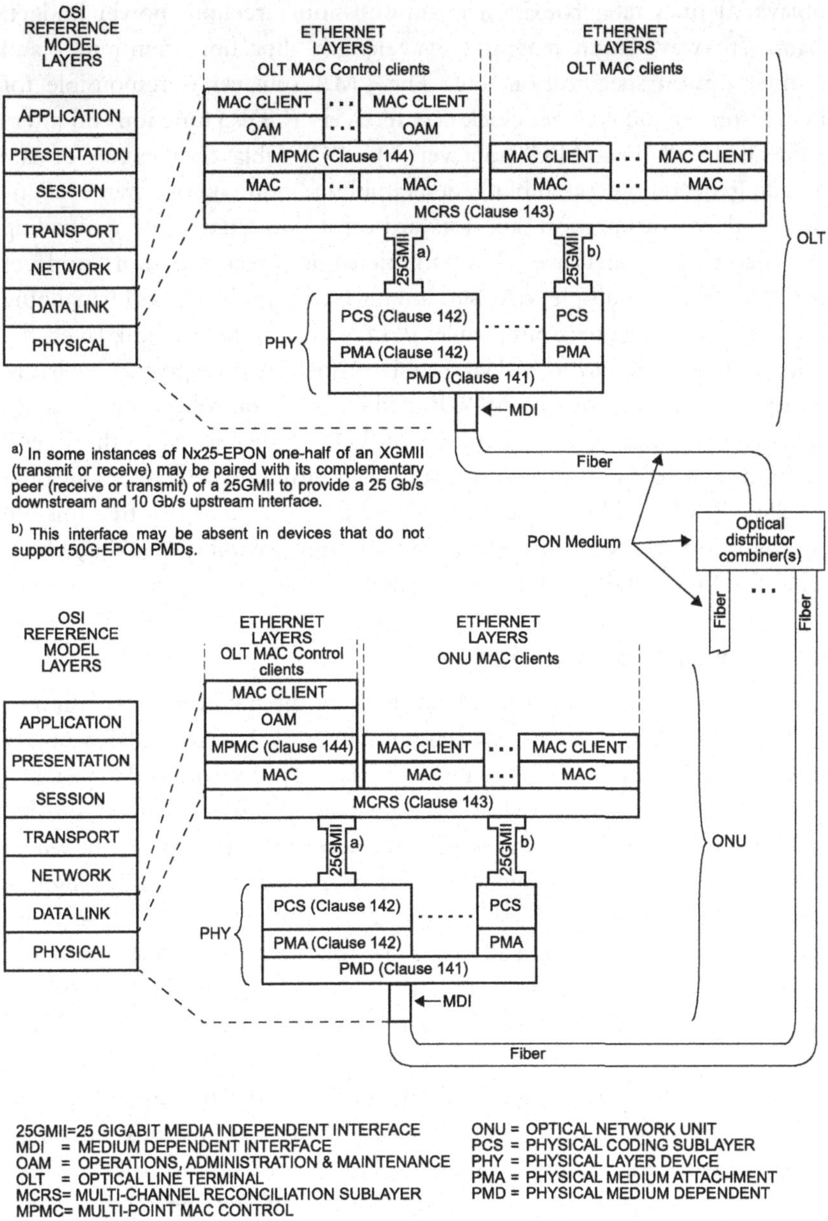

OSI REFERENCE MODEL LAYERS
APPLICATION
PRESENTATION
SESSION
TRANSPORT
NETWORK
DATA LINK
PHYSICAL
ETHERNET LAYERS
OLT MAC Control clients
ETHERNET LAYERS
OLT MAC clients
MAC CLIENT
OAM
MPMC (Clause 144)
MAC
MCRS (Clause 143)
25GMII
a)
b)
OLT
PHY
PCS (Clause 142)
PMA (Clause 142)
PMD (Clause 141)
PCS
PMA
MDI
Fiber
a) In some instances of Nx25-EPON one-half of an XGMII (transmit or receive) may be paired with its complementary peer (receive or transmit) of a 25GMII to provide a 25 Gb/s downstream and 10 Gb/s upstream interface.
b) This interface may be absent in devices that do not support 50G-EPON PMDs.
PON Medium
Optical distributor combiner(s)
ETHERNET LAYERS
OLT MAC Control clients
ETHERNET LAYERS
ONU MAC clients
ONU
25GMII=25 GIGABIT MEDIA INDEPENDENT INTERFACE
MDI = MEDIUM DEPENDENT INTERFACE
OAM = OPERATIONS, ADMINISTRATION & MAINTENANCE
OLT = OPTICAL LINE TERMINAL
MCRS= MULTI-CHANNEL RECONCILIATION SUBLAYER
MPMC= MULTI-POINT MAC CONTROL
ONU = OPTICAL NETWORK UNIT
PCS = PHYSICAL CODING SUBLAYER
PHY = PHYSICAL LAYER DEVICE
PMA = PHYSICAL MEDIUM ATTACHMENT
PMD = PHYSICAL MEDIUM DEPENDENT

Fig. 3.3 Relationship between 25/50G-EPON sublayers and ISO/IEC OSI reference model [19]. *(Courtesy of IEEE.)*

sublayer defines rate, coexistence, transmission direction, power budget, channel-to-wavelength mapping, wavelength allocation, transmitter and receiver optical specification, etc. The PMA sublayer is responsible for the transmitter and receiver clock data recovery (CDR) function and interconnects the PMD and PCS sublayer. The PCS sublayer takes care of data encoding/decoding, scrambling/descrambling, transcoding, frame alignment and synchronization, and forward error correction (FEC) encoding and decoding functions. The multichannel reconciliation sublayer (MCRS) enables multiple MACs to interface with multi-PHYs by mapping the MAC serial data to/from parallel PHY paths. In the data link layer, the multipoint MAC control (MPMC) sublayer includes two protocols, that is, multipoint control protocol (MPCP) and channel control protocol (CCP), which are responsible for arbitration of TDM-based access to the P2MP medium and querying and control of multiple channels within the N×25G-EPON PHY, respectively. Specifically, it includes transmission time allocation, ONU discovery and registration, enabling/disabling channel at the ONU, DBA (Dynamic Bandwidth Allocation), etc.

3.2.2 Physical layer design

There are two important and difficult parts in the physical layer design or PMD specification. One is wavelength plan, and the other one is power budget specification. There are a few important aspects that need to be taken into consideration. First of all, significant part of the usable spectrum has been occupied by the previous generation PON standards as shown in Fig. 3.4. There, the coexistence plan became very important in assigning wavelength plan for 25/50G-EPON. For example, if 25/50G-EPON was expected to coexist with all major previous generation PON standards, including 1G-EPON/G-PON, 10G-EPON/XGS-PON, and NG-PON2, the remaining usable spectra were very limited, mostly between 1330 and 1480 nm, and part of the red C-band between 1545 and 1575 nm (excluding PtP WDM channels specified in NG-PON2). On the other hand, if 25/50G-EPON was only expected to coexist with one previous generation PON standard, for example, either 1G-EPON/G-PON or 10G-EPON/XGS-PON, it opened up significant portion of the spectra for reuse.

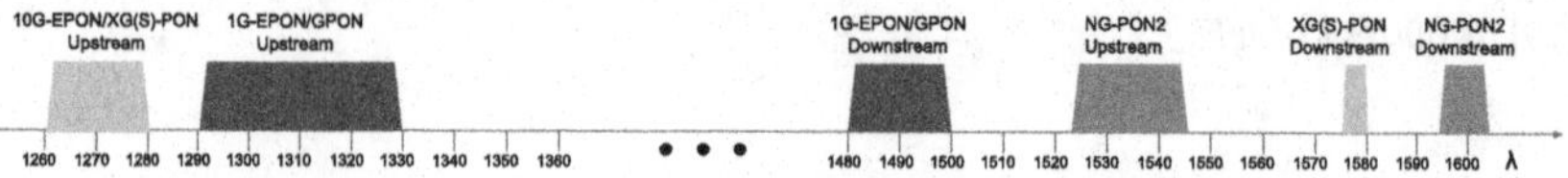

Fig. 3.4 Wavelength plan for previous generation PON standards.

Secondly, low-cost 25G directly modulated lasers (DMLs) were widely available in O-band (from 1270 to 1330 nm) due to the wide adoption in intradata center applications, for example, 100GE CWDM-4 [12]. Lastly, there were a few practical optical system and component limitations. From system level, dispersion penalty increases by a factor of 6 [20] as the data rate increases from 10 to 25 Gbps, which bonds the wavelength plan to low dispersion O-band rather than high dispersion C-band unless significant dispersion compensation solution is considered. From component level, low-cost bidirectional optical subassembly (BOSA) design prefers wide separation between downstream and upstream wavelength, for example, 45 nm or beyond [21]. In addition, low-cost DMLs without thermal electric cooler (TEC) for temperature control require a 20-nm wavelength window for operation. In a short summary, the selection of wavelength plan for 25/50G-EPON became an optimization problem to balance between performance and cost, with some constrains from system and component perspective. As a matter of fact, the significant difficulty in assigning 4 × 25 Gbps channels to support 100 Gbps operation was one of the major reasons to drop the objective of 100 Gbps operation from IEEE 802.3ca [18].

Taking all aforementioned aspects into considerations, the IEEE P802.3ca task force agreed with the following recommendations for wavelength plan. First of all, 25/50G-EPON should coexistence with either 1G-EPON/G-PON or 10G-EPON/XGS-PON, but not both, as there are significant challenges in supporting all generations of PON standards. Moreover, 1G-EPON/G-PON is expected to only work with 25G-EPON but not the 50G version. The assumption is that by the time ISPs start to migrate toward 50G-EPON, all previous generation 1G-EPON/G-PON customers should have already migrated into 25G-EPON. Furthermore, as shown in

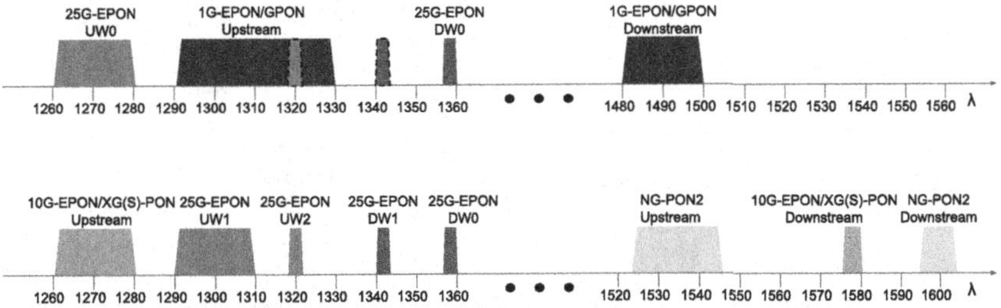

Fig. 3.5 Wavelength plan for 25/50G-EPON in coexistence with 1G-EPON/GPON *(top)* and 10G-EPON/XGS-PON *(bottom).*

Fig. 3.5, the initial 25/50G-EPON channel will be either DW0 for downstream and UW0 for upstream (1G-EPON/G-PON coexistence) or DW0 for downstream and UW1 for upstream (coexistence with 10G-EPON/ XGS-PON), where both 10 and 25 Gbps upstream are supported. Lastly, as traffic grows, the second 25/50G-EPON channel will be added with DW1 for downstream and UW2 for upstream for both coexistence cases, which increases the total system capacity to 50 Gbps with channel bonding as illustrated in Fig. 3.5.

Although IEEE standards do not mandate specific optical component technologies, the standard consensus buildup process inevitably takes feasibility and cost of optical component technologies into considerations. Specifically, for 25/50G-EPON, it has been assumed that cooled external modulated laser (EML) with wavelength tolerance of 4 nm within commercial temperature range between 0 and 70°C (also known as C-temp) will be adopted as the OLT transmitter, noted as DW0 (1358 ± 2 nm) and DW1 (1342 ± 2 nm) in Fig. 3.6. This assumption was mainly based on two reasons. From system performance perspective, cooled EML has better extinction ratio (ER) and smaller chirp, hence smaller transmitter and dispersion penalty (TDP). Therefore, DW0 and DW1 are deliberately chosen to be close to E-band (1360 to 1460 nm). Although E-band has higher dispersion than O-band, TDP of the OLT transmitter is expected to be on par if not better than the ONU transmitter. From cost perspective, the higher cost of cooled EML is offset by the fact that the OLT will be shared by a large number of ONUs. On the other hand, uncooled DML with wavelength tolerance of 20 nm within C-temp range will be adopted as the ONU transmitter, noted as UW0 (1270 ± 10 nm) and UW1 (1300 ± 2 nm) in Fig. 3.6. Therefore, UW0 and UW1 were chosen to be at O-band (1260 to 1360 nm), especially around typical zero dispersion wavelength, i.e., 1310 nm for standard

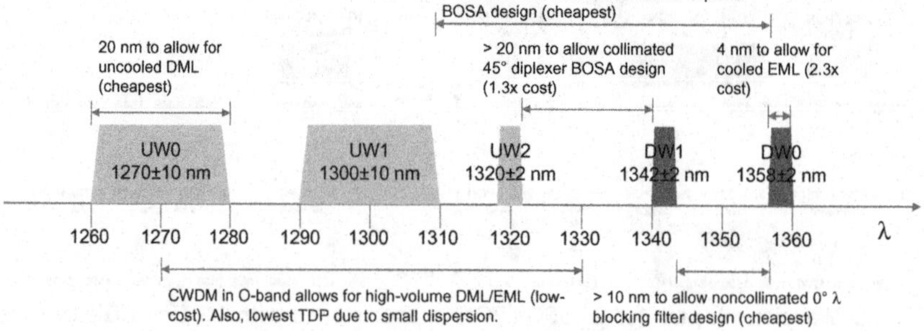

Fig. 3.6 Major considerations for 25/50G-EPON wavelength plan.

single-mode fiber. Negligible or even negative dispersion interacts with low ER and higher chirp and results in an acceptable TDP for uncooled DML in the upstream direction. It is worth noting that the second 25 Gbps channel noted as UW2 (1320 $\pm$ 2 nm) is expected to be implemented with cooled EML or DML to limit its wavelength tolerance within 4 nm, and this is mainly due to the limited available spectrum. In addition, 50/50 Gbps mode is likely to be adopted by business rather than residential applications, which are more tolerant to higher cost. The cost of cooled EML is >2 $\times$ of uncooled DML [22], and cooled DML is expected to be >1.5 $\times$ of uncooled DML due to complexity in TEC packaging in BOSA [23]. In addition to laser wavelength tolerance, there are two additional factors that were taken into consideration for the wavelength plan. The first one is that the separation between DW and UW should be >40 nm apart to allow noncollimated 45° diplexer-based BOSA design, which is the cheapest option [24]. If this requirement cannot be met, the next option is to allow >20 nm of separate and it supports collimated 45° diplexer-based BOSA design, which is about 1.3 $\times$ the cost of the former case [25]. As illustrated in Fig. 3.6, the DW and UW separation between the first 25 Gbps channel is beyond 40 nm to adopt the lowest cost design, and the second 25 Gbps channel maintains beyond 20 nm separation to utilize the slightly higher cost collimator-based BOSA design. The second aspect that needs to be considered for wavelength plan is the blocking filter design at the ONU receiver side, and > 10 nm separation allows for a noncollimated 0° blocking filter design, which has the lowest cost [24]. This is also confirmed from Fig. 3.6 that DW0 and DW1 are beyond 10 nm (about 12 nm) apart.

For the power budget specification, the IEEE P802.3ca task force agreed to specify two power classes: a medium class with channel insertion loss between 10 and 24 dB (noted as D2 or U2), and a high class with channel insertion loss between 15 and 29 dB (noted as D3 or U3), which is in compliance with the previous generation 10G-EPON [26]. This made sure that 25/50G-EPON is compatible with the previously constructed optical distribution network (ODN) and enables the reuse of 10 Gbps upstream PMD specification from 10G-EPON in the 25/50G-EPON specification in terms of transmitter launch power, receiver sensitivity, TDP, etc. Here, we will use the first 25 Gbps channel, high power class (D3 and U3), and 10G-EPON/XGS-PON coexistence scenario as an example to illustrate the power budget specification. Characteristics of OLT transmitter, ONU receiver, ONU transmitter, and ONU receiver for 25GBASE-PQX-D3 and 25GBASE-PQX-U3 are summarized in Table 3.3, Table 3.4, Table 3.5,

Table 3.3 OLT transmitter characteristics, 25GBASE-PQX-D3 [19].

Parameter	25GBASE-PQX-D3	Unit
Signaling rate (range)	25.78125 ± 100 ppm	GBd
Channel wavelength ranges	1356 to 1360	nm
Side-mode suppression ratio (SMSR) (min)	30	dB
Average launch power, each channel (max)	7.8	dBm
Optical modulation amplitude (OMA), each channel (min)	4.9	dBm
Difference in launch power between any two channels (OMA) (max)	3	dB
Launch power in OMA minus TDP, each channel (min)		
for extinction ratio ≥9 dB	4.8	dBm
for extinction ratio < 9 dB	4.9	dBm
Transmitter and dispersion penalty (TDP), each channel (max)	1.5	dB
Extinction ratio (min)	8	dB

and Table 3.6, respectively. The standard adopts the method of optical modulation amplitude (OMA) minus TDP as the transmitter launch power. This provides flexibility in specifying different optical component technologies. First, it offers the opportunity to trade off between OMA and TDP based on transmitter technologies. For example, DML with higher TDP can still meet specification by increasing OMA. For the downstream, reference TDP of 0 dB is chosen based on a best-case EML transmitter for the OLT, which corresponds to the minimum OMA requirement at 4.9 dBm as shown in Fig. 3.7 (left). This is a normative value, which has to be met even with

Table 3.4 ONU receiver characteristics, 25GBASE-PQX-U3 [19].

Parameter	25GBASE-PQX-U3	Unit
Signaling rate (range)	25.78125 ± 100 ppm	GBd
Channel wavelength ranges	1356 to 1360	nm
Bit error ratio (max)	10^{-2}	
Damage threshold	−6.2	dBm
Average receive power, each channel (max)	−7.2	dBm
Receiver sensitivity (OMA), each channel (max)	−24.1	dBm
Detect threshold, each channel (min)	−40	dBm
Stressed receiver sensitivity (OMA), each channel (max)	−22.6	dBm

Table 3.5 ONU transmitter characteristic, 25GBASE-PQX-U3 [19].

Parameter	25GBASE-PQX-U3	Unit
Signaling rate (range)	25.78125 ± 100 ppm	GBd
Channel wavelength ranges	1290 to 1310	nm
Side-mode suppression ratio (SMSR) (min)	30	dB
Average launch power, each channel (max)	9	dBm
Optical modulation amplitude (OMA), each channel (min)	4.7	dBm
Difference in launch power between any two channels (OMA) (max)	3	dB
Launch power in OMA minus TDP, each channel (min)		
for extinction ratio ≥6 dB	4	dBm
for extinction ratio < 6 dB	4.2	dBm
Transmitter and dispersion penalty (TDP), each channel (max)	2	dB
Extinction ratio (min)	5	dB
Turn-on time (max)	128	ns
Turn-off time (max)	128	ns

Table 3.6 OLT receiver characteristic, 25GBASE-PQX-D3 [19].

Parameter	25GBASE-PQX-D3	Unit
Signaling rate (range)	25.78125 ± 100 ppm	GBd
Channel wavelength ranges	1290 to 1310	nm
Bit error ratio (max)	10^{-2}	
Damage threshold	−5	dBm
Average receive power, each channel (max)	−6	dBm
Receiver sensitivity (OMA), each channel (max)	−24.3	dBm
Signal detect threshold, each channel (min)	−40	dBm
Stressed receiver sensitivity (OMA), each channel (max)	−22.8	dBm
Receiver settling time (max)	800	ns

negative TDP to ensure transmitter compliance. The corresponding ONU receiver OMA sensitivity at TDP of 0 dB is regarded as the unstressed receiver OMA sensitivity (URS), and it is an informative value. The other ONU receiver OMA sensitivity at TDP of 1.5 dB (maximum TDP allowed) is considered as the stressed receiver OMA sensitivity (SRS), and it is a normative value. Similarly, for the upstream, reference TDP of 0.5 dB is chosen based on

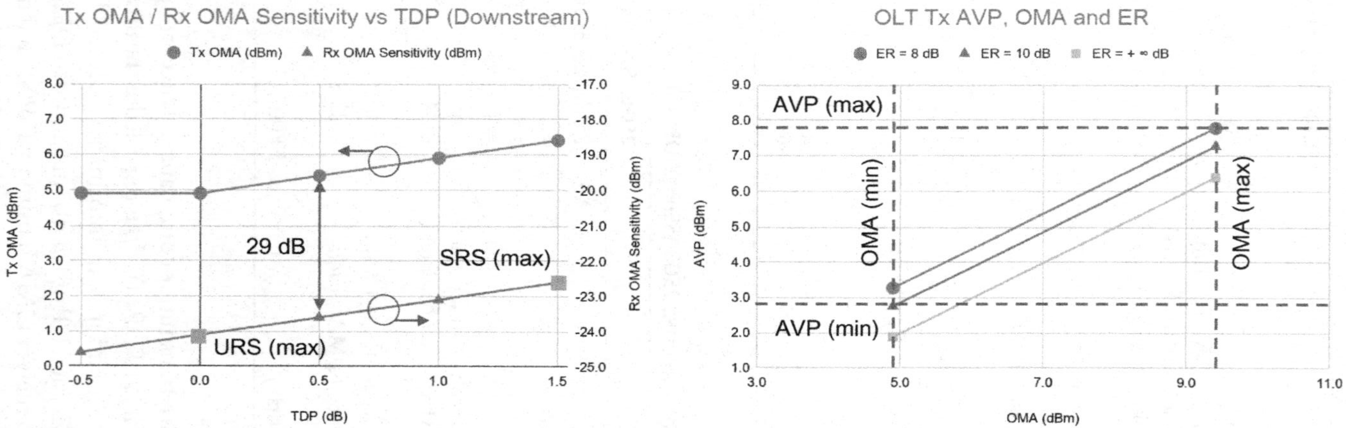

Fig. 3.7 25GBASE-PQX-D3/U3 downstream Tx OMA, Rx OMA sensitivity w.r.t. TDP *(left)*. 25GBASE-PQX-D3/U3 downstream OLT transmitter AVP vs. OMA *(right)*.

a best-case DML transmitter for the ONU, which corresponds to the minimum OMA requirement at 4.7 dBm as shown in Fig. 3.8 (left). This is a normative value, which has to be met even with TDP < 0.5 dB to ensure transmitter compliance. The corresponding OLT receiver OMA sensitivity at TDP of 0.5 dB is regarded as the URS, which is an informative value. The other OLT receiver OMA sensitivity at TDP of 2.0 dB (maximum TDP allowed) is considered as the SRS, which is a normative value. The other benefit of OMA minus TDP methodology is to trade off between average launch power (AVP) and ER. For example, as shown in Table 3.3, an OLT transmitter with ER ≥9 dB needs 0.1 dB lower OMA than an OLT transmitter with ER < 9 dB to maintain the same ONU receiver sensitivity. As shown in Fig. 3.7 (right), AVP (min) is an informative value based on OMA (min) and ER of 10 dB. AVP (max) is a normative value, and it is calculated by converting OMA (max), which includes maximum TDP and difference in launch power between any two channels (OMA). AVP (max) defines the average receiver power (max) based on minimum specified channel loss.

Both transmitter launch power and receiver sensitivity are quite challenging to meet. Therefore, quasicyclic low-density parity check (QC-LDPC) code (16,952, 14,392) is adopted to support a pre-FEC BER threshold of 10^{-2} (where receiver sensitivity is measured) and maintain post-FEC BER of $<10^{-12}$. The selected QC-LDPC code has an overhead of 15.2% and a net effective code gain (NECG) of >9.6 dB, which is about 2.5 dB better than the Reed-Solomon (255, 223) code adopted in 10G-EPON and XGS-PON. This also translates into about 1.5 to 2.0 dB optical gain in receiver sensitivity or the overall power budget [27]. In addition, applying an SOA as a booster or preamplifier may also be considered to meet transmitter and receiver performance. However, its cost may be too high, especially for residential applications.

3.2.3 Channel bonding

Another innovative approach adopted in 25/50G EPON is dynamic channel bonding, which allows two 25G channels to be bonded into one 50G channel at the unit of Envelope Quantum (EQ), which includes 64-bit data and 8-bit control symbols, in a total 72-bit block. As shown in Fig. 3.9, the MCRS encapsulates data transmitted by a MAC instance in transmission envelop, which is similar to transmission frame design in ITU-PONs such as G-PON and XGS-PON, etc. In systems with multiple channels, envelops may overlap and a frame can be striped over multiple channels with each

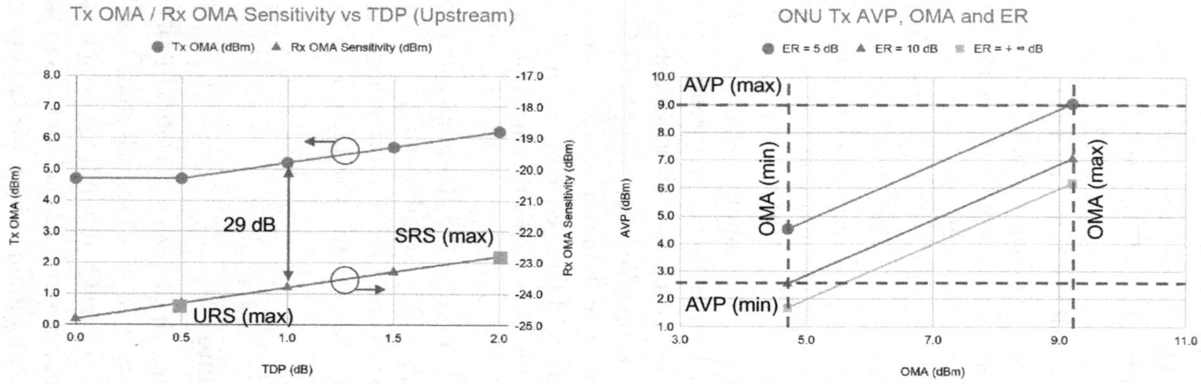

Fig. 3.8 25GBASE-PQX-D3/U3 upstream Tx OMA, Rx OMA sensitivity w.r.t. TDP *(left)*. 25GBASE-PQX-D3/U3 upstream ONU transmitter AVP vs. OMA *(right)*.

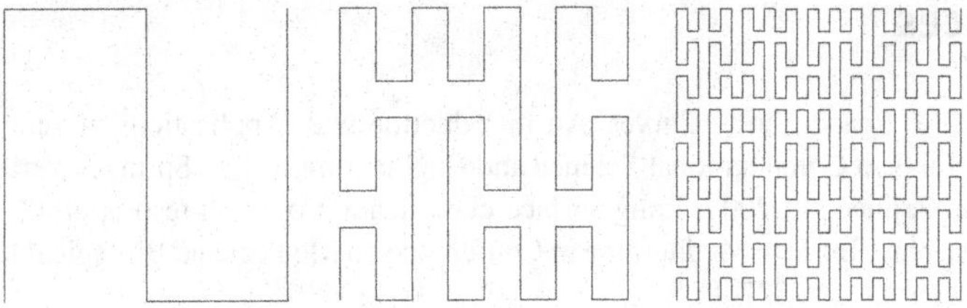

Figure 1.6 First iterations of the Peano curve (source: Wikimedia Commons).

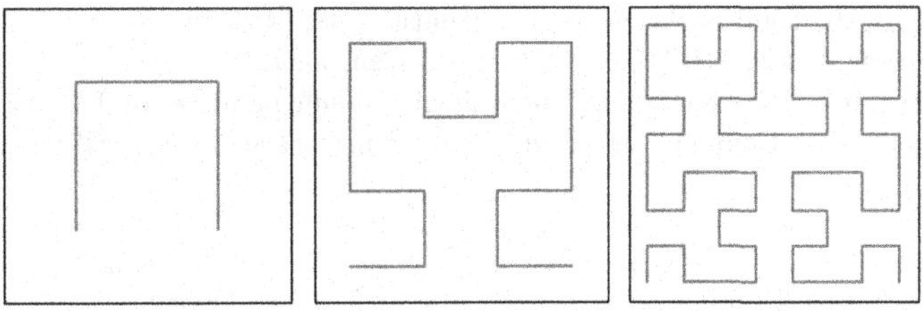

Figure 1.7 First iterations of the Hilbert curve (source: Wikimedia Commons).

Similar to the Koch snowflake, there are other curves constructed as a series of iterations that, while contained in a limited space, possess a length that tends to infinity. Some examples of this kind of space filling curves are the Peano curve (displayed in Fig. 1.6, created by António Miguel de Campos) or the Hilbert curve (Fig. 1.7, adapted from a work created by Zbigniew Fiedorowicz). Both images are distributed with Creative Commons Attribution-Share Alike 3.0 Unported licence, available at https://creativecommons.org/licenses/by-sa/3.0/deed.en. When used, for instance, in nanotechnology and the investigation of new materials, these structures could be employed for storing large quantities of information in very tiny amounts of space. For more information on this type of curves, see for example Bader (2013).

1.4 Conclusions

Limits of functions are a fundamental part of calculus, and as such they are used for example in the definition of the derivative of a function. However, limits have their own value as a mathematical tool, sometimes helping to provide results that, at a first glimpse, would seem to be counterintuitive.

In this chapter two examples of using limits in that kind of situations have been brought to the reader's attention: how to compute the continuous compound interest (a method that does not generate as much revenue as expected) and determining the perimeter and area of the Koch snowflake in its nth step as an approximation of those values for actual snowflakes (where the perimeter is infinite but the area is not).

References

Bader, M., 2013. Space-Filling Curves. An Introduction with Applications in Scientific Computing. Texts in Computational Science and Engineering, vol. 9. Springer-Verlag, Berlin.

Busayarat, S., Zrimec, T., 2007. Lung surface classification on high-resolution CT using machine learning. In: 11th Mediterranean Conference on Medical and Biological Engineering and Computing, pp. 822–825.

Cauchy, A.L., 1899. Cours d'analyse. Oeuvres complètes d'Augustin Cauchy 3 (19).

Churchill, G.E., 1969. Compound Interest Simplified. Pergamon Press Ltd.

von Koch, H., 1906. Une méthode géométrique élémentaire pour l'étude de certaines questions de la théorie des courbes planes. Acta Mathematica 30, 145–174.

Larson, R., Edwards, B.H., 2017. Calculus, 11th ed. Cengage Learning.

Sergeyev, Y.D., 2016. The exact (up to infinitesimals) infinite perimeter of the Koch snowflake and its finite area. Communications in Nonlinear Science and Numerical Simulation 1–3, 21–29.

Derivative: tool for approximation and investigation

2

Petr Habala
Czech Technical University in Prague, Prague, Czech Republic

2.1 Derivative: overview of theory

There are two basic motivations for the notion of derivative, physics and geometry. We start with the former. Imagine a car driving in a certain lane on a highway. This setting has the advantage that its position at time t can be given by one number $f(t)$, namely, how far along the highway it is (highways have convenient mileposts). We want to know its instantaneous velocity at time $t = a$.

One possible approach is to check what happens during a time interval $[a, x]$ for some $x > a$. Assuming that the car moves in the same direction, it covers the distance $f(x) - f(a)$, and the time it took was $x - a$. Thus its average speed over this interval was $\frac{f(x) - f(a)}{x - a}$. If we start taking x closer and closer to a, we expect that the corresponding average speeds will approach the instantaneous speed at time $t = a$. With minor modification, we can also apply this reasoning to time intervals $[x, a]$ for $x < a$ and eventually obtain the same approximating ratio.

The same approach can be applied to other quantities, and the independent variable can stand for other parameters than time. For instance, $T(h)$ can describe temperature as it depends on elevation h; then the above procedure tells us how fast the temperature changes as we change elevation.

In mathematical language, this rate of change is called the derivative. However, the procedure for determining it described above does not work for all functions, but only for those with a special quality that we call differentiability. We now state a formal definition, where we also show a version that focuses on the variable shift $h = x - a$, so depending on the sign of h we work with intervals $[a, a + h]$ or $[a + h, a]$.

Definition. Let f be a function defined on some neighborhood of a point $a \in \mathbb{R}$. We say that it is **differentiable** at a if the limit $\lim\limits_{x \to a} \left(\frac{f(x) - f(a)}{x - a} \right)$ converges.

If it is so, then we define the **derivative** of f at a as

$$f'(a) = \lim_{x \to a} \left(\frac{f(x) - f(a)}{x - a} \right) = \lim_{h \to 0} \left(\frac{f(a + h) - f(a)}{h} \right).$$

The geometric interpretation is that $f'(a)$ represents the slope of the tangent line to the graph of f at point a. We approximate it using the slopes of secant lines going through points $(a, f(a))$ and $(x, f(x))$. Common sense suggests that if $x = a + h$ is really close to a, then the secant line s should approximate the tangent line t well.

Calculus for Engineering Students. https://doi.org/10.1016/B978-0-12-817210-0.00009-6

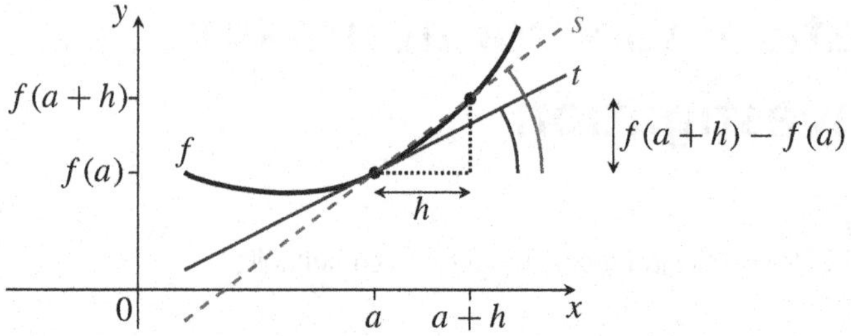

In order for this idea to work, the graph of f must have a nice-enough shape around a. A well-known example of this going wrong is the absolute value function $f(x) = |x|$ that is not differentiable at $a = 0$. Indeed, we cannot assign a reasonable tangent line there as the graph breaks sharply.

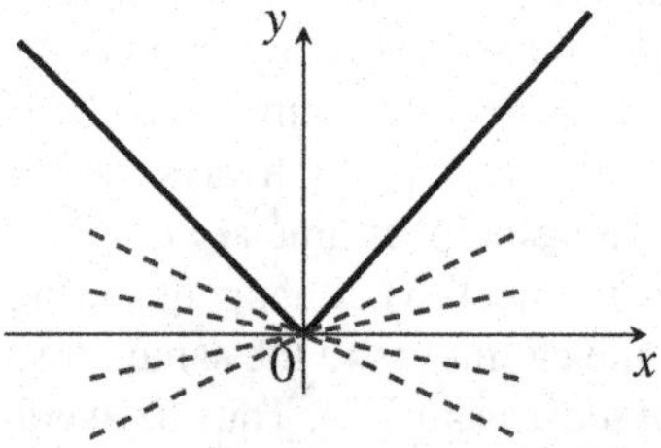

Just like with continuity, the true strength of differentiability is revealed when we consider it on intervals. We say that a function f is differentiable on some open set G if it is defined on G and differentiable at all points of G.

What kind of quality is this differentiability? It is well known that differentiable functions are automatically continuous; however, as the above example with absolute value shows, being continuous is not enough to guarantee differentiability. Intuitively, continuous functions (on intervals) have graphs that are not interrupted, but they can bend sharply. Differentiable functions are not permitted to bend sharply; their graphs must change shape gradually.

How do we determine derivatives? At school students typically encounter functions given by formulas, and for those there is a standard procedure. It relies on two basic pillars: a list of known derivatives (vocabulary) and five basic rules that allow us to differentiate more interesting formulas (grammar):

$$\begin{aligned} [cf]' &= cf', \\ [f \pm g]' &= f' \pm g', \\ [f \cdot g]' &= f'g + fg', \\ \left[\frac{f}{g}\right]' &= \frac{f'g - fg'}{g^2} \quad \text{on sets where } g \neq 0, \\ [g(f)]' &= g'(f) \cdot f'. \end{aligned}$$

Students usually quickly learn how to apply these formulas, so differentiation is not a problem to worry about. Since this is routine, we leave this topic to standard calculus texts.

Derivative is a very powerful tool. Most results that people use are based on the following key statement.

Theorem 2.1 (Mean value theorem). *Let f be a function that is continuous on some interval $[a, b]$ of real numbers. Assume also that f is differentiable on its interior (a, b). Then there must be a number $c \in (a, b)$ such that*

$$f'(c) = \frac{f(b) - f(a)}{b - a}.$$

Geometrically, if we connect two points on a graph of a suitably smooth function, then there must be a tangent line to this graph that is parallel to this secant line.

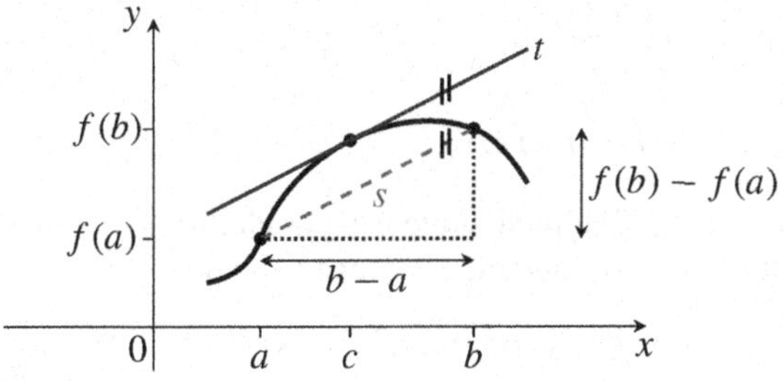

This statement has many useful consequences. For instance, it offers convenient tools for investigating shapes of functions. In particular, the following tests are known to all students of calculus.

Theorem 2.2. *Let f be a function that is continuous on some interval $[a, b]$ of real numbers and differentiable on its interior (a, b). Then the following are true:*

(i) If $f' > 0$ on (a, b), then f is strictly increasing on $[a, b]$.
(ii) If $f' < 0$ on (a, b), then f is strictly decreasing on $[a, b]$.

Let f be a function that is continuous on some interval $[a, b]$ of real numbers and twice differentiable on its interior (a, b). Then the following are true:

(i) If $f'' > 0$ on (a, b), then f is concave up on $[a, b]$.
(ii) If $f'' < 0$ on (a, b), then f is concave down on $[a, b]$.

This theorem forms the basis for a popular algorithm that can determine monotonicity, local extrema, and concavity of a given function. We also find these results in the basic toolbox for optimization, which is a very useful field of applied mathematics.

However, much more can be derived from the mean value theorem. For instance, the following statement can be very useful for proving inequalities when ordinary algebra fails.

Corollary 2.3. *Let f, g be functions continuous on an interval $[a, b]$ of real numbers that are differentiable on its interior (a, b). Assume that the following two conditions are met:*

(i) $f(a) \leq g(a)$,
(ii) $f' \leq g'$ on (a, b).

Then $f \leq g$ on $[a, b]$.

Another popular application of derivatives is found when evaluating limits. We, of course, refer to the famous L'Hôpital rule.

Theorem 2.4 (L'Hôpital's theorem). *Let f, g be functions defined and differentiable on some reduced neighborhood of a point a, where $a \in \mathbb{R} \cup \{-\infty, \infty\}$. Assume that one of the following two conditions is true:*

(i) $\lim_{x \to a} (f(x)) = \lim_{x \to a} (g(x)) = 0,$

(ii) $\lim_{x \to a} (|g(x)|) = \infty.$

Then

$$\lim_{x \to a} \left(\frac{f(x)}{g(x)} \right) = \lim_{x \to a} \left(\frac{f'(x)}{g'(x)} \right),$$

assuming that the limit on the right exists.

Evaluating limits using L'Hôpital's rule is a popular sport in introductory calculus courses, so we will leave it to standard calculus textbooks.

Applied mathematics makes heavy use of another benefit of derivatives: the Taylor polynomial. It can be proved that if we want to approximate some function f by a polynomial of degree n around some point a, then the best bet is the **Taylor polynomial** centered at a, i.e.,

$$\begin{aligned} T_n(x) &= f(a) + f'(a)(x-a) + \frac{1}{2!} f''(a)(x-a)^2 + \cdots + \frac{1}{n!} f^{(n)}(a)(x-a)^n \\ &= \sum_{k=0}^{n} \frac{f^{(k)}(a)}{k!}(x-a)^k. \end{aligned}$$

Obviously, one can form such a Taylor polynomial only if the function f has all the necessary derivatives at a. In applications we often prefer another form of approximation that is obtained by denoting $x = a + h$, where now h is the variable. We have

$$f(a+h) \approx f(a) + f'(a)h + \frac{1}{2!} f''(a)h^2 + \cdots + \frac{1}{n!} f^{(n)}(a)h^n.$$

It can be obtained directly by deriving the Taylor polynomial for $f(a+h)$, where the center for the variable h is 0.

Replacing functions by their Taylor polynomials has long been a basic tool for engineers and scientists. Some popular formulas include

$$e^x : T_n(x) = 1 + x + \frac{1}{2!}x^2 + \frac{1}{3!}x^3 + \cdots + \frac{1}{n!}x^n = \sum_{k=0}^{n} \frac{x^k}{k!},$$

$$\sin(x) : T_{2n+1}(x) = x - \frac{1}{3!}x^3 + \frac{1}{5!}x^5 - \frac{1}{7!}x^7 + \cdots = \sum_{k=0}^{n} (-1)^k \frac{x^{2k+1}}{(2k+1)!},$$

$$\cos(x): T_{2n}(x) = 1 - \frac{1}{2!}x^2 + \frac{1}{4!}x^4 - \frac{1}{6!}x^6 + \cdots = \sum_{k=0}^{n} (-1)^k \frac{x^{2k}}{(2k)!},$$

$$\arctan(x): T_{2n+1}(x) = x - \frac{1}{3}x^3 + \frac{1}{5}x^5 - \frac{1}{7}x^7 + \cdots = \sum_{k=0}^{n} (-1)^k \frac{x^{2k+1}}{2k+1},$$

$$\ln(x+1): T_n(x) = x - \frac{1}{2}x^2 + \frac{1}{3}x^3 - \frac{1}{4}x^4 + \cdots = \sum_{k=1}^{n} (-1)^{k+1} \frac{x^k}{k}.$$

When we want to replace a function $f(x)$ by its Taylor polynomial $T_n(x)$ of a certain degree around a, we need to have some control over the error that we make. This error $f(x) - T_n(x)$ is called the remainder R_n in this context. There are several formulas available for it; the most practical is the **Lagrange form of the remainder**. In the statement we have to be careful to cover both the case $x > a$ and the case $x < a$.

Theorem 2.5. *Let $a, x \in \mathbb{R}$, and denote by I the closed interval with endpoints a and x. Assume that a function f is defined on I, having continuous derivatives up to order n on I, and $f^{(n+1)}$ exists on the interior of I. Let T_n be the Taylor polynomial of f of degree n with center a.*

Then there is some number c in I such that

$$f(x) - T_n(x) = \frac{f^{(n+1)}(c)}{(n+1)!}(x-a)^{n+1}.$$

In situations when we prefer the $a + h$ form we can write

$$f(a+h) - \sum_{k=0}^{n} \frac{f^{(k)}(a)}{k!} h^k = \frac{f^{(n+1)}(c)}{(n+1)!} h^{n+1}.$$

It seems that this theorem does not help much in practical problems given that we do not actually know what c is, but we can avoid this problem by replacing $f^{(n+1)}(c)$ with the worst possible case. Then we no longer get an equality, but an upper estimate for the error, namely,

$$|f(x) - T_n(x)| \le \frac{1}{(n+1)!} \max_{t \in I} |f^{(n+1)}(t)| \cdot |x-a|^{n+1}.$$

However, in applications this is exactly what we need, which makes this estimate very useful.

Proper treatment of this background can be found in any good text on introductory calculus (e.g., Stroud, 1970).

2.2 Derivative in applications

- The basic interpretation of derivative explains it as the rate of change of some quantity. In the most natural setting, when $y(t)$ specifies a position at time t, then

$y'(t)$ provides velocity and $y''(t)$ acceleration. The derivative thus serves an important role in modeling of natural phenomena, because it allows us to access rate of change symbolically and thus include it when we try to capture natural laws mathematically. We then obtain so-called differential equations, for instance, $y' = ry$ describes a phenomenon where the change in quantity y is directly proportional to its current value. Having described our phenomenon mathematically, we may hope to extract some information.
Differential equations played a key role in advances in natural sciences and engineering in the past several hundred years. They belong to the basic toolbox of every engineer and we have special chapters on them in this book.
It is interesting that interpretation of a derivative usually stops with the second order. One may be excused to think that derivatives of higher order do not have a meaningful real-life interpretation (I thought so as well for a long time). However, there are some uses for higher-order derivatives. For instance, recently I learned that people who design control algorithms for vehicles worry a lot about the third derivative. They call it "jerk" and indeed, the third derivative of position tells us exactly how much passengers are jerked while a vehicle moves. We do not feel speed (first derivative) and we readily adjust to reasonably small acceleration (second derivative), it is the change of acceleration (i.e., the third derivative) that affects us most.

- Approximation formulas play key roles in numerical mathematics. They allow us to perform computations, because very few functions can be evaluated precisely by computers (human or electronic). But more than that, approximating formulas are the basis for several key numerical methods, and they also allow us to analyze errors associated with numerical methods.
- Optimization theory deals with the problem of finding the optimal configuration for a given system. In a typical setup we are given a set of possible solutions to some problem and we are expected to identify the one that best satisfies some criterion. This useful branch of mathematics is also known under other names, for instance operation research. Many problems can be reduced to the problem of finding a global extreme of some function, which naturally calls for derivatives.
- Tools based on the mean value theorem are indispensable when we want to analyze behavior of functions. The reader is undoubtedly familiar with basic analysis of monotonicity and concavity, identification of local extrema, and application of the derivative in optimization (global extrema). Here we will look at some less traveled approaches.

2.3 Exploring derivative

2.3.1 Related rates

Example. Imagine sand being poured from a conveyor belt onto a heap, forming a cone. Under ideal conditions, the sides of this cone – regardless of its size – will always have the same slope, determined by granular properties of the sand, humidity, and other factors. Assuming that the sand is poured at a steady rate, how fast is the height of the heap growing?

We start by setting up some mathematics. The sand coming from the conveyor belt is increasing the volume of the heap, let us call it $V(t)$. The fact that the sand comes at a constant rate means that the rate of change in volume is constant, mathematically speaking, $\frac{\mathrm{d}V}{\mathrm{d}t} = c$ for some $c > 0$.

The cone has dimensions h for height and r for the radius of the base; both change in time. The question asks for $\frac{\mathrm{d}h}{\mathrm{d}t}$. To sum it up, we have three functions of time, $V(t)$, $h(t)$, and $r(t)$, we are given $V'(t) = c$ and we want to find $h'(t)$.

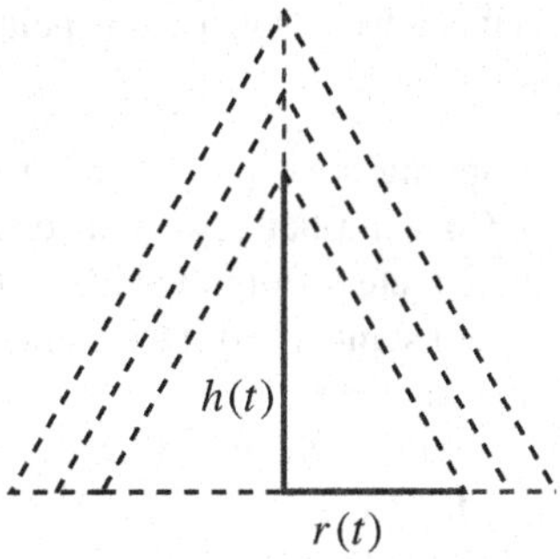

To find some connection between $V'(t)$ and $h'(t)$ we first relate $V(t)$ with $h(t)$, and for that we turn to geometry. We remember or deduce or find the formula for volume: $V = \frac{1}{3}\pi r^2 h$, in our case actually $V(t) = \frac{1}{3}\pi r(t)^2 h(t)$. Unfortunately, there is an extra unknown r in this picture, which we need to eliminate. This is where the last unused assumption from the question comes in. The slope of the side of the cone is supposed to be constant, which means that $\frac{r(t)}{h(t)} = k$ for some $k > 0$. From this we get $r(t) = h(t)k$ and hence $V(t) = \frac{1}{3}\pi k^2 h(t)^3$. This is the desired connection between V and h.

Now we differentiate both sides of the equality, not forgetting to use the chain rule. Then we obtain

$$V'(t) = \frac{1}{3}\pi k^2 3h(t)^2 h'(t) \implies h'(t) = \frac{V'(t)}{\pi k^2 h(t)^2}.$$

We obtained a general formula for the rate of growth of the height of the sand heap. Since the sand is pouring at a steady rate, we obtain

$$h'(t) = \frac{c}{\pi k^2 h(t)^2}.$$

Lumping all constants into one, we can write $h' = \frac{K}{h^2}$, so the heap is getting taller at a decreasing rate; the taller it is, the slower it grows. Qualitatively this fits in with our childhood experience when we were building sand castles, but now as grown ups we are able to deduce a quantitative result as well: the rate of growth is proportional to $\frac{1}{h^2}$.

This example shows typical features of "related rates" problems. When we translate such a problem into mathematical language, we end up with two functions $f(t)$, $g(t)$ that describe the situation; we know the rate of change (derivative) of one of these functions and we are asked to determine the derivative of the other. The general strategy is simple; first we find some connection between f and g in the form of an equation, and then we differentiate this equation symbolically to find the desired relation between their derivatives, that is, we relate the two rates.

Exercise 2.3.1.1. A balloon that is always spherical is being blown at a constant rate. Find the rate at which the radius grows.

Exercise 2.3.1.2. A radar is standing next to a straight road, measuring velocity of cars passing by. A certain car (for simplicity we will disregard its dimensions and consider it to be a point) travels 2 meters from the side of the road (which is also the nearest it gets to the radar) at a constant speed. Determine what speed the radar will show depending on the position of the car.

Note: The device measures the rate of change of the distance between the radar and the car, not the actual velocity of the car.

Hint: As the origin in your coordinate system take the point where the car is nearest to the radar (the segment between the car and the radar is then perpendicular to the road, as you can surely tell from the picture you have drawn). Point the main axis along the road, so the car travels along this axis.

Exercise 2.3.1.3. A lamppost is 6 meters high. I start walking away from this lamppost and experience tells us that the shadow that appears in front of me grows longer and longer.

Assuming that I walk at a steady pace, find the rate at which the shadow's length increases depending on how far I am from the lamppost. For the purposes of this problem we may assume that I am 3 meters tall.

For more problems of this type, see Math Tutor (2019).

2.3.2 Approximating derivative of a function

In engineering we often work with functions that are not given by a formula, but we are able to determine their values at specific points (for instance by running an experiment). How do we determine the derivative then? This topic actually belongs to numerical analysis (see, e.g., Burden and Faires, 2011).

Example. The definition offers a natural approach. If we take a really small h, we may approximate

$$f'(a) \approx \frac{f(a+h) - f(a)}{h}.$$

We expect that taking smaller h will result in a better approximation. Thus h serves as an indicator of quality, and it is traditional to take just $h > 0$. Then $a + h$ is to the right of a, so we look ahead in order to approximate the derivative. The resulting formula is a baseline method for a numerical estimation of derivative, called the **forward difference** (also forward divided difference).

How large is the error? In numerical mathematics, by the absolute error E we mean the difference $x - \hat{x}$ between some exact value x and its approximation $\hat{x}$. We then have $x = \hat{x} + E$. To judge the seriousness of the error we typically check its magnitude $|E|$. So how large is $|E|$ for our approximation?

To see this, we start with the Taylor expansion for $f(a+h)$:

$$f(a+h) = f(a) + f'(a)h + \frac{1}{2}f''(a)h^2 + \frac{1}{6}f'''(a)h^3 + \cdots.$$

We easily isolate $f'(a)$ from this to obtain

$$f'(a) = \frac{f(a+h) - f(a)}{h} + \Big(-\frac{1}{2}f''(a)h - \frac{1}{6}f'''(a)h^2 - \cdots\Big).$$

Comparing both sides we see that the expression in parentheses is exactly the error of approximation for the forward difference. When h is small, then higher powers of h become negligible compared to h and the error estimate becomes $E \approx -\frac{1}{2}f''(a)h$, that is, $|E| \approx \frac{1}{2}|f''(a)|h$. In other words, the error decreases more or less linearly as h approaches zero. This is nice (error goes to zero), but not really nice (linear speed is not all that great).

To see how this works in real life we choose $f(x) = e^x$ and $a = 0$. Then $f'(0) = 1$ and the forward difference is $\dfrac{e^h - e^0}{h}$. The error should be (for small h) approximately $-\frac{1}{2}h$. Note that since the exponential is concave up, by taking secants going forward we will always overestimate the slope just like in the first picture of this chapter, so the error should be negative, indeed.

So much for the theory. Now we look at some results. In the following chart we show several approximations of the derivative $f'(0) = 1$ and their errors. For comparison we also show the estimate of error as we derived it above.

h	**Approximation**	$\lvert E_h \rvert$	$\lvert -\frac{1}{2}h \rvert$
0.5	1.30...	0.30...	0.25
0.1	1.052...	0.052...	0.5
0.01	1.0050...	0.0050...	0.005
0.001	1.00050...	0.00050...	0.0005

This seems to confirm that taking smaller h leads to better approximations; it also seems that for smaller h the error estimate closely matches the actual error. It does not work so well for larger h, but that is to be expected, because then the higher powers of h in the error estimate are not small enough to be safely ignored.

Now we look at this globally. In the graph on the left the reader can see the actual error in red (gray line in print version) and the linear estimate in blue (dashed line in print version). As expected, the two curves do not agree when h is close to 1. However, as we get closer to zero, the higher powers in the error estimate become negligible compared to the linear power and the error seems to match the linear estimate.

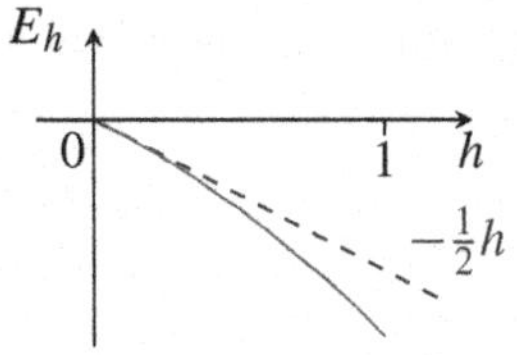

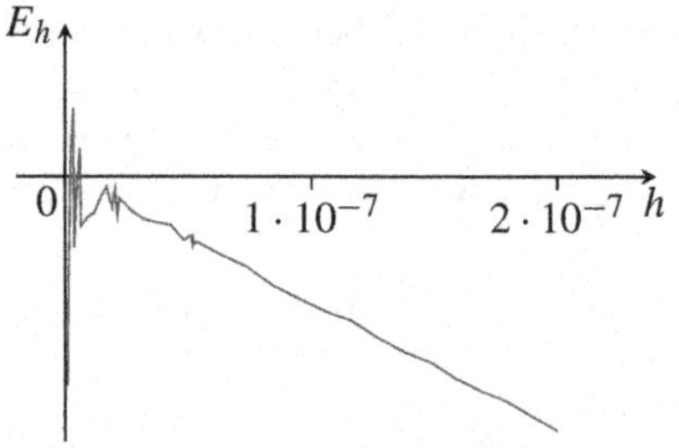

In the picture on the right we focused on really small values of h. Note that at first the error follows a straight line, but as we get to about $0.2 \cdot 10^{-7}$ it suddenly starts behaving erratically, growing in size and sometimes it even becomes positive. This is the result of numerical errors in actual calculations. Computers and calculators never work precisely; there are always small errors involved, and once we try to make our calculations too precise, these errors take over and obliterate our results. This is a major factor in all numerical calculations and engineers have to be really careful when solving real-life problems numerically (which is almost always).

Exercise 2.3.2.1. The forward difference approximates the derivative by looking ahead. This creates a bias; for instance, it will always overestimate for functions that are concave up. However, in the definition of the derivative we are also allowed to look back, when we take $x < a$ in the definition.

(a) Draw a new picture for secant approximation of the derivative (see the first picture in this chapter), but this time with $x < a$. Denote $x = a - h$, where $h > 0$, and deduce an approximating formula for the derivative. This is called the **backward difference** (or backward divided difference).

(b) Use the Taylor expansion of $f(a - h)$ to deduce an error estimate for the backward difference.

(c) Use Maple, Mathematica, MATLAB®, or some other system of your choice to compare theoretical forecasts with real-life performance, like we did in the example above. Try small h to find where the numerical error kicks in. You can also try other functions f and points a.

Exercise 2.3.2.2. The backward difference is also biased. Biased estimators are dangerous in numerical calculations, because they encourage numerical errors to accumulate. However, the forward and backward differences are biased in opposite ways. If we take their average, we can hope that the errors will cancel each other out (not completely, that would be too much to ask, but at least to some extent).

(a) Create a new approximation formula for the derivative by averaging the forward and backward difference. It is called the **central difference** (or central divided difference). Draw a picture describing its geometrical meaning.

(b) Estimate the error of this approximation.
Hint: Substitute Taylor expansions for $f(a+h)$ and $f(a-h)$ into your approximating formula for the derivative.

(c) Use a computer to compare the error estimate with the actual error for some function.

2.3.3 Approximating functions

Engineering calculations are done with decimal numbers. People can do (precisely) only addition, subtraction, and multiplication. Note that already division presents trouble, as the procedure that we all know can lead to a nonterminating process in the case the result has an infinite decimal expansion, and for other functions things get only worse. Generally we can perform and do precisely only those calculations that can be reduced to basic algebraic operations. Many people are not aware of the fact that calculators and computers are subject to the same limitations. Unfortunately, this leaves out most elementary functions (like $\sin(x)$, e^x, $\ln(x)$, roots, etc.); these cannot be evaluated using the basic algebraic operations. However, we routinely evaluate elementary functions on calculators or computers. Where do they get those values?

The answer is that they do not really supply us with those numbers, but with their approximations, which are obtained using basic algebraic operations. Fortunately, for practical applications of mathematics (like engineering) this is enough. Every machine works at a certain precision; for instance, my calculator works with 12-digit precision. The standard engineering precision as defined by IEEE is about 7 digits in decimal form, while double precision is about 16 digits (the specification is actually in binary representation).

Thus we need special procedures that can supply approximations of required precision. One popular tool for deriving approximation formulas is the Taylor polynomial.

Example. How do we calculate $e^{1.4}$? We can start with the Taylor polynomial for the function e^x with center a. We have

$$T_n(x) = e^a + e^a(x-a) + \frac{1}{2}e^a(x-a)^2 + \frac{1}{3!}e^a(x-a)^3 + \cdots + \frac{1}{n!}e^a(x-a)^n.$$

We can also use the version

$$e^{a+h} \approx e^a + e^a h + \frac{1}{2}e^a h^2 + \frac{1}{3!}e^a h^3 + \cdots + \frac{1}{n!}e^a h^n.$$

How do we choose a? We usually want a "nice" number close to the one we need, in our case close to 1.4. At the first glance, $a = 1$ looks inviting. We would get the formula

$$e^{1.4} \approx e + e(1.4-1) + \frac{1}{2}e(1.4-1)^2 + \frac{1}{3!}e(1.4-1)^3 + \cdots + \frac{1}{n!}e(1.4-1)^n.$$

However, note that it includes the number e, which is irrational, and we cannot calculate with it precisely. We could supply some approximation of it into our calculations, but that would be just another source of error. It is therefore better to choose $a = 0$, obtaining

$$e^{1.4} \approx 1 + 1.4 + \frac{1}{2}(1.4)^2 + \frac{1}{3!}(1.4)^3 + \cdots + \frac{1}{n!}(1.4)^n.$$

This is an expression that can be calculated by hand, and hence also by processors in calculators and computers. What degree of polynomial should we choose? It depends on the precision we need.

Let us say that we want the error to be at most $0.01 = 10^{-2}$. Theory helps us here, namely, the Lagrange form of the remainder tells us that when we use Taylor polynomial of degree n, the error will be

$$E_n = \frac{1}{(n+1)!}e^c(1.4)^{n+1},$$

where c is some number between 0 and 1.4. We pass to an upper estimate for the error as explained above, replacing e^c by the largest possible value. Since e^x is increasing, e^c is at most $e^{1.4}$. Thus we obtain the estimate

$$|E_n| \le \frac{1}{(n+1)!}e^{1.4}(1.4)^{n+1} \le \frac{1}{(n+1)!}10 \cdot (1.5)^{n+1} = e_n.$$

What happened at the second inequality? We have to keep in mind that we can only do calculations by hand, so numbers like $e^{1.4}$ are not good for us. It is better to use a less precise estimate in exchange for nicer numbers; in particular we know that $e < 3$ and hence $e^{1.4} < e^2 < 3^2 < 10$. I like 10. And when it comes to multiplication by hand, I much prefer $1.5 = \frac{3}{2}$ to 1.4.

So we have our error estimate and we return to our question: what n should we choose so that the error does not exceed 10^{-2}? If we are lucky, we can solve the equation $e_n \le 10^{-2}$ for n, but here this cannot be done. So we will experiment, evaluating the estimate e_n for various n.

To make our life easier we rewrite it as a recurrent sequence, namely,

$$e_2 = \frac{1}{3!}10 \cdot \left(\frac{3}{2}\right)^3 = \frac{45}{8}$$

and

$$e_n = \frac{1}{n+1}1.5e_{n-1} = \frac{3}{2(n+1)}e_{n-1}.$$

Let us evaluate some errors:

$$e_2 = \tfrac{45}{8} = 5.625, \qquad e_5 = \tfrac{81}{512} = 0.16...,$$
$$e_3 = \tfrac{135}{64} = 2.1..., \qquad e_6 = \tfrac{243}{7168} = 0.03...,$$
$$e_4 = \tfrac{81}{128} = 0.6..., \qquad e_7 = \tfrac{729}{114688} = 0.006....$$

We see that if we take $T_7(1.4)$, then we are guaranteed to obtain an approximation of $e^{1.4}$ with sufficient precision. Note that since the estimate for the error is an increasing function in x for $x > 0$, it follows that T_7 would be good for approximating e^x for all x from $[0, 1.4)$ with the given precision.

This whole procedure can be done by hand (in fact I did, and only then I checked the results in Maple), and calculations like that were done by people called computers in the days before electronic computers (and there being no computers, internet, or TV, people had lots of time for such calculations).

Similar procedures are wired into processors, so the next time you press the button $\boxed{e^x}$ on your calculator, you will know that there are some heavy duty calculations going on before it shows you a number on display. Of course, the procedures used in calculators and computers are much more sophisticated so that you do not have to wait seconds (or minutes) for the answer, but the basic idea is analogous.

A curious reader may wonder how much we lost when deriving the upper estimate e_n, as we were rather generous then. Most likely a lower degree of Taylor polynomial would already be sufficient. Using computer one can find that the true Lagrange estimate $\frac{1}{(n+1)!}e^{1.4}(1.4)^{n+1}$ becomes smaller than 10^{-2} for $n = 6$, so we did not loose that much in our rougher estimate.

Exercise 2.3.3.1. Consider the function $f(x) = \sqrt{x}$.

(a) Find Taylor polynomials of degrees 2, 3, and 4 for f that can be used to approximate $\sqrt{x}$ for numbers near 1. Use them to approximate $\sqrt{1.2}$.

(b) Use the Lagrange form of the remainder to estimate the errors $E_n(1.2)$ of the three approximations. Then modify the estimates to obtain some estimates that can be evaluated by hand, and evaluate them (you can use a calculator or a computer for that).

Exercise 2.3.3.2. Consider the function $f(x) = \ln(1 + x)$.

(a) Determine its Taylor polynomial T_n of degree n that can be used for approximating f for small x.

(b) Use the Lagrange form of the remainder to estimate the error $E_n(x)$. Find some upper estimate $e_n(x)$ for the error that can be evaluated efficiently by hand assuming that x satisfies $|x| \le 0.5$.

(c) Determine n so that the polynomial $T_n(x)$ provides approximation for $\ln(1 + x)$ for all $x \in [-0.5, 0.5]$ with error at most 10^{-6}.

Exercise 2.3.3.3. Consider the function $f(x) = \sin(x)$.

(a) Determine its Taylor polynomial T_n of degree n that can be used for approximating sine for small x.

(b) Use the Lagrange form of the remainder to estimate the error $E_n(x)$. Find some upper estimate e_n for the error that can be evaluated efficiently by hand assuming that x satisfies $|x| \le 1$.

(c) Determine n so that the polynomial $T_n(x)$ provides approximation for $\sin(x)$ for all $x \in \left[-\frac{\pi}{4}, \frac{\pi}{4}\right]$ with error at most $10^{-3} = 0.001$.

Note that due to properties of the sine, once we know how to evaluate it on this interval, we are able to evaluate it for all real numbers.

2.3.4 Approximating formulas

We saw how Taylor approximation can be useful in numerical approximations. They can be helpful also in other fields. Here we will look at some examples from physics.

Example. The capacitor is one of the basic components in electrical circuits. There are several popular configurations; one of them is two parallel plates.

If two parallel plates of area A are used and their distance is d, then the capacity is given by the formula $C = \varepsilon\frac{A}{d}$, assuming that d is very small compared to the dimensions of the plates. (Here ε is the permittivity of the environment between plates.)

Another popular type is a spherical capacitor. Imagine a sphere of radius R inside a larger sphere of radius R_2. The capacity is then given by the formula

$$C = \frac{4\pi\varepsilon}{\frac{1}{R} - \frac{1}{R_2}}.$$

If we introduce the distance d between the two spheres and write $R_2 = R + d$, the formula becomes

$$C = \frac{4\pi\varepsilon}{\frac{1}{R} - \frac{1}{R+d}} = \frac{4\pi\varepsilon}{\frac{d}{R(R+d)}} = \frac{4\pi\varepsilon}{d}R(R+d).$$

If d is very small compared to R, that is, the spheres are really close compared to their size, then $R + d$ is almost equal to R and our formula can be approximated as

$$C \approx \frac{4\pi\varepsilon}{d}R^2 = \varepsilon\frac{4\pi R^2}{d} = \varepsilon\frac{A}{d}.$$

Here A is the surface of the inner sphere, which is about the same as the surface of the outer sphere (since d is negligible compared to R). We arrived at the same formula as for the parallel-plate capacitor.

Another popular type is the coaxial cable type (some people call it a cylinder capacitor). It consists of an inner cable of radius R, inserted in an outer cylinder that has radius $R_2 = R + d$, both of length l. Then the formula for capacity is

$$C = \frac{2\pi\varepsilon l}{\ln(R_2/R)} = \frac{2\pi\varepsilon l}{\ln(1 + \frac{d}{R})}.$$

Here algebra does not help with simplification, and we have to turn to the Taylor polynomial. We know that

$$\ln(1 + t) = t - \frac{1}{2}t^2 + \frac{1}{3}t^3 - \frac{1}{4}t^4 + \cdots.$$

For very small t we can therefore approximate $\ln(1 + t) \approx t$. If d is very small compared to R, we can apply our approximation with $t = \frac{d}{R}$ and obtain

$$C = \frac{2\pi\varepsilon l}{\ln(1 + \frac{d}{R})} \approx \frac{2\pi\varepsilon l}{\frac{d}{R}} = \varepsilon\frac{2\pi Rl}{d} = \varepsilon\frac{A}{d}.$$

Now A is the surface of the cylinder, that is, of the active part in this capacitor.

Remarkably, all three types have capacity given (approximately) by the same formula if the active parts are very close compared to their dimensions.

Exercise 2.3.4.1. Physics tells us that the potential energy of a radial gravitational field when at the distance r from its source is given by the formula $U(r) = -\kappa \dfrac{mM}{r}$.

Assuming that the earth is a perfect ball, we can take its radius R and calculate the gravitational acceleration on the surface as $g = \kappa \frac{M}{R^2}$. We isolate κ from this formula, substitute it into the formula for the potential energy $U(r)$, and obtain the following fact. If an object is at distance r from the earth's center, then its potential energy is $U(r) = -\dfrac{mgR^2}{r}$.

(a) Determine potential energy gained when we lift a certain object from ground level (distance R) up by h meters.

(b) Form the Taylor expansion of $\frac{1}{r}$ with center $a = R$. Use it as a basis for linear approximation of $U(r)$, assuming that r is close to R.
Use it to obtain a linear approximation of $U(R+h)$.
Alternative approach: Find the Taylor expansion of $\frac{1}{R+h}$, where h is now considered the variable, with center $h = 0$. Use it to deduce a linear approximation, assuming that h is small relative to R.
Bonus question: Seeing the expansion, can you think of some other way of deducing it?

(c) Use the linear approximation of $U(R+h)$ to deduce an approximate formula for the potential energy $U(R+h) - U(R)$.

Exercise 2.3.4.2. Generally, when an object moves at velocity v in an environment where its potential energy is constant, then it has a relativistic energy given by the formula $E = \gamma m_0 c^2$, where $\gamma = \dfrac{1}{\sqrt{1-\frac{v^2}{c^2}}}$.

It can be rewritten as $E = m_0c^2 + (\gamma - 1)m_0c^2$, where $E_0 = m_0c^2$ is the rest mass energy, that is, the basic energy of the object when in rest. The other term must then be the kinetic energy $E_K = (\gamma - 1)m_0c^2$.

(a) Consider the function $f(t) = \dfrac{1}{\sqrt{1-t^2}} - 1$. Find its Taylor polynomial of degree 4 with center $t = 0$.
Can you see the connection between f and E_K?

(b) Assuming that t is small, the Taylor polynomial offers an interesting approximation. Apply it with $t = \frac{v}{c}$, assuming that v is very small compared to c. Deduce an approximating formula for the kinetic energy E_K.

For the first and third exercise I am indebted to V. Kříha of the Department of physics, FEE, CTU in Prague.

2.3.5 Investigating solutions of differential equations

Differential equations are an essential part of the language of natural sciences and engineering (see Chapter 11 and 12). Unfortunately, most equations coming from the real world cannot be solved analytically. However, often we can deduce some properties of such solutions using tools of mathematical analysis. Here we will look at some simpler examples.

Example. Consider one of the simplest initial value problems: $y' = y$, $y(0) = 1$. It can be easily solved by separation; the solution is $y(x) = e^x$. We will try to pretend that we do not know this.

The theory of ordinary differential equations (ODE) guarantees that such a problem has a solution $y(x)$ on $[0, \infty)$, so it makes sense to work with it. This solution is by necessity continuous and differentiable, as we should be able to substitute it into y'.

(a) We will prove that $y(x) > 0$ on $[0, \infty)$.

Assuming the opposite, consider the set $M = \{x \geq 0;\ y(x) \leq 0\}$ that should not be empty by this assumption. As a subset of $[0, \infty)$ this set is bounded from below, and hence it has a finite infimum. Since it is a preimage of a closed set $(-\infty, 0]$ under the continuous function y, M is closed, and hence it actually has a minimum x_M. Since $y(0) = 1 > 0$, we have $0 \notin M$ and thus $x_M > 0$. Consequently $y > 0$ on $[0, x_M)$. We also know that $y(x_M) \leq 0$ as $x_M \in M$.

However, if $y(x_M) < 0$ were true, then by continuity of y there would be some neighborhood U of x_M such that $y < 0$ on U. In particular there would be some $x < x_M$, $x \geq 0$, such that $y(x) < 0$, contradicting minimality of x_M. This proves that $y(x_M) = 0$.

Now we apply the mean value theorem to y on the interval $[0, x_M]$. There must be $c \in (0, x_M)$ such that

$$y'(c) = \frac{y(x_M) - y(0)}{x_M - 0} = \frac{0-1}{x_M} = -\frac{1}{x_M}.$$

It follows that $y'(c)$ is negative, and because y solves the given ODE, we have $y(c) = y'(c) < 0$. On the other hand, $c \notin M$ and thus $y(c) > 0$, a contradiction. This proves that our original assumption (y is zero or negative somewhere) was wrong.

(b) Since $y'(x) = y(x)$ and $y(x) > 0$, it follows that $y(x)$ is increasing on $[0, \infty)$. In particular, $y(x) \geq y(0) = 1$ on $[0, \infty)$.

It also follows that $y(x)$ must have a limit at infinity, but we do not know whether it is finite or infinite.

(c) From $y' = y$ we get $y'' = y' > 0$, so the solution $y(x)$ must be concave up on $[0, \infty)$.

(d) We claim that $y(x) \to \infty$ as $x \to \infty$.

The proof will combine two facts. We know that $y(x)$ is increasing, so it must go up at a certain rate around the origin. And because it is concave up, it can never slow down. Now we do it mathematically.

We take any $\delta > 0$ and consider the slope $k = \dfrac{y(\delta) - y(0)}{\delta - 0}$. Since y is increasing, we know that $k > 0$. Since y is concave up, for any $x > \delta$ the definition implies

$\frac{y(\delta)-y(0)}{\delta-0} \le \frac{y(x)-y(\delta)}{x-\delta}$, that is, $y(x) \ge y(\delta) + k(x-\delta)$. Since $k > 0$, the expression on the right tends to infinity at infinity and by comparison, also $y(x) \to \infty$ there.

A nicer comparison $y(x) \ge kx + y(0)$ can be deduced if we use the alternative characterization of concavity $\frac{y(\delta)-y(0)}{\delta-0} \le \frac{y(x)-y(0)}{x-0}$.

Note that we did not use derivative in this proof, which means that it is more general, but also more complicated. A shorter argument using derivative is outlined after Lemma 2.7 below.

(e) How fast does y go to infinity? We claim that $y(x)$ goes to infinity incomparably faster than x.

Indeed, $y(x) \to \infty$, so we will be able to use L'Hôpital's rule in the following calculation:

$$\lim_{x\to\infty}\left(\frac{y(x)}{x}\right) \underset{\text{l'H}}{\overset{\frac{\infty}{\infty}}{=\!=}} \lim_{x\to\infty}\left(\frac{y'(x)}{1}\right) = \lim_{x\to\infty}\big(y(x)\big) = \infty.$$

We were able to replace y' by y since y solves the given ODE.

It is not hard to show by induction that, in fact, $\lim\limits_{x\to\infty}\left(\frac{y(x)}{x^k}\right) = \infty$ for any $k > 0$. Indeed, we just proved it for $k = 1$. Assuming it to be true for $k \in \mathbb{R}$, we deduce

$$\lim_{x\to\infty}\left(\frac{y(x)}{x^{k+1}}\right) \underset{\text{l'H}}{\overset{\frac{\infty}{\infty}}{=\!=}} \lim_{x\to\infty}\left(\frac{y'(x)}{(k+1)x^k}\right) = \frac{1}{k+1}\lim_{x\to\infty}\left(\frac{y(x)}{x^k}\right) = \infty.$$

The solution y thus grows to infinity faster than any power.

We learned quite a bit about our solution $y(x)$ without actually solving the equation.

In this example we learned several things that will come handy in exercises below. First we have the following statement.

Lemma 2.6. *If a function $y(x)$ is increasing and concave up on an interval $[K, \infty)$, then $\lim\limits_{x\to\infty}\big(y(x)\big) = \infty$.*

It will be also useful to be able to use knowledge of the derivative of a function to estimate its growth. We have the following result.

Lemma 2.7. *Let $y(x)$ be a differentiable function whose derivative satisfies $y(x) \ge A$ for some A on an interval $[K, \infty)$. Then $y(x) \ge y(K) + A(x-K)$ on $[K, \infty)$.*

Prove this as an additional exercise using Corollary 2.3 applied to $y(x)$ and $g(x) = y(K) + A(x-K)$. One can also apply the Mean value theorem to the interval $[K, x]$.

In our example we could use the result of part (b) to conclude that $y' = y \ge 1$ for $x \ge 0$, and therefore $y(x) \ge x + y(0) = x + 1$ on $[0, \infty)$ by this lemma. In particular this independently confirms that $\lim\limits_{x\to\infty}\big(y(x)\big) = \infty$.

In general, if the assumptions of the lemma were true and also $A > 0$, then we would get $y \to \infty$. This will be used later.

Exercise 2.3.5.1. Consider the initial value problem

$$y' = \frac{1}{y+1}, \qquad y(0) = 0.$$

(a) Apply the definition of the derivative to $y'(0)$ to show that there must be some right neighborhood $(0, \delta)$ on which $y(x) > 0$.

(b) Show that $y(x) > 0$ on $[0, \infty)$.
Note that when proving positivity of y in our example above we made use of the fact that $y(0) = 1 > 0$. Now we do not have this, instead the result of part (a) can be used.

(c) Show that $y(x)$ is increasing on $[0, \infty)$.
Note: It follows that $y(x)$ must have a limit at infinity, but we do not know now whether it is finite or infinite.

(d) Show that $y(x)$ is concave down on $[0, \infty)$.

(e) Show that $y'(x)$ converges to some $L \geq 0$ at infinity.

(f) Show that $y(x) \to \infty$ at infinity.
Hint: Use the previous part, consider cases $L > 0$ and $L = 0$. One of the cases makes use of Lemma 2.7.

(g) Show that $\lim\limits_{x\to\infty} \left(\frac{y(x)}{x}\right) = 0$.
This tells us that $y(x)$ is negligible compared to x at infinity, which can be written as $y(x) = o(x)$, $x \to \infty$. It can be also expressed by saying that the asymptotic rate of growth of $y(x)$ at infinity is smaller than that of $y(x) = x$.

Note: Solving this equation by separation is not straightforward, as it leads to the equation $\frac{1}{2}y^2 + y = x + C$. From this we can deduce two candidates for solution: $y(x) = \sqrt{2x + C} - 1$ and $y(x) = -\sqrt{2x + C} - 1$. Only one of them can satisfy our initial condition: $y(x) = \sqrt{2x + 1} - 1$. We see that it really fits with our observations above. It also explains why attempts to prove that $y(x) \geq x$ as in the previous example would fail: it is not true.

Exercise 2.3.5.2. Consider the initial value problem

$$y' = \frac{y}{y+1}, \qquad y(0) = 1.$$

(a) Show that $y(x) > 0$ on $[0, \infty)$.

(b) Show that $y(x)$ is increasing on $[0, \infty)$.
Consequently, it has a limit at infinity.

(c) Show that $y(x)$ is concave up on $[0, \infty)$.

(d) Show that $y(x) \to \infty$ at infinity.

(e) Prove that $y(x)$ has an oblique asymptote of the form $y = x + B$ at infinity.

Note: This differential equation cannot be solved by separation, as it leads to the equation $y + \ln(y) = x + C$ from which we cannot isolate y analytically. The given ODE is not linear either, so none of the methods that are typically covered in ODE courses can handle it. In fact, it simply cannot be solved analytically no matter what method is used, since it is known that its solution exists but it cannot be expressed as an algebraic formula. Thus it is nice that we can say something about that elusive solution.

Exercise 2.3.5.3. Consider the initial value problem

$$y' = \frac{1}{xy+1}, \qquad y(0) = 0.$$

(a) Show that $y(x) > 0$ on $(0, \infty)$.
(b) Show that $y(x)$ is increasing on $[0, \infty)$.
Thus $y(x)$ must have a limit at infinity. As far as we know right now, this limit can be infinity or a positive number by part (a).
(c) Show that $y(x)$ is concave down on $[0, \infty)$.
(d) Show that $y'(x) \to 0$ at infinity.
(e) Show that $\lim\limits_{x\to\infty}\left(\frac{y(x)}{x}\right) = 0$.
(f) Look at the proof that $y(x) \to \infty$ in Example 2.3.5.2. Find which step is not true when we try to adapt it to this ODE.

Note: This differential equation is not even separable, and obviously not linear, so again we are unable to solve it analytically using common tools. And again, the solution of this simple initial problem exists but cannot be expressed using an algebraic formula.

Answers to exercises

2.3.1.1: Setup: radius $r(t)$, volume $V(t)$, given V', and we need to know r'.

Connection: $V = \frac{4}{3}\pi r^3$. Then $V' = 4\pi r^2 r'$, and hence $r' = \frac{V'}{4\pi r^2}$. If we blow at a constant rate, then the radius of the balloon grows at the rate $r' = \frac{K}{r^2}$.

2.3.1.2: Setup: position $x(t)$ (in meters), distance $d(t)$, known $x'(t)$, and we need $d'(t)$.

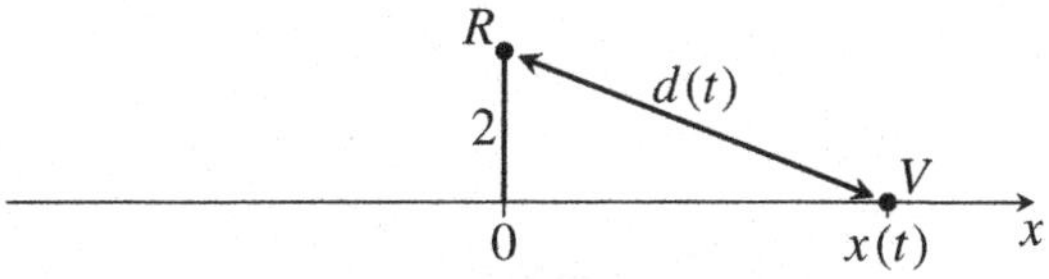

The relation we seek is $d = \sqrt{2^2 + x^2}$. Taking derivative (and canceling 2) we obtain $d' = \frac{x}{\sqrt{2^2+x^2}}x'$. Note that there is a difference between velocity, speed, and rate of change. In particular, if the car goes towards the radar, then d' is negative, while the speed of the car should be positive. We therefore introduce v_C for the true speed and v_A for apparent speed of the car and obtain the formula $v_A = \frac{|x|}{\sqrt{2^2+x^2}}v_C$.

Now we look at this formula. First, note that $\sqrt{4+x^2} > |x|$, so the factor $\frac{|x|}{\sqrt{4+x^2}}$ is always less than 1. In other words, the radar always shows less than the actual speed of the car. When the car is closest to the radar ($x = 0$), the apparent velocity is zero.

On the other hand, when the car is really far away ($x \approx \infty$), then $\frac{|x|}{\sqrt{4+x^2}} \approx \frac{|x|}{\sqrt{x^2}} = 1$. Thus the radar almost shows the actual velocity, which makes sense; the car is then approaching almost head on.

2.3.1.3: Setup: shadow length $l(t)$, distance from lamppost $d(t)$, we know d', and we want to know l'.

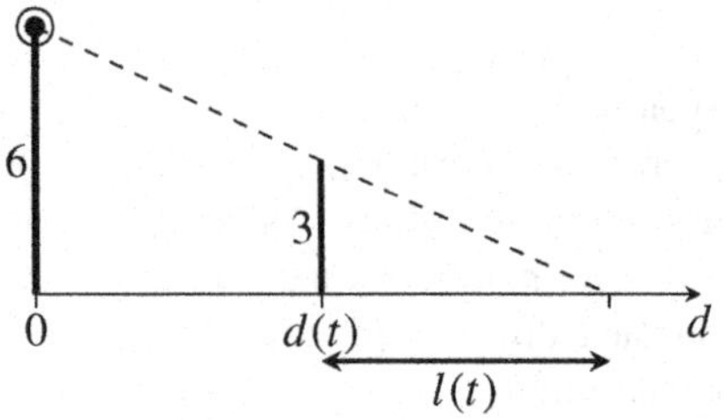

Relation: From similarity of triangles we get $\frac{l}{3} = \frac{l+d}{6}$. Hence $l = d$, that is, $l' = d' = c$ by assumption of steady walk.

Surprisingly enough, my shadow grows at a rate that does not depend on my distance from the lamppost, just on the speed at which I walk. I actually expected that the further I am, the faster the shadow grows, but my intuition was wrong this time.

2.3.2.1: (a) $f'(a) \approx \frac{f(a)-f(a-h)}{h}$.

(b) $E \approx \frac{1}{2}f''(a)h$. The error is linear and should be comparable to that of forward difference.

(c) In Maple I used the following code for the default example $f(x) = e^x$, $a = 0$:

```
f:=exp(x);a:=0; der:=eval(diff(f,x),x=a);
approx:=(1/2)*eval(diff(f,x,x),x=a)*h;
plot([der-(eval(f,x=a)-eval(f,x=a-h))/h,approx],
                h=0.0..0.9, color=[red,navy]);
```

2.3.2.2: (a) $f'(a) \approx \frac{f(a+h)-f(a-h)}{2h}$.

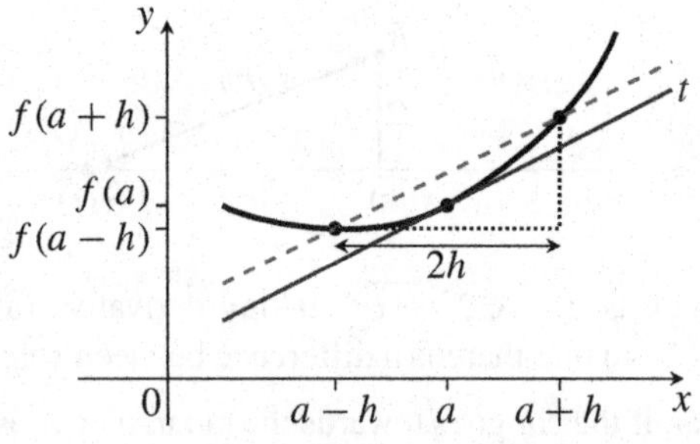

(b) $E \approx -\frac{1}{3!}f'''(a)h^2$. Quadratic error disappears much faster than linear error, so this approximation is much more effective than forward or backward difference. In particular, we can get very precise approximations while h is still relatively large, before the numerical errors kick in. But again, if we overdo it, they will show up eventually.

(c) In Maple we can use the following modification of the above code:

```
approx:=(-1/6)*eval(diff(f,x,x,x),x=a)*h^2;
plot([der-(eval(f,x=a+h)-eval(f,x=a-h))/(2*h),approx],
        h=0.0..0.9, color=[red,navy]);
```

2.3.3.1: (a) Choice $a = 1$, find T_4, the others are obtained by truncating it. We have $T_4(x) = \sqrt{1} + \frac{1}{2}\frac{1}{\sqrt{1}}(x-1) - \frac{1}{2}\frac{1}{4}\frac{1}{\sqrt{1}^3}(x-1)^2 + \frac{1}{6}\frac{3}{8}\frac{1}{\sqrt{1}^5}(x-1)^3 - \frac{1}{24}\frac{15}{16}\frac{1}{\sqrt{1}^7}(x-1)^4$.
$T_4(x) = 1 + \frac{1}{2}(x-1) - \frac{1}{8}(x-1)^2 + \frac{1}{16}(x-1)^3 - \frac{5}{128}(x-1)^4$.
Alternative: $T_4(1+h) = 1 + \frac{1}{2}h - \frac{1}{8}h^2 + \frac{1}{16}h^3 - \frac{5}{128}h^4$.
Approximations:
$T_2(1.2) = 1 + \frac{1}{2}\cdot 0.2 - \frac{1}{8}(0.2)^2 = 1 + \frac{1}{10} - \frac{1}{200} = 1 + \frac{1}{10} - \frac{5}{1000} = 1.095$,
$T_3(1.2) = 1 + \frac{1}{2}\cdot 0.2 - \frac{1}{8}(0.2)^2 + \frac{1}{16}(0.2)^3 = 1 + \frac{1}{10} - \frac{1}{200} + \frac{1}{2000} = 1.0955$,
$T_4(1.2) = 1 + \frac{1}{2}\cdot 0.2 - \frac{1}{8}(0.2)^2 + \frac{1}{16}(0.2)^3 - \frac{5}{128}(0.2)^4 = 1.0954375$.
(b) Lagrange form of the remainder: $E_2(1.2) \le \frac{1}{3!}\max\left|-\frac{3}{8}c^{-5/2}\right|\cdot(0.2)^3$, $E_3(1.2) \le \frac{1}{4!}\max\left|\frac{15}{16}c^{-7/2}\right|\cdot(0.2)^4$, $E_4(1.2) \le \frac{1}{5!}\max\left|-\frac{105}{32}c^{-9/2}\right|\cdot(0.2)^5$.
Since negative powers of c are decreasing for $1 \le c \le 1.2$, we can estimate
$|E_2(1.2)| \le \frac{1}{3!}\frac{3}{8}\cdot\frac{1}{5^3} = \frac{1}{2000} = 0.0005$,
$|E_3(1.2)| \le \frac{1}{4!}\frac{15}{16}\cdot\frac{1}{5^4} = \frac{1}{16000} = 0.0000625$,
$|E_4(1.2)| \le \frac{1}{5!}\frac{105}{32}\cdot\frac{1}{5^5} = \frac{7}{800000} = 0.00000875$.
Just to satisfy our curiosity, actual errors are $E_2 \approx 0.0004$, $E_3 \approx -0.00005$, and $E_4 \approx 0.000008$. It seems that the Lagrange estimates are very good here.

2.3.3.2: (a) Choose $a = 0$, $f(x) = \ln(1+x)$, then $f^{(n)} = (-1)^{n+1}\frac{(n-1)!}{(x+1)^n}$ for $n \ge 1$ (by induction). Therefore
$T_n(x) = 0 + \sum\limits_{k=1}^{n} \frac{1}{k!}(-1)^{k+1}\frac{(k-1)!}{1^k}x^k = \sum\limits_{k=1}^{n}(-1)^{k+1}\frac{1}{k}x^k$.
(b) Lagrange estimate:
$E_n(x) = \frac{1}{(n+1)!}(-1)^{n+2}\frac{n!}{(c+1)^{n+1}}x^{n+1} = (-1)^n\frac{1}{n+1}\frac{1}{(c+1)^{n+1}}x^{n+1}$.
We have c between 0 and x and $|x| \le 0.5$, so definitely $|c| \le 0.5$ as well. Since $\frac{1}{(c+1)^{n+1}}$ is decreasing and positive on this range, we can estimate
$|E_n(x)| \le \frac{1}{n+1}\frac{1}{(0.5)^{n+1}}\cdot(0.5)^{n+1} = \frac{1}{n+1} = e_n$.
(c) We have $e_n \le 10^{-6} \implies n+1 \ge 10^6$, so we take $n = 10^6$.
The corresponding T_n is a very long polynomial and it would take a long time to evaluate. In fact here the actual error is much smaller, so the error estimate is extremely bad, but the Lagrange form does not allow for a better one. The main problem is the term $\frac{1}{(c+1)^{n+1}}$. If we restricted our attention to $x \ge 0$, then $c \in [0, 0.5]$ and we get $|E_n(x)| \le \frac{1}{n+1}\frac{1}{1^{n+1}}(0.5)^{n+1} = \frac{1}{2^{n+1}(n+1)}$, which tends to zero very fast.

2.3.3.3: (a) Choose $a = 0$, $f(x) = \sin(x)$, all derivatives are sines or cosines with plus/minus: $f' = \cos(x)$, $f'' = -\sin(x)$, $f''' = -\cos(x)$, $f'''' = \sin(x)$, $f''''' = \cos(x)$, etc. Therefore
$T_n(x) = 0 + x - 0 - \frac{1}{3!}x^3 - 0 + \frac{1}{5!}x^5 + 0 - \frac{1}{7!}x^7 - \dots = x - \frac{1}{3!}x^3 + \frac{1}{5!}x^5 - \frac{1}{7!}x^7 + \dots$.
(b) Lagrange estimate: $E_n(x) = \frac{1}{(n+1)!}f^{(n+1)}(c)x^{n+1}$. The derivative is sines and cosines with alternating signs, but it is never more than 1. Thus, using $|x| \le 1$, we get

$|E_n(x)| \leq \frac{1}{(n+1)!} \cdot 1 \cdot 1^{n+1} = \frac{1}{(n+1)!} = e_n.$

(c) Since $\left[-\frac{\pi}{4}, \frac{\pi}{4}\right] \subseteq [-1, 1]$, we can use the above estimate. We want $e_n \leq 10^{-3}$, and this needs $(n+1)! \geq 1000$. By experimentation, $n = 6$ is enough.

2.3.4.1: (a) $E_p = U(R+h) - U(R) = -\frac{mgR^2}{R+h} + \frac{mgR^2}{R}$.

(b) $\frac{1}{r} = \frac{1}{R} - \frac{1}{R^2}(r-R) + \frac{1}{R^3}(r-R)^2 - \frac{1}{R^4}(r-R)^3 + \cdots$, so $\frac{1}{r} \approx \frac{1}{R} - \frac{1}{R^2}(r-R)$.
Thus $U(R+h) \approx -mg(R-h)$.
Alternative: $\frac{1}{R+h} = \frac{1}{R} - \frac{1}{R^2}h + \frac{1}{R^3}h^2 - \frac{1}{R^4}h^3 + \cdots$, so $\frac{1}{R+h} \approx \frac{1}{R} - \frac{1}{R^2}h$.
Bonus question: We can use the formula for summing up geometric series:

$$\frac{1}{R+h} = \frac{1}{R}\frac{1}{1-(-\frac{h}{R})} = \frac{1}{R} \cdot \left(1 + \left(-\frac{h}{R}\right) + \left(-\frac{h}{R}\right)^2 + \left(-\frac{h}{R}\right)^3 + \cdots\right).$$

Now we see why we can ignore higher powers when h is negligible compared to R, since then the higher powers of the very small number $-\frac{r}{R}$ go to zero fast.

Another possibility is to apply the binomial formula to $(R+h)^{-1}$, but that requires more work.

(c) $E_p = U(R+h) - U(R) \approx -mgR^2(\frac{1}{R} - \frac{h}{R^2}) + \frac{mgR^2}{R} = mgh$.

We obtained the formula for potential energy that we all remember from elementary school. Now we see that it is actually just an approximation of a more precise formula, and that it is valid only for relatively small values of h.

2.3.4.2: (a) $T_4 = \frac{1}{2}t^2 + \frac{1}{4!} \cdot 9t^4 = \frac{1}{2}t^2 + \frac{3}{8}t^4$.

(b) For t small we can approximate $\gamma - 1 \approx \frac{1}{2}(\frac{v}{c})^2$, so we get
$E_{\text{K}} = \frac{1}{2}\frac{v^2}{c^2}mc^2 = \frac{1}{2}mv^2$.

Note: Instead of using the Taylor polynomial, one can apply the binomial theorem to $(1-t^2)^{-1/2}$.

2.3.5.1: (a) $y'(0) = \frac{1}{0+1} = 1$, so $\lim\limits_{x\to 0^+}(\frac{y(x)-y(0)}{x}) = 1$. Choosing, $\varepsilon = 1$ we obtain $\delta > 0$ such that $\frac{y(x)-y(0)}{x} > 1 - \varepsilon = 0$ on $(0, \delta)$. Since $x > 0$ there, we get $y(x) - y(0) > 0$ and $y(0) = 0$.

(b, c) Similar to the first example.

(d) $y'' = \frac{-1}{(y+1)^2}y' < 0$.

(e) We know that $y' > 0$ on $[0, \infty)$ and it is a decreasing function there (since $[y']' = y'' < 0$), and therefore it must have a proper limit L that satisfies $L \geq 0$.

Another argument: We know that $y \to l$, where $l > 0$. Then $\lim(y') = \lim(\frac{1}{y+1}) = \frac{1}{l+1} = L$. Since $l > 0$, this cannot be infinity. Since $l = \infty$ is possible, L can be zero.

(f) Case $L > 0$: Using the definition of the limit with $\varepsilon = \frac{1}{2}L$ we find some $K > 0$ so that $y' \geq L - \varepsilon = \frac{1}{2}L > 0$ on $[K, \infty)$. By Lemma 2.7 it follows that $y(x) \to \infty$ at infinity.

Case $L = 0$: By the given ODE, $\frac{1}{y(x)+1} \to 0$ at infinity. This is possible only if $y(x) \to \infty$ there.

(g) Since $y(x) \to \infty$, L'Hôpital's rule can be used:

$$\lim_{x\to\infty}\left(\frac{y(x)}{x}\right) = \lim_{x\to\infty}\left(\frac{y'(x)}{1}\right) = \lim_{x\to\infty}\left(\frac{1}{y(x)+1}\right) = \frac{1}{\infty+1} = 0.$$

2.3.5.2: (a, b) Similar to the first example.

(c) $y'' = \frac{1}{(y+1)^2} y' > 0$.

(d) It follows by Lemma 2.6 from the fact that y is increasing and concave up on $[0, \infty)$.

(e) Since $y(x) \to \infty$ at infinity, we can use L'Hôpital's rule in evaluating $\lim_{x\to\infty} \left(\frac{y(x)}{x}\right) = \lim_{x\to\infty} \left(\frac{y'(x)}{1}\right) = \lim_{x\to\infty} \left(\frac{y(x)}{y(x)+1}\right) = \lim_{x\to\infty} \left(\frac{1}{1+1/y(x)}\right) = \frac{1}{1+0} = 1$. This proves the claim.

2.3.5.3: Solution combines approaches from previous exercises.

(c) $y'' = \frac{-1}{(xy+1)^2}(y + xy') < 0$.

(d) We have $\lim_{x\to\infty} \left(y'(x)\right) = \lim_{x\to\infty} \left(\frac{1}{xy(x)+1}\right) = \frac{1}{\infty \cdot L+1}$ by (b).
Since $L \cdot \infty = \infty$ is true for both $L > 0$ and $L = \infty$, we get $\frac{1}{\infty \cdot L+1} = \frac{1}{\infty} = 0$.

(e) Case L proper: Then $\lim_{x\to\infty} \left(\frac{y(x)}{x}\right)$ is of the type $\frac{L}{\infty} = 0$.
Case $L = \infty$: L'Hôpital's rule can be used:
$\lim_{x\to\infty} \left(\frac{y(x)}{x}\right) = \lim_{x\to\infty} \left(\frac{y'(x)}{1}\right) = 0$ by part (d).

(f) Before, the only way to conclude $\frac{1}{y(x)+1} \to \frac{1}{L+1} = 0$ was to have $L = \infty$.
Now we have $\frac{1}{xy(x)+1} \to \frac{1}{\infty \cdot L+1} = 0$ and this is true also for L that is not infinity.

References

Math Tutor, 2019. http://math.feld.cvut.cz/mt/. Derivatives → Exercises → Global extrema and optimization, related rates.

Burden, R.L., Faires, J.D., 2011. Numerical Analysis, 9th edition. Brooks/Cole, Cengage Learning.

Stroud, K.A., 1970. Engineering Mathematics. Macmillan. Later editions with Dexter J. Booth, Industrial Press, Inc., New York.

[illegible]

(c) [illegible] ... the only way to [illegible] ... [illegible]

References

[illegible]

[illegible] ... Numerical Analysis, [illegible] edition [illegible]

[illegible]

Complex numbers and some applications

3

Fatih Yılmaz
Ankara Hacı Bayram Veli University, Ankara, Turkey

3.1 Introduction

Complex numbers are very important in engineering and science. They have applications in many areas, including control theory, signal analysis, relativity, and fluid dynamics. Engineers use complex numbers in analyzing stresses and strains on beams and in studying resonance phenomena in structures such as tall buildings and suspension bridges. The solution of physical equations is often made simpler through the use of complex numbers. Electrical engineers benefit from complex numbers to deal with the fact that the current through a circuit element such as a capacitor or inductor is not in phase with the voltage across it. They are also needed to analyze the fluid or air flows around the object. Another particularly important application of complex numbers is in quantum mechanics, when representing wave functions of a quantum system.

3.1.1 *Complex arithmetic*

In elementary mathematics, complex numbers emerge under the conditions where simple algebraic equations have solutions that cannot be expressed in terms of real numbers alone. Although complex numbers occur in many branches of mathematics, they arise most directly when solving polynomial equations. Unfortunately not all equations have (real number) solutions. Consider the following general quadratic equation:

$$ax^2 + bx + c = 0,$$

where a, b, and c are real numbers. The quadratic formula

$$x_{1,2} = \frac{-b \pm \sqrt{b^2 - 4ac}}{2a}$$

gives the zeros of the polynomial. It has no real solution for $\Delta = b^2 - 4ac < 0$. For example,

$$x^2 \geq 0; \quad x \in \mathbb{R}$$

$$x^2 + 1 \geq 0 + 1 = 1 > 0; \quad x \in \mathbb{R}$$

Calculus for Engineering Students. https://doi.org/10.1016/B978-0-12-817210-0.00010-2

$$x^2 \neq -1; \quad x \in \mathbb{R}.$$

Around this problem, let us consider the function $f(x) = x^2 + 1$, and the main question is where the function $f(x)$ crosses the x-axis. As it can be seen, it has no solution. However, about 200 years ago, Carl Friedrich Gauss proved that every polynomial of degree n has exactly n roots, which was later called the *fundamental theorem of algebra*. Taking into account this theorem, the function $f(x) = x^2 + 1 = 0$ must have two roots.

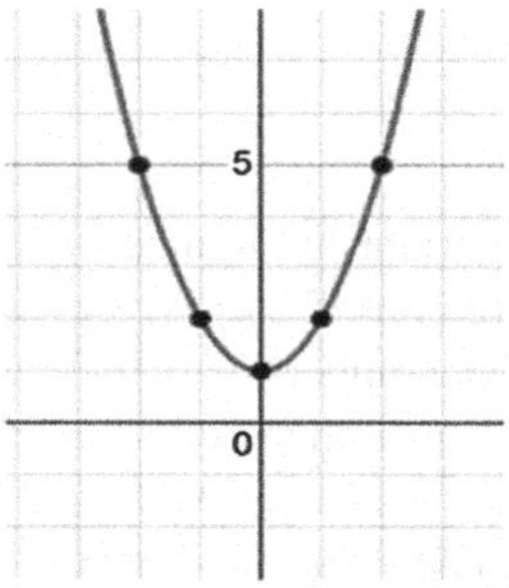

$f(x) = x^2 + 1$

Here, we introduce the symbol i, whose defining property is that it satisfies the equation

$$i^2 = -1.$$

From its definition, it is clear that $i \notin \mathbb{R}$. The notation i was first introduced by the Swiss mathematician and physicist Leonhard Euler (1707–1783).

A *complex number* is any symbol of the form $z = a + ib$, or $z = a + bi$, where a and b are any two real numbers. The real part of z is the number a, and the imaginary part of z is b; note that both parts are real numbers. One may define a complex number as nothing more than an ordered pair of two real numbers (a, b). The ordering is significant.

3.1.2 Properties of complex numbers

Let us give the basic definitions and terminology associated with equality, addition, subtraction, and multiplication of complex numbers.

The *addition/subtraction* of two complex numbers, z_1 and z_2, in general, gives another complex number. The real components and the imaginary components are added/subtracted separately and in a like manner to the familiar addition/subtraction of real numbers. In other words, if $z_1 = a_1 + ib_1$ and $z_2 = a_2 + ib_2$ are any complex numbers, then

$$z_1 \pm z_2 = (a_2 \pm a_2) + i(b_1 \pm b_2) = z_3.$$

Example 3.1. Let $z_1 = 3 + 4i$ and $z_2 = -1 + i$, then the addition and the subtraction, respectively, are obtained by

$$z_1 + z_2 = (3 + 4i) + (-1 + i) = 2 + 5i$$

and

$$z_1 - z_2 = (3 + 4i) - (-1 + i) = 4 + 3i.$$

The more difficult concept is *multiplication* of complex numbers, which is carried out as follows:

$$z_1 z_2 = (a_1 + ib_1)(a_2 + ib_2) = (a_1 a_2 - b_1 b_2) + i(a_2 b_1 + a_1 b_2) = z_3.$$

Example 3.2. Let $z_1 = 3 + 4i$ and $z_2 = -1 + i$ be any complex numbers. Then

$$z_1 z_2 = (3 + 4i)(-1 + i) = (-3 - 4) + (3i - 4i) = -7 - i.$$

Before we consider division of two complex numbers, let us introduce the number

$$z^* = a - ib \quad \left(\text{or } \overline{z} = z^*\right),$$

which is the complex *conjugate* to $z = a + ib$. The conjugate numbers are symmetric with respect to the real axis. Thus, z^* and z have the same real part and opposite signs in the imaginary part. (See Fig. 3.1.) Their multiplication is given as

$$zz^* = (a + ib)(a - ib) = a^2 + b^2.$$

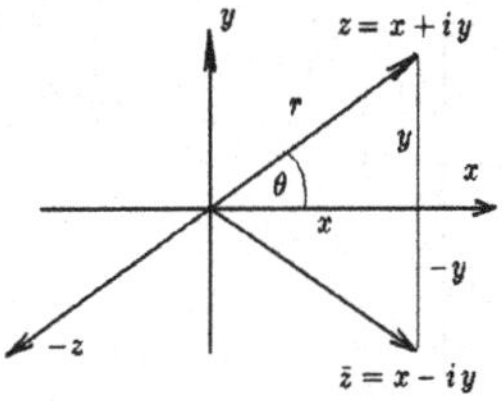

Figure 3.1 Complex z and its conjugate.

To perform *division* of two complex numbers,

$$z_3 = \frac{z_2}{z_1},$$

which can also be represented as

$$z_3 = \frac{z_2 z_1^*}{z_1 z_1^*} = \frac{(a_2 + ib_2)(a_1 - ib_1)}{(a_1 + ib_1)(a_1 - ib_1)} = \frac{(a_2 + ib_2)(a_1 - ib_1)}{(a_1^2 + b_1^2)}.$$

Example 3.3. If $z_1 = 3+4i$ and $z_2 = -1+i$ are any complex numbers, then

$$\frac{z_1}{z_2} = \frac{3+4i}{-1+i} = \frac{3+4i}{-1+i} \cdot \frac{-1-i}{-1-i} = \frac{1-7i}{2}.$$

Two complex numbers $z_1 = a_1 + ib_1$ and $z_2 = a_2 + ib_2$ are said to be *equal* when both their real and imaginary parts are equal, that is,

$$z_1 = z_2 \Leftrightarrow \begin{cases} \text{real}(z_1) = \text{real}(z_2) \text{ i.e. } a_1 = a_2 \\ \text{imag}(z_1) = \text{imag}(z_2) \text{ i.e. } b_1 = b_2, \end{cases}$$

where real(z_1) indicates the real part of z_1 and imag(z_1) indicates the imaginary part of z_1.

3.1.3 Geometric interpretation

One of the problems that mathematicians came across was the visualization of complex numbers. In the 19th century, the genius mathematician Jean Robert Argand solved this problem.

To every complex number $a+ib$, there corresponds exactly one point with coordinates (a, b).

Points for which $b = 0$, fill the so-called real axis and points for which $a = 0$, fill the so-called imaginary axis.

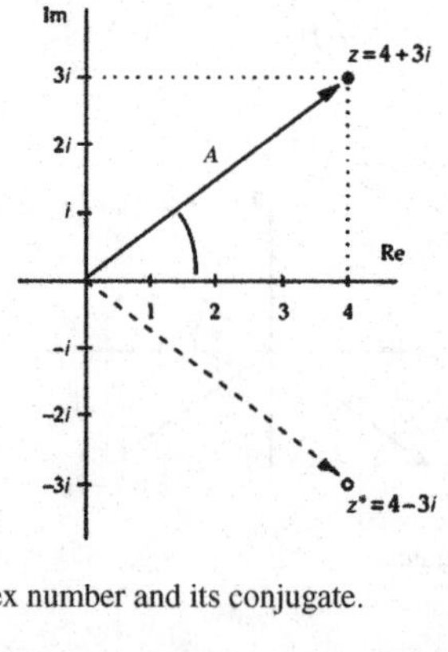

Figure 3.2 Interpretation a complex number and its conjugate.

Numbers bi are often called *imaginary numbers*. In Fig. 3.2, the geometric interpretation of the complex number $z = 4 + 3i$ and its conjugate $z^* = 4 - 3i$ can be seen. Similar to x and y in the Cartesian plane, the Argand diagram uses the real and imaginary parts.

The term "*imaginary*" was first used by the French mathematician René Descartes (1596–1650), who is known more as a philosopher. However, as a result of his contributions to mathematics and coordinate geometry, the name "*Cartesian coordinates*" is used.

The complex numbers also can be represented in polar coordinates. Let

$$x = r\cos\theta,$$
$$y = r\sin\theta,$$

where $r = \sqrt{x^2 + y^2}$.

Here r is called as the *modulus* of the complex z. Also, the angle θ is called the *argument* of the complex z. Then the *polar form* of a complex number is

$$\begin{aligned} z &= x + iy = r\cos\theta + ir\sin\theta \\ &= r(\cos\theta + i\sin\theta). \end{aligned}$$

In calculus, the exponential function is introduced together with the sine and cosine functions.

Note that the *modulus* tells us how far we go from the origin to get a complex number in the Argand diagram. The direction of a complex number is known as *argument*.

The Taylor series expansions are given as follows:

$$e^x = 1 + \frac{x}{1!} + \frac{x^2}{2!} + \cdots + \frac{x^n}{n!} + \cdots$$
$$\cos x = 1 - \frac{x^2}{2!} + \frac{x^4}{4!} + \cdots + (-1)^n \frac{x^{2n}}{(2n)!} + \cdots$$
$$\sin x = x - \frac{x^3}{3!} + \frac{x^5}{5!} + \cdots + (-1)^n \frac{x^{2n+1}}{(2n+1)!} + \cdots$$

By substituting $x = i\theta$ in the expansion of e^x,

$$\begin{aligned} e^{i\theta} &= 1 + \frac{i\theta}{1!} + \frac{(i\theta)^2}{2!} + \cdots + \frac{(i\theta)^n}{n!} + \cdots \\ &= \left[1 - \frac{\theta^2}{2!} + \frac{\theta^4}{4!} + \cdots\right] + i\left[\theta - \frac{\theta^3}{3!} + \frac{\theta^5}{5!} + \cdots\right]. \end{aligned}$$

We get $e^x = \cos\theta + i\sin\theta$. This is known as Euler's formula, which helps to express a complex number in the polar form as the following:

$$z = r\cos\theta + ir\sin\theta = r(\cos\theta + i\sin\theta) = re^{i\theta}.$$

Note 1: The German mathematician Carl Gauss (1777–1855), who is considered by many as the greatest mathematician ever, formally introduced the standard notation $a + ib$ for complex numbers.

In engineering texts, j is often used instead of i for the square root of -1, to avoid conflict with the notation for electrical current.

Let us consider the sine and cosine of a complex variable. (See Fig. 3.3.) Then we get

$$\sin z = \frac{e^{iz} - e^{-iz}}{2i}$$

and

$$\cos z = \frac{e^{iz} + e^{-iz}}{2}.$$

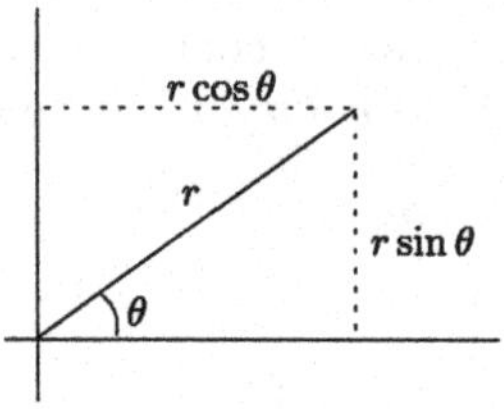

Figure 3.3 The sine and cosine of a complex variable.

By substituting $x = i\theta$ at the expansion of e^x, we get $e^x = \cos\theta + i\sin\theta$. This is known as Euler's formula, which helps to express a complex number in the polar form as follows:

$$\begin{aligned} z &= r\cos\theta + ir\sin\theta \\ &= r(\cos\theta + i\sin\theta) = re^{i\theta}. \end{aligned}$$

Note 2: To determine division of two complex numbers, the polar form can be used. That is,

$$\begin{aligned} z &= r(\cos\theta + i\sin\theta) = re^{i\theta} \\ z^* &= r(\cos\theta - i\sin\theta) = re^{-i\theta}. \end{aligned}$$

From here,

$$\frac{1}{z} = \frac{z^*}{zz^*} = \frac{e^{-i\theta}}{r}.$$

Thus,

$$\begin{aligned} \frac{z_1}{z_2} &= \frac{r_1 e^{i\theta_1} e^{-i\theta_2}}{r_2} \\ &= \frac{r_1}{r_2} e^{i(\theta_1 - \theta_2)}. \end{aligned}$$

Using the exponential representation, consider the sine and cosine functions, by substituting the argument x via the complex variable z. Then we have

$$\begin{aligned} e^{iz} &= \left(1 - \frac{z^2}{2!} + \frac{z^4}{4!} + \cdots\right) + i\left(z - \frac{z^3}{3!} + \frac{z^5}{5!} + \cdots\right) \\ &= \cos z + i\sin z. \end{aligned}$$

In a similar manner,

$$\cos z = \frac{1}{2}(e^{iz} + e^{-iz}) \qquad \sin z = \frac{1}{2i}(e^{iz} - e^{-iz})$$

The sine and cosine of an imaginary angle are

$$\cos(ib) = \frac{1}{2}(e^{-b} + e^{b}) = \cosh b,$$
$$\sin(ib) = \frac{1}{2i}(e^{-b} - e^{b}) = i \sinh b,$$

which we call *hyperbolic functions*. It is clear that the equation $\cosh^2\theta - \sinh^2\theta = 1$ holds, which reminds us of the equation of a hyperbola, $x^2 - y^2 = 1$. This is why they are called hyperbolic functions.

Note that $\cos\theta$ is an even (symmetric) function of θ, so changing the sign of θ should not change the definition; but $\sin\theta$ is an odd (antisymmetric) function, so changing the sign of θ should change the sign in the definition.

Taking into account all these definitions, it is easy to see that the following rules apply to arithmetic in $\mathbb{C}$:

$*z_1 + z_2 = z_2 + z_1$ $\quad$ $*(z_1 + z_2) + z_3 = z_1 + (z_2 + z_3)$

$*z_1 z_2 = z_2 z_1$ $\quad$ $*(z_1 z_2) z_3 = z_1(z_2 z_3)$

$*\overline{z_1 + z_2} = \overline{z_1} + \overline{z_2}$ $\quad$ $*\overline{z_1 z_2} = \overline{z_1}.\overline{z_2}$

$*|\overline{z}| = |z|$ $\quad$ $*|z_1 z_2| = |z_1||z_2|$

$*(z_1 + z_2) z_3 = z_1 z_3 + z_2 z_3$

By exploiting the properties above, it can be said that the set of complex numbers forms a field. Moreover, the set of all unit complex numbers forms a group under multiplication.

3.1.4 The complex logarithm

The logarithmic function is the inverse of the exponential function, meaning that if one acts on z by the logarithmic function $f(z) = \log z$, then by the exponential function one gets z, just as $e^{\log z} = z$. By exploitation of this property,

$$e^{\log z} = z = re^{i\theta} = e^{\ln r} e^{i\theta} = e^{(\ln r) + i\theta}.$$

In other words,

$$\left.\begin{aligned} e^{ix} &= \cos x + i \sin x \\ a + ib &= r[\cos x + i \sin x] \end{aligned}\right\} \Rightarrow re^{ix} \begin{aligned}[t] &= r[\cos x + i \sin x] \\ &= e^{\ln r}[\cos x + i \sin x] \\ &= e^{\ln r} e^{ix} \\ &= e^{\ln r + ix}. \end{aligned}$$

Thus,

$$\log\left(re^{ix}\right) = \log\left(r[\cos x + i\sin x]\right) = \ln r + ix.$$

Here, we see that

$$\log z = \ln r + i\theta.$$

The addition of an integer multiple of 2π to the argument of z changes the imaginary part of the logarithm in the same amount. Hence, for an integer n,

$$\log z = \ln r + i(\theta + 2\pi n).$$

This representation indicates the logarithm of a complex number in polar form. Since $r[\cos x + i\sin x] = r[\cos(x + 2\pi n) + i\sin(x + 2\pi n)]$, for any integer n,

$$\begin{aligned}\log\left(r[\cos x + i\sin x]\right) &= \log\left(r\left[\cos(x + 2\pi n) + i\sin(x + 2\pi n)\right]\right)\\ &= \ln r + i(x + 2\pi n).\end{aligned}$$

This means that the logarithm of a complex number has an infinite number of answers.

Example 3.4. We have $\log(i) = \log(1[\cos\frac{\pi}{2} + i\sin\frac{\pi}{2}]) = \ln(1) + i(\frac{\pi}{2}) = i\frac{\pi}{2}$.

Example 3.5. We have $i^i = [e^{\log(i)}]^i = e^{i\log(i)} = e^{i.i\pi/2} = e^{-\pi/2} = 0.201\ldots$.

Here it is interesting that a pure imaginary power of an imaginary number is completely equal to a real number.

Example 3.6. Compute $(1 + i)^8$.

Let us write the complex number in polar form. Then we have $1 + i = \sqrt{2}e^{i\pi/4}$. Then

$$(1 + i)^8 = \left(\sqrt{2}e^{i\pi/4}\right)^8 = 16e^{2\pi i} = 16.$$

Note 3: For $z = a + ib = re^{i\theta}$, $e^z = e^{a+ib} = e^a e^{ib}$. Therefore,

$$\log e^z = \ln e^a + i(b + 2\pi in) = a + ib + 2\pi in = z + 2\pi in.$$

This means that $\log(e^z)$ is not always equal to z. But as mentioned before, the equation $e^{\log z} = z$ always holds.

Note 4: It is clear that $e^{2\pi i} = 1$. Moreover, for any integer n, $e^{i(2\pi + 2\pi n)} = 1$ holds. Then

$$\left[e^{i(2\pi+2\pi n)}\right]^i = 1^i = 1, \quad \text{since } \left(e^{2\pi i}\right)^i = e^{-2\pi} = 1.$$

Here, we obtain $e^{-(2\pi+2\pi n)} = 1$. For $n = -1$; we verify the following well-known equation:

$$e^{-(2\pi-2\pi)} = e^0 = 1.$$

3.1.5 *Important theorems*

Here, we present some well-known theorems (Flanders and Price, 1975).

De Moivre's theorem

The ease with which two complex numbers in modulus-argument form can be multiplied together gives rise to a very important result known as

$$\left[r(\cos\theta + i\sin\theta)\right]^n = r^n(\cos n\theta + i\sin n\theta).$$

We shall now consider nth roots of complex numbers, where n is an arbitrary positive integer.

Fundamental theorem of algebra

Any polynomial $p(x) \in \mathbb{C}[x]$ of degree $n \geq 1$ has at least one root among the complex numbers. This means that the field of complex numbers is algebraically closed.

As a result, if $p(x)$ is a polynomial of degree $n \geq 1$, then $p(x)$ has precisely n roots among the complex numbers when a root of multiplicity k is counted k times. Note that if the factor $x - r$ appears k times, we say that r is a root of multiplicity k.

Linear factor theorem

A polynomial $p(x)$ of degree $n \geq 1$ can be written as the product of n linear factors. Then we have

$$P(x) = a(x - r_1)(x - r_2)\cdots(x - r_n).$$

Conjugate roots theorem

If $p(x)$ is a polynomial of degree $n \geq 1$ with real coefficients, and if $a + ib$, $b \neq 0$, is a root of $p(x)$, then the conjugate $a - ib$ is also a root of $p(x)$.

As a corollary of this theorem, we can say that if the degree of a real polynomial is odd, it must have at least one real root. We can prove this as follows. Complex roots come in conjugate pairs, so they are in even numbers. But a polynomial of odd degree has the roots in odd number. That is why some of them must be real.

Note 5: For arbitrary complex numbers $z_1$andz_2, we have the inequalities

$$|z_1 + z_2| \leq |z_1| + |z_2|\big||z_1| - |z_2|\big| \leq |z_1 - z_2|.$$

Note 6: Powers of complex numbers can be computed with the help of the binomial theorem, which tells how to compute the powers of a binomial like $x + y$. In other words,

$$(x+y)^n = \sum_{k=0}^{n} \binom{n}{k} x^{n-k} y^k = \sum_{k=0}^{n} \binom{n}{k} x^k y^{n-k}.$$

For example,

$$\begin{aligned}(a+ib)^4 &= a^4 + 4a^3bi + 6a^2b^2i^2 + 4ab^3i^3 + b^4i^4 \\ &= \left(a^4 - 6a^2b^2 + b^4\right) + i\left(4a^3b - 4ab^3\right).\end{aligned}$$

3.1.6 Roots of complex numbers

Let us consider $z = r(\cos\theta + i\sin\theta)$ and $u = \rho(\cos\alpha + i\sin\alpha)$. By exploitation of De Moivre's theorem,

$$r(\cos\theta + i\sin\theta) = \rho^n(\cos\alpha + i\sin\alpha)^n = \rho^n(\cos n\alpha + i\sin n\alpha).$$

Here, $\rho^n = r$, $n\alpha = \theta + 2k\pi$, and k is any integer. Then $\rho = r^{1/n}$, $\alpha = \theta/n + 2k\pi/n$. Thus, we get n distinct values for k. This means that every complex number has exactly n distinct nth roots.

The n distinct roots of $z^n = r(\cos\theta + i\sin\theta)$ are given by

$$z_k = \sqrt[n]{r}\left[\cos\left(\frac{\theta + 2\pi k}{n}\right) + i\sin\left(\frac{\theta + 2\pi k}{n}\right)\right],$$

where $k = 0, 1, 2, \ldots, n-1$ (Andreescu and Andrica, 2006).

Example 3.7. Find the roots of $z^3 = (-1+i)$. We have

$$u = (-1+i)^{\frac{1}{3}} = (\sqrt{2})^{\frac{1}{3}}\left[\cos\left(\frac{3\pi}{4}\cdot\frac{1}{3} + \frac{2k\pi}{3}\right) + i\sin\left(\frac{3\pi}{4}\cdot\frac{1}{3} + \frac{2k\pi}{3}\right)\right],$$

$$k = 0, 1, 2.$$

Then the roots are

$$(\sqrt{2})^{\frac{1}{3}}\left[\cos\left(\frac{\pi}{4}\right) + i\sin\left(\frac{\pi}{4}\right)\right] \quad (k = 0),$$

$$(\sqrt{2})^{\frac{1}{3}}\left[\cos\left(\frac{11\pi}{12}\right) + i\sin\left(\frac{11\pi}{12}\right)\right] \quad (k = 1),$$

$$(\sqrt{2})^{\frac{1}{3}}\left[\cos\left(\frac{19\pi}{12}\right) + i\sin\left(\frac{19\pi}{12}\right)\right] \quad (k = 2).$$

3.1.7 Matrix representation

Another way of considering complex numbers exploits matrices. Let us consider the following map:

$$\theta : (\mathbb{C}, +, \cdot) \to \left(M(2, \mathbb{R}), \oplus, \odot\right)$$

$$\theta(a + ib) \mapsto \begin{bmatrix} a & -b \\ b & a \end{bmatrix},$$

where $\oplus$, and $\odot$ denotes the matrix addition and multiplication, respectively, is an isomorphism (Ward, 1997).

Then one can write

$$(a+ib)(c+id) = (ac-bd)+i(ad+bc) = \begin{bmatrix} a & -b \\ b & a \end{bmatrix}\begin{bmatrix} c \\ d \end{bmatrix} = \begin{bmatrix} ac-bd \\ ad+bc \end{bmatrix},$$

i.e., $(a,b)(c,d) = (ac-bd, ad+bc)$. The real and imaginary part of a complex number can be represented as follows (Ward, 1997):

$$\left.\begin{matrix} a=1 \\ b=0 \end{matrix}\right\}; 1+i0 = \begin{bmatrix} 1 & 0 \\ 0 & 1 \end{bmatrix} \quad \text{and} \quad \left.\begin{matrix} a=0 \\ b=1 \end{matrix}\right\}; 0+i1 = \begin{bmatrix} 0 & -1 \\ 1 & 0 \end{bmatrix}.$$

To verify that $i^2 = -1$, we can use the matrix representation, i.e.,

$$\begin{bmatrix} 0 & -1 \\ 1 & 0 \end{bmatrix}\begin{bmatrix} 0 & -1 \\ 1 & 0 \end{bmatrix} = \begin{bmatrix} -1 & 0 \\ 0 & -1 \end{bmatrix} = -\begin{bmatrix} 1 & 0 \\ 0 & 1 \end{bmatrix} = -1.$$

As is well known, complex numbers are commutative under multiplication. We can verify it for the matrix representation as follows:

$$z_1 z_2 = z_2 z_1$$

$$\begin{bmatrix} a & -b \\ b & a \end{bmatrix}\begin{bmatrix} c & -d \\ d & c \end{bmatrix} = \begin{bmatrix} c & -d \\ d & c \end{bmatrix}\begin{bmatrix} a & -b \\ b & a \end{bmatrix}$$

Moreover the complex conjugate is

$$z_1 = a+ib \Rightarrow \overline{z_1} = a-ib \Rightarrow \begin{bmatrix} a & b \\ -b & a \end{bmatrix}.$$

Note that the determinant of the matrix form of a complex number gives the square of the length, i.e.,

$$z_1 = a+ib \Rightarrow \det\left(\begin{bmatrix} a & -b \\ b & a \end{bmatrix}\right) = a^2+b^2 = |z_1|^2.$$

By exploiting the polar form of a complex number, $e^{i\theta} = \cos\theta + i\sin\theta$, which is in the form of $a+ib$, we have

$$\exp\left(\begin{bmatrix} 0 & -\theta \\ \theta & 0 \end{bmatrix}\right) = \begin{bmatrix} \cos\theta & -\sin\theta \\ \sin\theta & \cos\theta \end{bmatrix},$$

where $i\theta = \begin{bmatrix} 0 & -\theta \\ \theta & 0 \end{bmatrix}$. This matrix is defined as a 2D rotation matrix.

Note that the determinant of the rotation matrix is equal to "1" and the inverse of the rotation matrix is its transpose.

Let us consider the complex number $z = a+ib$, which is the sum of

$$z = z_x + z_y,$$

where $z_x = (a,0)$ and $z_y = (0,b)$. Let z' be the rotated form of z. Then $z' = z'_x + z'_y$.

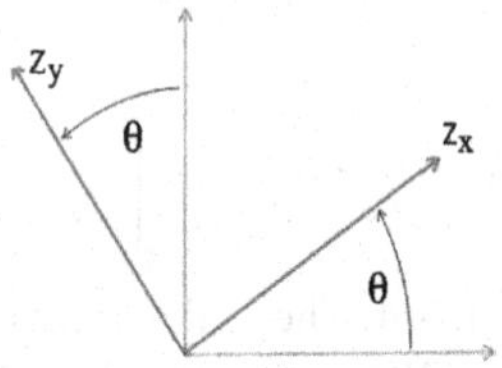

This means

$$\begin{aligned}(a,b)' &= (a,0)' + (0,b)' \\ &= (a\cos\theta, a\sin\theta) + (-b\sin\theta, b\cos\theta) \\ &= (a\cos\theta - b\sin\theta, a\sin\theta + b\cos\theta)\end{aligned}$$

which is the rotated form of (a,b).

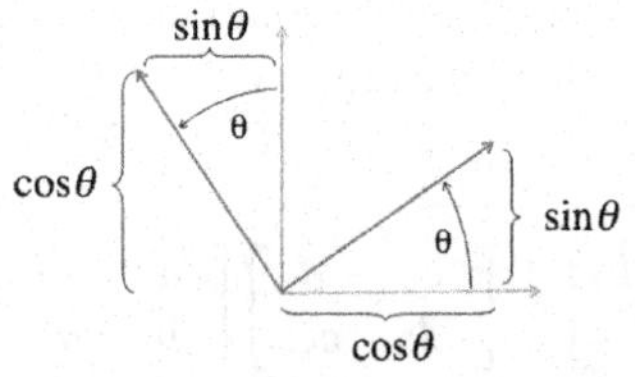

We know that $(a,b)(c,d) = (ac - bd, ad + bc)$, and let $c = \cos\theta$ and $d = \sin\theta$. Then

$$\begin{aligned}(a,b)(\cos\theta, \sin\theta) &= (\cos\theta, \sin\theta)(a,b) \\ &= (\cos\theta + i\sin\theta)(a + ib).\end{aligned}$$

The rotation is counterclockwise.

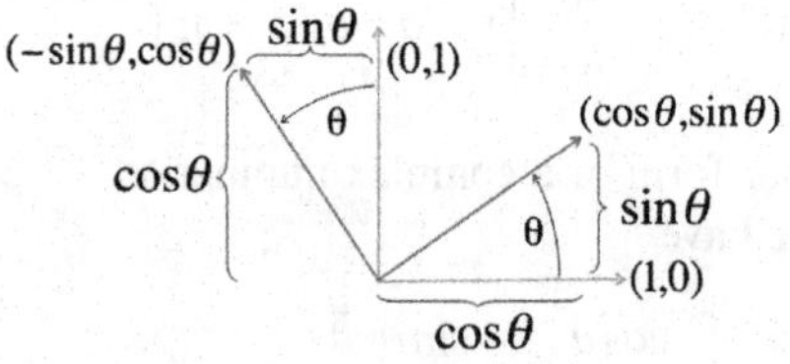

Namely, the rotated form of $z = a + ib = (a,b)$ is

$$\begin{aligned}(a+ib)' &= (\cos\theta + i\sin\theta)(a+ib) \\ &= e^{i\theta}(a+ib).\end{aligned}$$

Example 3.8. Let $r = 1$, $\theta = 30°$, and $\varphi = 30°$, so

$$\begin{aligned}x &= r\cos\varphi = \cos\varphi = \cos\theta = \cos 30° = \sqrt{3}/2, \\ y &= r\sin\varphi = \sin\varphi = \sin\theta = \sin 30° = 1/2.\end{aligned}$$

Then we can compute the values x' and y':

$$\begin{bmatrix} x' \\ y' \end{bmatrix} = \begin{bmatrix} \sqrt{3}/2 & -1/2 \\ 1/2 & \sqrt{3}/2 \end{bmatrix} \begin{bmatrix} \sqrt{3}/2 \\ 1/2 \end{bmatrix} = \begin{bmatrix} 1/2 \\ \sqrt{3}/2 \end{bmatrix}.$$

The result makes sense since $\theta + \varphi = 60°$.

3.2 Illustrations

Complex numbers are utilized by engineers in numerous branches of science, usually in applications related with physics or computer design, or in operations where a point can be considered as a complex number. We are giving some examples in the next sections, but also in Chapter 12, about higher-order ordinary differential equations (ODEs), because problems modeled with second-order ODEs usually require complex numbers in their resolution.

Example 3.9. Simplify $(2+i)^5$ in the form $a+ib$: By exploiting the binomial theorem, we have

$$\begin{aligned}(2+i)^5 &= 32 + 80i - 80 - 40i + 10 + i \\ &= -38 + 41i.\end{aligned}$$

Example 3.10. We have

$$\begin{aligned}\left(-\frac{1}{2}+\frac{\sqrt{3}}{2}i\right)^{17} &= \left(\cos\frac{2\pi}{3}+i\sin\frac{2\pi}{3}\right)^{17} = \cos\frac{34\pi}{3}+i\sin\frac{34\pi}{3} \\ &= \cos\frac{4\pi}{3}+i\sin\frac{4\pi}{3} = -\frac{1}{2}-i\frac{\sqrt{3}}{2}.\end{aligned}$$

Example 3.11. We have $(-\frac{1}{2}-\frac{\sqrt{3}}{2}i)^{-1} = (\cos\frac{4\pi}{3}+i\sin\frac{4\pi}{3})^{-1} = \cos\frac{4\pi}{3} - i\sin\frac{4\pi}{3} = -\frac{1}{2}+i\frac{\sqrt{3}}{2}$.

Example 3.12. Solve the equation $z^3 = 2i$.

Since $z = r(\cos\theta + i\sin\theta)$, $2i = 2(\cos\frac{\pi}{2}+i\sin\frac{\pi}{2})$, $z^3 = 2i$ means

$$r^3(\cos 3\theta + i\sin 3\theta) = 2\left(\cos\frac{\pi}{2}+i\sin\frac{\pi}{2}\right).$$

Taking into account that $r = \sqrt[3]{2}$ and $3\theta = \frac{\pi}{2} + 2k\pi$ for $k = 0, 1, 2$; the roots are

$$z_0 = \sqrt[3]{2}\left(\cos\frac{\pi}{6}+i\sin\frac{\pi}{6}\right) = \sqrt[3]{2}\left(\frac{\sqrt{3}}{2}+\frac{1}{2}i\right),$$

$$z_1 = \sqrt[3]{2}\left(\cos\frac{5\pi}{6} + i\sin\frac{5\pi}{6}\right) = \sqrt[3]{2}\left(-\frac{\sqrt{3}}{2} + \frac{1}{2}i\right),$$

$$z_2 = \sqrt[3]{2}\left(\cos\frac{3\pi}{2} + i\sin\frac{3\pi}{2}\right) = -i\sqrt[3]{2}.$$

Example 3.13. Find all roots of the equation $z^4 - i - 1 = 0$.
Here $z^4 = i + 1$, and $\tan\theta = 1 \Rightarrow \theta = \frac{\pi}{4}$, $r = \sqrt{2}$. By exploiting the theorem,

$$z_k = \sqrt[8]{2}\left[\cos\left(\frac{\pi}{16} + \frac{\pi k}{2}\right) + i\sin\left(\frac{\pi}{16} + \frac{\pi k}{2}\right)\right],$$

where $k = 0, 1, 2, 3$. Then

$$z_0 = \sqrt[8]{2}\left[\cos\left(\frac{\pi}{16}\right) + i\sin\left(\frac{\pi}{16}\right)\right],$$

$$z_1 = \sqrt[8]{2}\left[\cos\left(\frac{9\pi}{16}\right) + i\sin\left(\frac{9\pi}{16}\right)\right],$$

$$z_2 = \sqrt[8]{2}\left[\cos\left(\frac{17\pi}{16}\right) + i\sin\left(\frac{17\pi}{16}\right)\right],$$

$$z_3 = \sqrt[8]{2}\left[\cos\left(\frac{25\pi}{16}\right) + i\sin\left(\frac{25\pi}{16}\right)\right].$$

Alternative: We have $z^4 = i + 1 \Rightarrow$ let $z^4 = z_1$ where $z_1 = i + 1$. So we have $z_1 = |z_1|e^{i\theta}$ where $|z_1| = \sqrt{2}$ and $\theta = \frac{\pi}{4}$. So,

$$z^4 = \sqrt{2}e^{i\pi/4} \Rightarrow z_k = (\sqrt{2})^{\frac{1}{4}} e^{i\left(\frac{\pi/4 + 2k\pi}{4}\right)}, \text{ where } k = 0, 1, 2, 3.$$

Example 3.14. Calculate $\sqrt{i}$.
Firstly, we have $i = e^{i\pi/2}$, from Euler's formula. Then

$$\sqrt{i} = \left(e^{i\pi/2}\right)^{1/2} = e^{i\pi/4} = \cos(\pi/4) + i\sin(\pi/4) = \frac{1}{\sqrt{2}} + i\frac{1}{\sqrt{2}}.$$

Let us verify the result. We have

$$\left(\frac{1}{\sqrt{2}} + i\frac{1}{\sqrt{2}}\right)^2 = \frac{1}{2} + 2i\frac{1}{\sqrt{2}}\frac{1}{\sqrt{2}} + i^2\left(\frac{1}{\sqrt{2}}\right)^2 = i.$$

Example 3.15. Consider the rotation of $z = a + ib$ with $\theta = \frac{\pi}{2}$. We have

$$e^{\pi i/2}(a + ib) = i(a + ib) = ai + i^2 b = -b + ai = (-b, a).$$

Exercises. 1. If $z_1 = 3+4i$ and $z_2 = -1+i$; find the values:

$$\text{a) } z_1 + z_2, \qquad \text{b) } z_1 z_2, \qquad \text{c) } \frac{z_1}{z_2}, \qquad \text{d) } z_1^3.$$

2. Prove that every complex number with modulus equal to 1 is a quotient of two conjugate numbers.

3. Express the following terms in the form $a+ib$:

$$\text{a) } \frac{1}{2+i}, \qquad \text{b) } \frac{-1}{i}, \qquad \text{c) } \frac{i}{1+i}, \qquad \text{d) } \frac{2}{2+3i}.$$

4. Find a quadratic equation whose roots are $1-2i$ and $1+2i$.

5. Express the following terms in the form $a+ib$:

$$\text{a) } \frac{i}{2+i}, \qquad \text{b) } \frac{i-1}{i+3}, \qquad \text{c) } \frac{2i}{1+i}, \qquad \text{d) } \frac{i}{1+3i}.$$

6. Solve the following equations:

$$\text{a) } x^2+9=0, \qquad \text{b) } x^2+x+1=0, \qquad \text{c) } x^2+8x+20=0, \qquad \text{d) } x^2+27=0.$$

7. Simplify the following expressions in the form $a+ib$:

$$\text{a) } (3+4i)+(1+i), \qquad \text{b) } (3+4i)(1+i),$$
$$\text{c) } (3-4i)+(1+i), \qquad \text{d) } (3+4i)(1-i).$$

3.3 Quaternions

As an extension of the complex number system, quaternions are comprised of a scalar and three imaginary numbers, and are generally represented in the form of their linear combinations, which were introduced by Hamilton in 1843. It is an interesting mathematical concept with a deep relationship with the foundations of algebra and number theory. Quaternions are widely used as attitude representation parameter of rigid bodies and in control theory, signal processing, computer graphics, and physics.

As is well known, complex numbers describe a 2D space (the xy-plane) and rotations in the xy-plane are commutative. Also, the multiplication of complex numbers is commutative. However, to describe rotations in 3D space, we are immediately faced with a difficulty. The quaternion is most useful to us as a means of representing orientations.

A quaternion has four components (Hamilton, 1866), $q = [q_0\ q_1\ q_2\ q_3]$, where one of the four components is a real scalar number, and the other three form a vector in the imaginary $\{i, j, k\}$-space, i.e.,

$$q = q_0 + q_1 i + q_2 j + q_3 k,$$

where $i^2 = j^2 = k^2 = ijk = -1$, $i = jk = -kj$, $j = ki = -ik$, and $k = ij = -ji$.

Sometimes, it is denoted as the combination of a scalar value s and a vector value v,

$$q = \langle s, v \rangle,$$

where $s = q_0$ and $v = [q_1 \;\; q_2 \;\; q_3]$. In other words,

$$q = s_q + v_q,$$

where s_q denotes the scalar part and v_q is the vector part ($s_q = a$ and $v_q = q_1 i + q_2 j + q_3 k$).

In a similar manner with complex numbers, the *conjugate* of q is

$$\overline{q} = s_q - v_q.$$

The *sum of quaternions* is the same as the vector sum, i.e., for $q = q_0 + q_1 i + q_2 j + q_3 k$ and $p = p_0 + p_1 i + p_2 j + p_3 k$,

$$q + p = [q_0 + p_0 \;\; q_1 + p_1 \;\; q_2 + p_2 \;\; q_3 + p_3].$$

The *product of quaternions* follows a special rule: for $q = s_q + v_q$ and $p = s_p + v_p$,

$$q * p = (s_q s_p - \langle v_q v_p \rangle, s_q v_p + s_p v_q + v_q \times v_p),$$

where $\langle v_q v_p \rangle$ denotes the *dot product* and $v_q \times v_p$ denotes the *cross product* of vectors v_q and v_p, (Baek et al., 2017). The product of two quaternions can be derived by writing the quaternions in extended complex form, i.e.,

$$\begin{aligned} q * p = &(q_0 p_0 - q_1 p_1 - q_2 p_2 - q_3 p_3) + (q_1 p_0 + q_0 p_1 - q_3 p_2 + q_2 p_3) i \\ &+ (q_2 p_0 + q_3 p_1 + q_0 p_2 - q_1 p_3) j + (q_3 p_0 - q_2 p_1 + q_1 p_2 + q_0 p_3) k. \end{aligned}$$

In general, quaternion multiplication does not commute. Since the quaternion algebra is associative, they can be considered in terms of matrices. Also, the product of two quaternions can be also written as *matrix product.* In other words, the following map is an isomorphism:

$$\theta : (\mathrm{H}, +, \cdot) \to \big(M(4, \mathbb{R}), \oplus, \odot\big)$$

$$\theta(q_0 + q_1 i + q_2 j + q_3 k) \mapsto \begin{bmatrix} q_0 & -q_1 & -q_2 & -q_3 \\ q_1 & q_0 & -q_3 & q_2 \\ q_2 & q_3 & q_0 & -q_1 \\ q_3 & -q_2 & q_1 & q_0 \end{bmatrix},$$

where $\mathrm{H} = \{q_0 + q_1 i + q_2 j + q_3 k : q_0, q_1, q_2, q_3 \in \mathbb{R}\}$, and $\oplus$, and $\odot$ denote the matrix addition and multiplication, respectively (Ward, 1997). Via this property, the multiplication of two quaternions can be written as

$$q * p = (q_0 + q_1 i + q_2 j + q_3 k)(p_0 + p_1 i + p_2 j + p_3 k)$$
$$= \begin{bmatrix} q_0 & -q_1 & -q_2 & -q_3 \\ q_1 & q_0 & -q_3 & q_2 \\ q_2 & q_3 & q_0 & -q_1 \\ q_3 & -q_2 & q_1 & q_0 \end{bmatrix} \begin{bmatrix} p_0 \\ p_1 \\ p_2 \\ p_3 \end{bmatrix},$$

which corresponds to the definition of product of quaternions.

The *length of a quaternion* is the same as the norm of the corresponding vector, i.e.,

$$|q| = \sqrt{q\overline{q}} = \sqrt{q_0^2 + q_1^2 + q_2^2 + q_3^2}.$$

The *inverse of* a non-zero quaternion q with respect to the quaternion product is given as

$$q^{-1} = \frac{\overline{q}}{|q|^2}.$$

Let us compute some powers of $q = (0, v_q)$, i.e.,

$$q^2 = (0, v_q)(0, v_q) = (-v_q v_q) = -|v_q|^2.$$

$$q^3 = q^2 q = -|v_q|^2 v_q.$$

$$q^4 = q^2 q^2 = |v_q|^4.$$

Then consider the Taylor series expansion of the exponential function for $q = (0, v_q)$. We have

$$e^q = 1 + q + \frac{q^2}{2!} + \frac{q^3}{3!} + \frac{q^4}{4!} + \cdots$$
$$= 1 + v_q - \frac{|v_q|^2}{2!} - \frac{|v_q|^2 v_q}{3!} + \frac{|v_q|^4}{4!} + \frac{|v_q|^4 v_q}{5!} - \frac{|v_q|^6}{6!} \cdots$$

We need to find what function has this Taylor expansion. The scalar and vector parts have the following forms, respectively:

$$1 - \frac{|v_q|^2}{2!} + \frac{|v_q|^4}{4!} - \frac{|v_q|^6}{6!} \cdots$$
$$\left[1 - \frac{|v_q|^2}{3!} + \frac{|v_q|^4}{5!} - \frac{|v_q|^6}{7!} + \cdots\right] v_q.$$

The vector part can be characterized by

$$\frac{\sin(v_q)}{|v_q|} v_q$$

and the scalar part looks like a cosine function for $x = |v_q|$. By combining these results, we have

$$e^{q_1 i+q_2 j+q_3 k} = \cos\big(|v_q|\big) + \frac{\sin(|v_q|)}{|v_q|} v_q,$$

where $|v_q| = \sqrt{q_1^2 + q_2^2 + q_3^2}$. Consequently, the *exponential of quaternion* $q = s_q + v_q$ is given as

$$\exp(q) = \exp(s_q)(\cos(|v_q|) + \frac{v_q}{|v_q|} \sin\big(|v_q|\big)).$$

By exploiting this formula, we verify that

$$e^{ix} = \cos(x) + \frac{\sin(x)}{x}(ix) = \cos(x) + i\sin(x).$$

Example 3.16. We have $e^{1+i+j} = e.e^{i+j} = e[\cos(\sqrt{2}) + \frac{\sin(\sqrt{2})}{\sqrt{2}}(i + j)] = e[\cos(\sqrt{2}) + \frac{\sin(\sqrt{2})}{\sqrt{2}} i + \frac{\sin(\sqrt{2})}{\sqrt{2}} j)]$.

Note: For quaternions, the statement $e^{i+j} = e^i e^j$ is not valid.

Note: We have $e^{\pi i} = e^{\pi j} = e^{\pi k} = e^{\frac{\pi i}{\sqrt{2}} + \frac{\pi j}{\sqrt{2}}} = e^{\frac{\pi i}{\sqrt{3}} + \frac{\pi j}{\sqrt{3}} + \frac{\pi k}{\sqrt{3}}} = -1$.

Example 3.17. We have $e^{3j+4k} = [\cos(5) + \frac{\sin(5)}{5}(3i + 4j)] \approx 0,284 - 0,575j - 0,767k$.

The *logarithm of quaternion* $q = s_q + v_q$ is given as

$$\log(q) = \log\big(|q|\big) + \frac{v_q}{|v_q|} \arccos\left(\frac{s_q}{|q|}\right).$$

A *rotation* from one coordinate frame A to another B is given by the conjugation operation (Parent, 2002).

$$q_B = q_R q_A q_R^{-1}.$$

For the real unit quaternion, let us obtain the matrix that corresponds to the rotation. If $q = q_0 + q_1 i + q_2 j + q_3\, k \in \mathbb{H}$, then

$$q_0^2 + q_1^2 + q_2^2 + q_3^2 = 1 \Rightarrow q_0^2 + q_1^2 = 1 - q_2^2 - q_3^2, \quad \text{and}$$

(i)

$$\begin{aligned} q1q^{-1} &= (q_0 + q_1 i + q_2 j + q_3 k) * 1 * (q_0 - q_1 i - q_2 j - q_3 k) \\ &= q * \overline{q} = q_0^2 + q_1^2 + q_2^2 + q_3^2 = 1, \end{aligned}$$

(ii)

$$
\begin{aligned}
qiq^{-1} &= (q_0 + q_1 i + q_2 j + q_3 k) * \mathrm{i} * (q_0 - q_1 i - q_2 j - q_3 k) \\
&= (q_0 i - q_1 - q_2 k + q_3 j) * (q_0 - q_1 i - q_2 j - q_3 k) \\
&= q_0^2 i + q_0 q_1 - q_0 q_2 k + q_0 q_3 j \\
&\quad - q_1 q_0 + q_1^2 i + q_1 q_2 j + q_1 q_3 k \\
&\quad - q_2 q_0 k + q_2 q_1 j - q_2^2 i - q_2 q_3 \\
&\quad + q_3 q_0 i + q_3 q_1 k + q_3 q_2 - q_3^2 i \\
&= i\left(q_0^2 + q_1^2 - q_2^2 - q_3^2\right) + j(2q_0 q_3 + 2q_1 q_2) + k(2q_1 q_3 - 2q_2 q_0) \\
&= i\left(1 - 2\left(q_2^2 + q_3^2\right)\right) + j(2q_0 q_3 + 2q_1 q_2) + k(2q_1 q_3 - 2q_2 q_0),
\end{aligned}
$$

(iii)

$$
\begin{aligned}
qjq^{-1} &= (q_0 + q_1 i + q_2 j + q_3 k) * \mathrm{j} * (q_0 - q_1 i - q_2 j - q_3 k) \\
&= (q_0 j + q_1 k - q_2 - q_3 i) * (q_0 - q_1 i - q_2 j - q_3 k) \\
&= q_0^2 j + q_0 q_1 k - q_0 q_2 - q_0 q_3 i \\
&\quad + q_1 q_0 k - q_1^2 j + q_1 q_2 i + q_1 q_3 \\
&\quad - q_2 q_0 + q_2 q_1 i + q_2^2 j - q_2 q_3 k \\
&\quad - q_3 q_0 i - q_3 q_1 + q_3 q_2 k - q_3^2 j \\
&= i(2q_1 q_2 - 2q_0 q_3) + j\left(q_0^2 - q_1^2 + q_2^2 - q_3^2\right) + k(2q_0 q_1 + 2q_2 q_3) \\
&= i(2q_1 q_2 - 2q_0 q_3) + j\left(1 - 2\left(q_1^2 + q_3^2\right)\right) + k(2q_0 q_1 + 2q_2 q_3),
\end{aligned}
$$

(iv)

$$
\begin{aligned}
qkq^{-1} &= (q_0 + q_1 i + q_2 j + q_3 k) * \mathrm{k} * (q_0 - q_1 i - q_2 j - q_3 k) \\
&= (q_0 k - q_1 j + q_2 i - q_3) * (q_0 - q_1 i - q_2 j - q_3 k) \\
&= q_0^2 k - q_0 q_1 j + q_0 q_2 i + q_0 q_3 \\
&\quad - q_1 q_0 j - q_1^2 k - q_1 q_2 + q_1 q_3 i \\
&\quad + q_2 q_0 i + q_2 q_1 - q_2^2 k + q_2 q_3 j \\
&\quad - q_3 q_0 + q_3 q_1 i + q_3 q_2 j + q_3^2 k \\
&= i(2q_2 q_0 + 2q_1 q_3) + j(2q_2 q_3 - 2q_0 q_1) + k\left(q_0^2 - q_1^2 - q_2^2 + q_3^2\right) \\
&= i(2q_2 q_0 + 2q_1 q_3) + j(2q_2 q_3 - 2q_0 q_1) + k\left(1 - 2\left(q_1^2 + q_2^2\right)\right).
\end{aligned}
$$

Then the rotation matrix is

$$
Q = \begin{bmatrix}
1 & 0 & 0 & 0 \\
0 & 1 - 2(q_2^2 + q_3^2) & 2(q_1 q_2 - q_0 q_3) & 2(q_2 q_0 + q_1 q_3) \\
0 & 2(q_0 q_3 + q_1 q_2) & 1 - 2(q_1^2 + q_3^2) & 2(q_2 q_3 - q_0 q_1) \\
0 & 2(q_1 q_3 - q_2 q_0) & 2(q_0 q_1 + q_2 q_3) & 1 - 2(q_1^2 + q_2^2)
\end{bmatrix}.
$$

Quaternions have a number of advantages over matrices as a means of representing rotations, since only four numbers, the minimum possible, are needed to represent a rotation, fast and correctly.

Example 3.18. Consider a security camera whose position is given by a real quaternion $\hat{x}_b = \frac{\sqrt{2}}{2}i + \frac{\sqrt{2}}{2}j$. Find new position of the camera with rotation matrix, after rotation around the pillar whose position is $q = \frac{\sqrt{2}}{2} + \frac{\sqrt{2}}{2}j$. (See Fig. 3.4.)

Figure 3.4 Rotation of security camera.

Solution: By exploiting the rotation matrix, we get

$$q_0 = \frac{\sqrt{2}}{2}, \qquad q_1 = 0, \qquad q_2 = \frac{\sqrt{2}}{2}, \qquad q_3 = 0,$$

$$Q = \begin{bmatrix} 1 & 0 & 0 & 0 \\ 0 & 0 & 0 & 1 \\ 0 & 0 & 1 & 0 \\ 0 & -1 & 0 & 0 \end{bmatrix} \begin{bmatrix} 0 \\ \frac{\sqrt{2}}{2} \\ \frac{\sqrt{2}}{2} \\ 0 \end{bmatrix}$$

$$= \begin{bmatrix} 0 \\ 0 \\ \frac{\sqrt{2}}{2} \\ -\frac{\sqrt{2}}{2} \end{bmatrix} := \frac{\sqrt{2}}{2}j - \frac{\sqrt{2}}{2}k.$$

Rotation matrices, which can be characterized as orthogonal matrices whose determinant is 1, are square matrices with real entries.

Example 3.19. Find the image of $(1, -1, 2)$ with rotation by 60° on an axis in the yz-plane that is curved with 60° to the positive y-axis.

Solution: Let **u** be the unit vector in the direction of the axis of rotation $\cos 60^\circ j + \sin 60^\circ k = \frac{1}{2}j + \frac{\sqrt{3}}{2}k$. The vector corresponding to the point $(1, -1, 2)$ is $p = i - j + 2k$. To obtain the image of p under the rotation, we need to compute qpq^{-1}, where q is the quaternion $\cos\frac{\theta}{2} + \sin\frac{\theta}{2}\boldsymbol{u}$ and θ is the rotation angle of the rotation (here 60°). We have

$$q = \frac{\sqrt{3}}{2} + \frac{1}{2}\mathrm{u}$$

$$= \frac{\sqrt{3}}{2} + \frac{1}{4}j + \frac{\sqrt{3}}{4}k = \frac{1}{4}(2\sqrt{3} + j + \sqrt{3}k).$$

Since q is a unit quaternion, its inverse is also its conjugate, i.e., $q^{-1} = \frac{1}{4}(2\sqrt{3} - j - \sqrt{3}k)$. Then

$$qp = \frac{1}{4}[(1 - 2\sqrt{3}) + (2 + 3\sqrt{3})i - \sqrt{3}j + (4\sqrt{3} - 1)k]$$

and

$$qpq^{-1} = \frac{1}{8}[(10 + 4\sqrt{3})i + (1 + 2\sqrt{3})j + (14 - 3\sqrt{3})k].$$

That point is

$$\left(\frac{10 + 4\sqrt{3}}{8}, \frac{1 + 2\sqrt{3}}{8}, \frac{14 - 3\sqrt{3}}{8}\right).$$

Exercises. **1.** Every transformation of a rigid body or a coordinate system with respect to a reference coordinate system can be expressed by a screw displacement, which is a translation along a λ-axis with a rotation by an angle θ about the same axis. This description of transformation is the basis of the screw theory. Describe screw motions by means of quaternions.

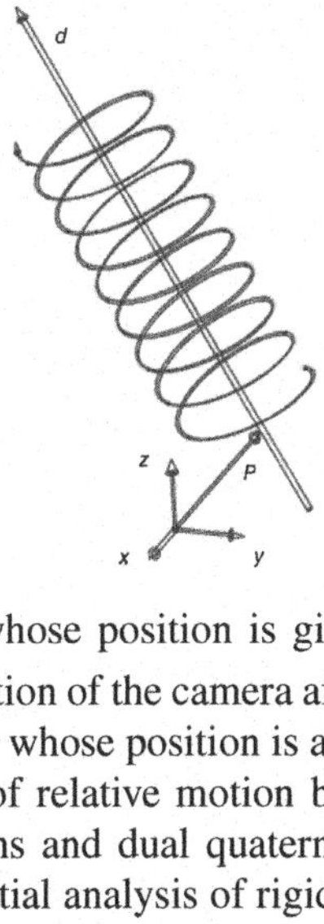

2. Consider a security camera whose position is given by a real quaternion $\hat{x}_b = \frac{\sqrt{2}}{2}i + \frac{\sqrt{2}}{2}j$. Find the new position of the camera after rotation with angle $\frac{\pi}{2}$ counterclockwise, around the pillar whose position is also a real quaternion $\hat{q}_b = j$.

3. The mathematical definition of relative motion between human body segments is a complex task. Quaternions and dual quaternion interpolation are powerful mathematical tools for the spatial analysis of rigid body motions. By using these tools, present the kinematic modeling of human joints.

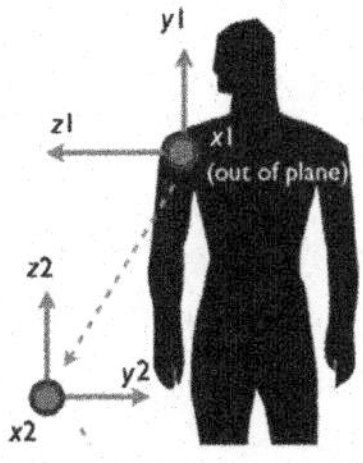

References

Andreescu, T., Andrica, D., 2006. Complex Numbers From A to ... Z. Birkhauser, Boston.
Baek, J., Jeon, H., Kim, G., Han, S., 2017. Visualizing Quaternion Multiplication, vol. 5. IEEE, pp. 8948–8955.
Flanders, H., Price, J.J., 1975. Algebra and Trigonometry. Academic Press, Inc.
Hamilton, W., 1866. Elements of Quaternions. Cambridge Univ. Press, Cambridge, U.K.
Parent, R., 2002. Computer Animation Algorithms and Techniques. M. Kaufmann Publishers.
Ward, J.P., 1997. Quaternions and Cayley Numbers. First Press, Springer.

Sequences and series: a tool for approximation

4

Petr Habala
Czech Technical University in Prague, Prague, Czech Republic

4.1 Sequences and series: overview of theory

The traditional mathematical definition of a sequence uses the notion of mapping. However, for practical purposes we usually quickly change our point of view and also notation to something more natural.

Informally, by a sequence we mean any countable ordered set of real or complex numbers. We can index it using successive integers starting from some n_0, and we then write such a sequence as $\{a_n\}_{n=n_0}^{\infty}$ or as a_n, $n = n_0, n_0 + 1, n_0 + 2, \ldots$.

The most popular starting points for indexing are $n_0 = 1$ and $n_0 = 0$. The former is natural; we count the given numbers as the first, second, third, and so on. The latter is not natural (even professional mathematicians rarely count things as zeroth, first, second,...), but it makes good sense in some settings, most notably when it comes to series.

Note that indexing is just a technical tool, the substance of every sequence is the numbers in it. For instance, the formulas $\{n+1\}_{n=0}^{\infty}$ and $\{n-1\}_{n=2}^{\infty}$ are formally different, but they represent the same sequence $\{1, 2, 3, 4, 5, 6, \ldots\}$. It is hard to do any meaningful mathematical work with sequences without having them captured in a formula, but we should be aware that every such expression is just one of many possible mathematical guises for the underlying object.

The key notion related to sequences is that of a limit.

Definition. Consider a sequence $\{a_n\}_{n=n_0}^{\infty}$. We say that $L \in \mathbb{R}$ is the **limit** of this sequence, denoted $\lim\limits_{n\to\infty}(a_n) = L$ or $a_n \to L$, if the following condition is true:

- For every $\varepsilon > 0$ there exists $N \in \mathbb{N}$ so that $|a_n - L| < \varepsilon$ whenever $n \geq N$.

If such L exists, we say that the sequence $\{a_n\}$ converges to L or that it is **convergent**; we may also say that the limit $\lim\limits_{n\to\infty}(a_n)$ converges.

Otherwise we say that this sequence or the limit $\lim\limits_{n\to\infty}(a_n)$ is **divergent**.

Note that from the notation for limit we cannot tell where the indexing of our sequence starts. There is a good reason for it: Limit is decided at the "end" of the sequence; hence where (and how) it starts is irrelevant.

Note that the definition also applies to sequences of complex numbers.

There are many popular sequences that obviously go somewhere, but not necessarily to a limit in the above sense, for instance the sequence $a_n = 13n$. As n grows,

Calculus for Engineering Students. https://doi.org/10.1016/B978-0-12-817210-0.00011-4

numbers a_n grow beyond any bound, intuitively we would say that they go to infinity. This is a useful idea that is worth capturing mathematically.

Definition. Consider a sequence $\{a_n\}_{n=n_0}^{\infty}$ of real numbers.

We say that ∞ is the limit of this sequence, denoted $\lim\limits_{n\to\infty}(a_n)=\infty$ or $a_n\to\infty$, if the following condition is true:

• For every $K>0$ there exists $N\in\mathbb{N}$ so that $a_n>K$ whenever $n\geq N$.

We say that $-\infty$ is the limit of this sequence, denoted $\lim\limits_{n\to\infty}(a_n)=-\infty$ or also $a_n\to-\infty$, if the following condition is true:

• For every $K>0$ there exists $N\in\mathbb{N}$ so that $a_n<-K$ whenever $n\geq N$.

A sequence can have a **proper limit** L (a real number) or an **improper limit** $\pm\infty$; in both cases we would say that the limit $\lim\limits_{n\to\infty}(a_n)$ exists. When a sequence does not have any limit at all, proper or improper, we say that $\lim\limits_{n\to\infty}(a_n)$ does not exist. One can expect to see some oscillation in such a sequence.

If we work with complex sequences, the improper limit has to be done differently, since there is a fundamental difference in topology of real and complex numbers. We will leave this topic to complex analysis.

How do we determine limits of sequences? There is no reliable algorithm, but there are some useful strategies.

• We usually start with the "substitute and see" step. If a_n is given by an algebraic formula featuring elementary functions, we can "substitute" infinity for n; if it leads to an answer, then this answer is correct. The "evaluation" uses special limit algebra valid for extended real numbers that includes formulas like $L^\infty=\infty$ for $L>1$, $\frac{L}{\infty}=0$, and others.

This strategy fails when evaluation leads to expressions that the limit algebra cannot handle, for instance indeterminate expressions like $\frac{\infty}{\infty}$ and 1^∞, or limits where we substitute infinity into sine and cosine. We then have to use another approach. The failed outcome of this step then serves as a guide which approach is likely to work, because it identified problems.

• A very popular and powerful approach is to use tools coming from the world of functions. When a sequence a_n is given by a formula, then the same formula usually also defines a function $f(x)$. For instance, when dealing with the sequence $a_n=\frac{n}{2^n}$, we may instead consider the function $f(x)=\frac{x}{2^x}$. Then we can investigate $\lim\limits_{x\to\infty}\big(f(x)\big)$ using powerful techniques from Chapter 1, in particular L'Hôpital's rule and procedures for dealing with other indeterminate forms. If this yields a limit L, then also $\lim\limits_{n\to\infty}(a_n)=L$.

Unfortunately, this approach has its limitations. First, if $\lim\limits_{x\to\infty}\big(f(x)\big)$ does not exist, then we have no information about $\lim\limits_{n\to\infty}(a_n)$. The second problem is that some sequences like $(-13)^n$ or $n!$ cannot be turned into functions.

• If the given sequence is composed of powers and related expressions, we can often successfully apply intuitive evaluation based on **scale of powers**. It is known that some expressions dominate others at infinity; informally we say that a_n dominates b_n

if $\frac{a_n}{b_n} \to \infty$ or, equivalently, if $\frac{b_n}{a_n} \to 0$. For instance, we know that powers dominate logarithms. The most popular groups in this hierarchy are:

(i) power n^n;
(ii) factorial $n!$;
(iii) exponentials (or geometric sequences) a^n for $a > 1$;
(iv) powers x^a for $a > 0$;
(v) logarithms $\ln^a(n)$ for $a > 0$.

We wrote these groups in the order of domination. For instance, every geometric sequence q^n dominates every power n^a regardless of the choice of $q > 1$, $a > 0$. There is also hierarchy within each of the groups (iii) through (v); there the value of the parameter decides. For instance, n^3 dominates n^2, which we easily check by calculating $\lim\limits_{n\to\infty}\left(\frac{n^3}{n^2}\right) = \infty$.

Knowing the scale of powers allows one to quickly determine the limit of many expressions. In particular, the scale of powers allows one to disregard unimportant terms in sums, as only the dominant terms determine behaviour at infinity. For instance, we may argue as follows:

$$\frac{\sqrt{n^2 + \ln(n)} + 2^n}{n! + n^3} \sim \frac{\sqrt{n^2} + 2^n}{n!} = \frac{n + 2^n}{n!} \sim \frac{2^n}{n!} \to 0.$$

However, one has to understand limitations of this approach, because there are situations when it would lead to incorrect answers.

When a limit is found using this approach, the correctness of the answer can usually be confirmed formally by factoring dominant terms out of the investigated expression.

• There is a technique of "equivalent infinitesimals" (no relation to infinitesimals from early calculus days or hyperreal numbers) that allows replacement of some functions near zero. It can help with sequences at infinity when a term like $\frac{1}{n}$ appears in a function. Some possible and useful replacements include $\sin(\frac{1}{n}) \sim \frac{1}{n}$, $\ln\left(1 + \frac{1}{n}\right) \sim \frac{1}{n}$, and many others. However, here one really needs to be careful, as it is easy to get a wrong answer by incorrect application.

• While we cannot differentiate sequences, L'Hôpital's rule actually does have its counterpart in the world of sequences.

Theorem 4.1 (Stolz–Cesàro). *Let $\{a_n\}$ and $\{b_n\}$ be sequences of real numbers. If $\{b_n\}$ is strictly monotone and unbounded, then*

$$\lim_{n\to\infty}\left(\frac{a_n}{b_n}\right) = \lim_{n\to\infty}\left(\frac{a_{n+1} - a_n}{b_{n+1} - b_n}\right),$$

assuming that the limit on the right exists.

• When a sequence involves sines and cosines, we can often deal with it using a suitable version of the comparison theorem.

We often study properties of sequences. Among the most popular are **monotone** sequences. There are four kinds of these: a sequence $\{a_n\}$ is (strictly) increasing, nondecreasing, (strictly) decreasing, or nonincreasing, if it satisfies $a_{n+1} > a_n$, $a_{n+1} \geq a_n$,

$a_{n+1} < a_n$, or $a_{n+1} \le a_n$, respectively, for all n. Note that some people prefer slightly different terminology, but all agree on what "monotone" means. Monotone sequences are well behaved, for instance they always have a limit. Moreover, if a monotone sequence is bounded, then it always converges to a proper limit.

Sequences are the underlying notion for series. What is a series? Here the answer is not so simple. It is an abstract object that expresses our intention to add infinitely many numbers. Formally, given a sequence $\{a_n\}_{n=n_0}^{\infty}$ of real (or complex) numbers, we may want to add them (if this makes any sense) and we express this intention by talking about series $\sum\limits_{n=n_0}^{\infty} a_n$.

Series as abstract objects can be formally manipulated.

Definition. Consider series $\sum\limits_{n=n_0}^{\infty} a_n$, $\sum\limits_{n=n_0}^{\infty} b_n$ of real (or complex) numbers. Let c be a real (or complex) constant. We define the series $c \sum\limits_{n=n_0}^{\infty} a_n$ as $\sum\limits_{n=n_0}^{\infty} ca_n$ and the series $\sum\limits_{n=n_0}^{\infty} a_n + \sum\limits_{n=n_0}^{\infty} b_n$ as $\sum\limits_{n=n_0}^{\infty} (a_n + b_n)$.

These operations follow all the usual rules (like commutativity). Thus, when we fix an integer n_0 and consider the set of all series of real numbers of the form $\sum\limits_{n=n_0}^{\infty} a_n$, then it becomes a linear space when equipped with the two operations. Similar statement applies to series of complex numbers.

Note that these operations are purely formal; we created a linear space of symbols that do not have an actual meaning yet (and many will never have, as not all attempts to add infinitely many numbers succeed). It is time to give this notion some contents.

Definition. Consider a series $\sum\limits_{n=n_0}^{\infty} a_n$ of real or complex numbers.

We say that the series converges to a (real or complex) number A, written as $\sum\limits_{n=n_0}^{\infty} a_n = A$, if

$$\lim_{N\to\infty}\left(\sum_{n=n_0}^{N} a_n\right) = A.$$

If such an A exists, then we say that the series $\sum\limits_{n=n_0}^{\infty} a_n$ is convergent.

Otherwise we say that it is divergent.

We write that $\sum\limits_{n=n_0}^{\infty} a_n = \infty$, respectively, $\sum\limits_{n=n_0}^{\infty} a_n = -\infty$ if

$$\lim_{N\to\infty}\left(\sum_{n=n_0}^{N} a_n\right) = \infty, \text{ respectively, } \lim_{N\to\infty}\left(\sum_{n=n_0}^{N} a_n\right) = -\infty.$$

Again, we talk of proper and improper sums of a series. Note that unlike the limit of a sequence, here we do care about indexing, since by changing the starting index n_0 we add or remove numbers in the (infinite) summation, thus obviously influencing its outcome.

The fact that a series converges to some A can be equivalently expressed as

$$\lim_{N\to\infty}\left(A-\sum_{n=n_0}^{N} a_n\right)=0.$$

We may think of a series as a representation of a number A. This representation is true if the error of approximation (the expression in the limit) goes to zero as we take longer and longer sums.

We learned how to add series and now we relate this process to operations. Things behave as expected.

Theorem 4.2. *Consider convergent series* $\sum\limits_{n=n_0}^{\infty} a_n = A$ *and* $\sum\limits_{n=n_0}^{\infty} b_n = B$. *Let* $c \in \mathbb{R}$. *Then the series* $c\sum\limits_{n=n_0}^{\infty} a_n$ *and* $\sum\limits_{n=n_0}^{\infty} a_n + \sum\limits_{n=n_0}^{\infty} b_n$ *also converge and*

$$\sum_{n=n_0}^{\infty} ca_n = cA, \quad \sum_{n=n_0}^{\infty}(a_n+b_n) = A+B.$$

Note that finite sums can be viewed as series, say, $2+3=2+3+0+0+0+\cdots$. It is easy to observe that when we turn a finite sum into a series by extending it with zeros, then the sum of the series according to the definition above agrees with the usual addition.

Having this in mind, we can see the first rule in the theorem as a generalization of the distributive law, while the second one is a special case of the commutative and associative law. In a summation of the form

$$(a_{n_0}+a_{n_0+1}+a_{n_0+2}+\cdots)+(b_{n_0}+b_{n_0+1}+b_{n_0+2}+\cdots)$$

we first remove parentheses, then reorganize the order, and finally add parentheses to obtain

$$(a_{n_0}+b_{n_0})+(a_{n_0+1}+b_{n_0+1})+(a_{n_0+2}+b_{n_0+2})+\cdots,$$

obtaining the same sum according to the theorem. Note that only this very special form is generally true, we are not allowed to permute terms in a series in general.

In fact, our freedom to manipulate series is quite limited. For instance, there are series that converge to certain numbers, but when we permute the order in which their terms are summed up, the outcome changes! This (and similar surprising properties) are not welcome in applications. If we want to get rid of them, we have to restrict our attention to series whose convergence is more robust.

Definition. Consider a series $\sum\limits_{n=n_0}^{\infty} a_n$. We say that it **converges absolutely** if $\sum\limits_{n=n_0}^{\infty} |a_n|$ converges.

The relationship between the two notions is captured in the following well-known theorem.

Theorem 4.3. *If a series converges absolutely, then it converges.*

Since there are series that are convergent but not absolutely convergent, it shows that absolute convergence indeed improves on the general notion of convergence. This fact is then highlighted by further theorems confirming some pleasant properties: An absolutely convergent series can be rearranged (we can change the order in which the terms are added) without changing the total sum, and we can split such a series into a sum of two subseries in any way we wish without affecting the outcome.

The definition shows that in order to recognize absolute convergence it is enough to master series with nonnegative real numbers. However, even under such a restriction it is not easy to recognize convergence. In fact, we do not have a general test that would tell apart convergent and divergent series. Instead, we have a wide range of tests that can confirm convergence. The following two criteria seem to be very popular in applications.

Theorem 4.4. *Consider a series* $\sum\limits_{n=n_0}^{\infty} a_n$ *of real or complex numbers.*

(i) Root test: Assume that the limit $\varrho = \lim\limits_{n\to\infty}\big(\sqrt[n]{|a_n|}\big)$ *exists.*

If $\varrho < 1$, *then* $\sum\limits_{n=n_0}^{\infty} a_n$ *converges absolutely.*

If $\varrho > 1$, *then* $\sum\limits_{n=n_0}^{\infty} a_n$ *does not converge absolutely.*

(ii) Ratio test: Assume that the limit $\lambda = \lim\limits_{n\to\infty}\big(\frac{|a_{n+1}|}{|a_n|}\big)$ *exists.*

If $\lambda < 1$, *then* $\sum\limits_{n=n_0}^{\infty} a_n$ *converges absolutely.*

If $\lambda > 1$, *then* $\sum\limits_{n=n_0}^{\infty} a_n$ *does not converge absolutely.*

The problem is that when such a test fails, say, when we get $\lambda = 1$ in the above test, then they leave us without information. There are many series for which we still do not know whether they converge or diverge, for instance the Flint Hills series $\sum \frac{1}{n^3 \sin^2(n)}$, and in fact neither it is known whether the sequence $\{\frac{1}{n^2 \sin(n)}\}$ tends to zero. Another simple yet so far victorious series is $\sum \frac{(-1)^n n}{p_n}$, where p_n is the nth prime number. Remarkably, one of the most famous (yet) unsolved problems in mathematics, the Riemann hypothesis, can be also formulated as a question on convergence of certain series.

Things get really interesting when the terms in a series depend on some parameter. Then the series can converge (absolutely) or diverge depending on the choice of parameter. If we interpret this parameter as a free variable, we in fact add infinitely many functions. However, this is a topic for Chapter 5.

If recognizing convergence can get tricky, then finding the actual sum of a convergent series is almost impossible. In fact, we can do it only for a very restricted group of series, probably the most popular (and most useful) result concerns the geometric series. If $|q| < 1$, then

$$\sum_{n=N}^{\infty} q^n = \frac{q^N}{1-q}.$$

For more information see e.g. Stroud (1970) or another text on elementary calculus.

4.2 Sequences and series in applications

- Sequences play a crucial role in (discrete) signal processing. Many phenomena, for instance digitalized music, are in fact sequences of values, and thus their mathematical treatment (transformations, filters, etc.) is based on tools for manipulating and investigating sequences. We actually meet transforms of sequences in our everyday lives, for instance every time we listen to mp3-coded music.
- Another interesting application of sequences comes from the field of algorithm analysis. A typical algorithm accepts input whose size (or complexity) can be described by a natural number n. When assessing an algorithm, people would ask good questions along these lines: How long will it take for the algorithm to do its work (time complexity)? How many operations will have to be performed (arithmetic complexity)? How much memory will it need (space complexity)? For each of these questions, the analysis typically yields some sequence a_n that goes to infinity and we would like to know how fast it goes there.
- In practical applications we often know that some quantity exists, but we cannot determine it directly. For instance, we know for sure that there is a number x satisfying $\cos(x) = x$ (draw a picture), but we are unable to provide it as a formula. One popular approach to such a situation is to use iterative methods that start with a rough guess and then repeatedly improve it, creating a sequence if we let this procedure run indefinitely.
  Since different applications have different requirements on precision, we only care about procedures that are capable of approximating the desired number with arbitrary precision (eventually). When we capture this requirement in mathematical language, we end up with the definition of a limit. Indeed, one of the basic requirements on approximating procedures in numerical mathematics is that the sequences that they produce converge to the desired quantities (see Chapter 8).
  Many interesting quantities (π, e, etc.) are thus related to certain sequences that allow for their approximate evaluation.
  Another tool for approximating interesting quantities is series. If a certain number can be expressed as a sum of a series, we can use partial sums as approximations of increasing precision. We will see this in our projects below.

- Sequences and series are a part of the basic toolbox of anyone who uses mathematics to explore nature and technology. While they seldom feature as leading mathematical tools in applications, they do appear naturally in a supporting role in many settings. For instance, when investigating passage of light through a slab of material, one encounters an interesting phenomenon when a portion of light reflects from the second (rear) surface back to the first surface, then a portion of this light reflects back to the second surface, where a portion of this portion reflects back to the first surface, and so on. If we want to make some sense of what is happening with this bouncing light, we necessarily have to apply series.

4.3 Exploring sequences and series

4.3.1 Asymptotic growth at infinity

Many numerical approaches in engineering lead to enormous systems of linear equations. In their course on linear algebra, students typically learn several possible methods, among them the Cramer rule and the Gaussian elimination. Which one is better for really large systems?

Usually the most important factor is arithmetic complexity, which is closely related to how long we will wait for the answer. It is well known that if we want to solve an $n \times n$ system of linear equations, it will require more than $n^2 n!$ operations using the Cramer rule with direct determinant evaluation (here by operations we mean the four basic algebraic operations). Gaussian elimination will cost us $n^3 + \frac{1}{2}n^2 - \frac{1}{2}n$ operations if we reduce the matrix to identity matrix, obtaining the solution directly (GJM), and $\frac{2}{3}n^3 + \frac{3}{2}n^2 - \frac{7}{6}n$ operations if we first reduce the matrix to an upper-triangular form and then work out the solution using back substitution (GEM+BS). (If the reader wonders about the fractions, one can check that both expressions yield integer values for $n \in \mathbb{N}$.)

Obviously, all three expressions determine sequences that tend to infinity. The question is how fast, as we want to choose an algorithm that requires the least work.

The key tool in judging how fast sequences tend to infinity is the notion of dominance and scale of powers that we introduced above. In this context the notion of "little o" is often used.

Consider sequences $a_n, b_n > 0$ that go to infinity. We say that $b_n = o(a_n)$, we read it "b_n is little oh a_n" if $\lim\limits_{n\to\infty}\left(\frac{b_n}{a_n}\right) = 0$, that is, $\lim\limits_{n\to\infty}\left(\frac{a_n}{b_n}\right) = \infty$.

Informally we would say that the sequence a_n dominates the sequence b_n. The scale of powers, that is, the relationship between the five groups listed above, can now be expressed for instance as follows: $\ln(n) = o(n^2)$ (square power dominates the logarithm).

How does it help us with our three algorithms? We know that factorial dominates powers, so the Cramer rule will definitely need significantly, even incomparably longer time to run compared to elimination as n grows large. We confirm this by evaluating appropriate limits, for instance comparison to GJM goes as follows:

$$\lim_{n\to\infty}\left(\frac{n^3+\frac{1}{2}n^2-\frac{1}{2}n}{n^2 n!}\right)=\lim_{n\to\infty}\left(\frac{n}{n!}+\frac{1}{2n!}-\frac{1}{2n\cdot n!}\right)=0.$$

In the last step we used our knowledge of the scale of powers.

So the Cramer rule is out, how about the two versions of Gauss elimination? Experience with limits at infinity tells us that near infinity, polynomials behave like their leading terms. Thus, for large n, we are comparing n^3 versus $\frac{2}{3}n^3$. The conclusion is that the version GEM+BS should be faster.

However, here the conclusion is not so clear-cut. Indeed, the difference is not as pronounced as when comparing $n^2\cdot n!$ versus n^3. GJM is slower than GEM-BS by a factor that is even less than 2, which in computer science is not considered crucial; such a difference can be easily remedied by buying a better hardware. It may happen that a bit slower algorithm turns out to be better because of other factors, for instance it may be less susceptible to numerical errors.

This brings us to another notion. It allows us to specify that two sequences are not in domination relationship, that they grow at comparable rate.

Consider sequences $a_n, b_n > 0$ that go to infinity. We say that $b_n = \Theta(a_n)$ if there are constants $c, C > 0$ such that $c \le \frac{b_n}{a_n} \le C$ for all n.

An easy way to confirm this is to show that $\lim\limits_{n\to\infty}\left(\frac{a_n}{b_n}\right)$ converges to a positive number. Note that this is just a sufficient condition, it is not an equivalent statement. However, it shows how this is related to dominance. We investigate the ratio $\frac{a_n}{b_n}$ around infinity. If it is 0 or infinity, then one of the two sequences dominates the other. If it is a positive number, then none dominates the other, they grow at comparable rates. Of course, this leaves out the case when the limit does not exist, which is exactly the reason why this limit condition is not equivalent to $b_n = \Theta(a_n)$.

We could show that $n^3+\frac{1}{2}n^2-\frac{1}{2}n=\Theta\left(\frac{2}{3}n^3+\frac{3}{2}n^2-\frac{7}{6}n\right)$. However, a more typical use of this notion is to characterize the rate of growth of the given sequence at infinity by some simple expression. Here we would say that $n^3+\frac{1}{2}n^2-\frac{1}{2}n=\Theta(n^3)$ and $\frac{2}{3}n^3+\frac{3}{2}n^2-\frac{7}{6}n=\Theta(n^3)$. The proof of the latter is as follows:

$$\lim_{n\to\infty}\left(\frac{\frac{2}{3}n^3+\frac{3}{2}n^2-\frac{7}{2}n}{n^3}\right)=\lim_{n\to\infty}\left(\frac{2}{3}+\frac{3}{n}-\frac{7}{6n^2}\right)=\frac{2}{3}.$$

Since the result is a positive number, our assertion is confirmed.

Thus, from this point of view, both versions of Gauss elimination look comparable. How do we decide then? A closer inspection shows that both versions use analogous computations and have the same memory requirements, so there is no other factor that could influence our decision. Thus it makes sense to choose the GEM+BS version, as it will save a third of our time for large systems.

It should be noted that there are more advanced methods for solving massively large systems, for instance matrix decompositions (they typically also have complexity n^3)

or iterative methods (see, e.g., the Gauss–Seidel iteration). However, this is a topic beyond this chapter.

We will now explore the scale of powers and investigation of sequences.

Example. (a) Where on the scale of powers do we find $\sqrt{4^n+n}+13\ln(n)$?

Our question is a bit informal but descriptive. A proper way to ask could be as follows: Investigate the asymptotic rate of growth of $\sqrt{4^n+n}+13\ln(n)$. Practically speaking, we want to find a simple representative for the given expression.

First of all, we know that powers are dominated by geometric sequences. This means that when n is very large, then n is negligible compared to 4^n and can be safely ignored. Thus $\sqrt{4^n+n}$ is essentially $\sqrt{4^n}=2^n$. We obtained a representative for the whole square root and compare it with the rest of the expression. Since logarithms are dominated by geometric sequences, $13\ln(n)$ can be disregarded.

We conclude that $\sqrt{4^n+n}+13\ln(n)=\Theta(2^n)$.

Confirmation: We use the limit approach. We have

$$\begin{aligned}\lim_{n\to\infty}\left(\frac{\sqrt{4^n+n}+13\ln(n)}{2^n}\right)&=\lim_{n\to\infty}\left(\sqrt{\frac{4^n+n}{(2^n)^2}}+\frac{13\ln(n)}{2^n}\right)\\&=\lim_{n\to\infty}\left(\sqrt{1+\frac{n}{4^n}}+13\frac{\ln(n)}{2^n}\right)=\sqrt{1+0}+0=1.\end{aligned}$$

Since the limit is a positive number, the conclusion is confirmed. In the calculations we used what we already know about the scale of powers.

(b) Where on the scale of powers do we find $\log_2(n)$?

Since $\log_2(n)=\frac{1}{\ln(2)}\ln(n)$, this sequence is in fact just a multiple of $\ln(n)$. Thus $\log_2(n)=\Theta(\ln(n))$, and it belongs to the group represented by $\ln(n)$.

This is the reason why we did not discuss the base of the logarithm when introducing the scale of powers; it is irrelevant.

(c) We will investigate the asymptotic rate of growth of $n\ln(n)$.

Incidentally, it is the complexity of the most efficient general algorithm for sorting n numbers by their size (or n names by alphabet).

Since $\dfrac{n\ln(n)}{n}=\ln(n)\to\infty$, we see that $n=o(n\ln(n))$, that is, $n\ln(n)$ dominates n. Could it be that this expression actually grows at the same rate as some power? This power would have to be larger than 1, so we will test $n\ln(n)$ against n^{1+a} for $a>0$. Then we have

$$\frac{n\ln(n)}{n^{1+a}}=\frac{\ln(n)}{n^a}\to 0,$$

since logarithms are dominated by (positive) powers, however small. The conclusion is that $n\ln(n)$ dominates n^1, but is dominated by any higher power. Thus it is a new type. As you can see, the power scale as given above features only the most popular expressions, but one can also meet other types in applications.

Exercise 4.3.1.1. Investigate the asymptotic growth of $a_n=n^2+\sqrt{n^6+13n}$.

Exercise 4.3.1.2. **(a)** Investigate the asymptotic growth of $a_n = \ln(n^2)$.
(b) Investigate the asymptotic growth of $a_n = \ln(n^2 + 1)$.

Exercise 4.3.1.3. Investigate the asymptotic growth of $a_n = n2^n$.

Exercise 4.3.1.4. **(a)** Prove that $\ln(n) = o(n^a)$ for any $a > 0$.
Hint: Investigate $\lim\limits_{x\to\infty}\left(\frac{\ln(x)}{x^a}\right)$. Note the change of variable; this is now a limit of a function, thus L'Hôpital's rule is available.
(b) Prove that $\ln^2(n) = o(n^a)$ for any $a > 0$.
(c) Prove that $\ln^m(n) = o(n^a)$ for any $a > 0$ and positive integer m.
(d) Prove that $\ln^b(n) = o(n^a)$ for any $a, b > 0$.
Hint: Make use of the number $B = \lceil b \rceil$, that is, B is b rounded up to the nearest integer.

4.3.2 Decimal expression of a number

Engineering calculations are done with decimal numbers. Most of them have infinite decimal expansions, which presents a problem when we want to store them in a computer with limited memory. How do computers solve this problem?

Every nonnegative number can be expressed in the form $0.a_1a_2a_3... \times 10^E$, where $a_n \in \{0, 1, 2, \ldots, 9\}$ and we want $a_1 \neq 0$ for positive numbers. This actually represents the number

$$\sum_{n=1}^{\infty} a_n \cdot 10^{-n} \cdot 10^E = \sum_{n=1}^{\infty} a_n \cdot 10^{E-n}.$$

We then decide how many digits we want to use to express this number, we will call it N here. There are two popular ways to obtain those N digits for a given number. The more precise but also more complicated approach is rounding off. A simpler way is truncation, when we simply keep the first N digits and discard the rest. For instance, if we decided to keep $\pi = 3.141592...$ with 5-digit precision, then rounding off leads to 0.31416×10^1, while truncating leads to 0.31415×10^1. In this way we introduce error to our calculations, but this is inevitable in the world of computing. Here we will look at this topic closer.

Example. (a) What is the largest number that can be expressed using the decimal notation $\sum\limits_{n=1}^{\infty} a_n 10^{-n}$?

Since $a_n \leq 9$, we can estimate

$$\sum_{n=1}^{\infty} a_n 10^{-n} \leq \sum_{n=1}^{\infty} 9 \cdot 10^{-n} = 9 \sum_{n=1}^{\infty} \left(\frac{1}{10}\right)^n = 9 \frac{\frac{1}{10}}{1 - \frac{1}{10}} = 9 \frac{\frac{1}{10}}{\frac{9}{10}} = 1.$$

We see that we can express numbers from the interval $[0, 1]$. We used the standard formula for the sum of a geometric series, and we had to be careful as our summation started only from $n = 1$.

Note the following. If we set $a_n = 9$ for all $n \geq 1$, then according to the above calculation $\sum_{n=1}^{\infty} 9 \cdot 10^{-n} = 1$. That is, $0.9999... = 1$. In other words,

$$1.00000000... = 0.99999999...$$

We see that the number 1 has two distinct decimal representations. This may be surprising for some readers, but it is true. If you are not convinced, think about this: What is $a = 1 - 0.999999...$? The fact is that every real number with a finite decimal expansion can be written in two ways. For instance, $0.13 = 0.1299999...$ While it may sound profound, it actually does not have any noticeable impact on the way we work with real numbers.

(b) What is the error if we truncate a real number from the interval $[0, 1)$ to N decimal digits?

When we approximate a number x by another number $\hat{x}$, the error is defined as $E = x - \hat{x}$. In our case $x - \hat{x} \geq 0$, as truncating can never make a number larger, so we easily estimate

$$|E| = E = x - \hat{x} = \sum_{n=1}^{\infty} a_n 10^{-n} - \sum_{n=1}^{N} a_n 10^{-n} = \sum_{n=N+1}^{\infty} a_n 10^{-n}$$

$$\leq \sum_{n=N+1}^{\infty} 9 \cdot 10^{-n} = 9 \frac{1}{10^{N+1}} \frac{1}{1 - \frac{1}{10}} = \frac{1}{10^N}.$$

We obtained a nice upper bound for the error. We can try an experiment: If instead of 0.1299999 we keep 0.12 (2 digits), then we made the error $0.0099999 \approx 0.01 = \frac{1}{10^2}$. This seems to fit.

Exercise 4.3.2.1. (a) Show that $0.3333333...$ is in fact $\frac{1}{3}$ using the series approach.

(b) Find some fractional expression for $0.13131313...$ using the series approach. Hint: Express this number not with powers 10^{-n}, but with powers 100^{-n}.

Exercise 4.3.2.2. Consider the binary representation $\sum_{n=1}^{\infty} a_n 2^{-n}$, now a_n can be only 0 or 1.

(a) Show what is the largest possible number that one can obtain in this way.

(b) Find the largest error that can appear if we truncate a binary decimal number to N decimal digits.

Remark: When entering numbers into calculator or computer, people feel that computations with numbers like 0.3 are more precise compared to numbers like $\frac{1}{3} = 0.3333....$, since the latter involve rounding errors.

However, this would be true only if the processor also worked in the decimal system. Typically all numbers that we enter are immediately converted to binary code. Since $0.3 = (0.0\overline{1001}1)_2$, it cannot be stored in binary form precisely and we encounter rounding errors.

4.3.3 Expressing numbers as sequences and series

In mathematics one meets interesting constants like π and e. We often use them symbolically, but we understandably also want to know what numbers these actually represent. Mathematicians proved that both numbers are transcendental, which means that that they cannot be expressed by any formula featuring basic algebraic operations and roots. Still, one easily finds in books or on the internet approximate values of amazing precision, going to millions of digits. Where do these numbers come from?

Example. In their introductory calculus course, students learn that e is the limit of the sequence $a_n = \left(1 + \frac{1}{n}\right)^n$. This sequence thus provides a possible way to determine e with increasing precision. To try it out, I started up my calculator (12-digit precision) and obtained the following:

n	a_n	E_n
1300	2.717...	0.001...
13000	2.7182...	0.0001...
130000	2.718272...	0.0000096...
1300000	2.718278...	0.0000038...
13000000	2.71817...	0.0001...
130000000	2.7175...	0.0008...

Note how the error decreases at first, but then starts growing again. This is the outcome of numerical errors that appear in the calculation. Although we work with 12-digit precision, we will never be able to get more than 6 digits right in e. This shows that while theoretically, the sequence can approximate e arbitrarily well, in the world of computer computing things are different. Given how much precision we lost, it may be better to try another approach.

Taylor series theory offers the famous expansion

$$e^x = 1 + x + \frac{x^2}{2} + \frac{x^3}{3!} + \frac{x^4}{4!} + \cdots.$$

Substituting $x = 1$ we obtain

$$e = 1 + 1 + \frac{1}{2} + \frac{1}{3!} + \frac{1}{4!} + \cdots.$$

This offers another possibility for approximation:

$$e \approx \sum_{k=0}^{n} \frac{1}{k!}.$$

For convenience we will rewrite it as a recurrent sequence, i.e.,

$$s_1 = 2, \qquad s_n = s_{n-1} + \frac{1}{n!} \text{ for } n \geq 2.$$

Firing up my calculator I found the following:

n	s_n	E_n
5	2.7167...	0.0016...
7	2.71825...	0.00003...
9	2.7182815...	0.0000003...
11	2.718281826...	0.000000002...
13	2.71828182845	0.00000000001

One distinct disadvantage of sums (or recurrent sequences) is that one cannot easily calculate the value for, say, $n = 100000$, just by plugging it into some formula. If I had to add that many numbers on a calculator (or by hand as people used to), I would be deeply unhappy. But as you can see, I was spared; the sequence s_n converges to e amazingly fast. This is definitely a good way to approximate e.

Since we obtained a good approximation for relatively small values of n, we did not reach the moment when we would have to start worrying about numerical errors, which is another advantage.

Can we estimate the precision of the sum approximation? If we use s_n to approximate e, then the error is

$$|E| = e - s_n = \sum_{k=0}^{\infty} \frac{1}{k!} - \sum_{k=0}^{n} \frac{1}{k!} = \sum_{k=n+1}^{\infty} \frac{1}{k!}.$$

Summing up this series requires very advanced tools, but we will be happy with an upper estimate. Note that for $k \geq 2$ the inequality $k! \geq 2^k$ is true, so we can compare and estimate. We have

$$|e - s_n| = \sum_{k=n+1}^{\infty} \frac{1}{k!} \leq \sum_{k=n+1}^{\infty} \frac{1}{2^k} = \frac{1}{2^{n+1}} \frac{1}{1 - \frac{1}{2}} = \frac{1}{2^n}.$$

We see that the error goes to zero. We can use this estimate to guess what n to take in order to get a desired precision, but it would not be very good as our estimate was very rough. Better estimates can be obtained, but we want to show here what can be achieved just using basic tools.

Exercise 4.3.3.1. The first scientific approach to approximate π dates to Archimedes. His idea was to take a circle of radius $\frac{1}{2}$ and approximate its circumference (which is π) using inscribed and circumscribed regular polygons. To avoid evaluations of sine or cosine, Archimedes employed a clever iterative scheme where at each step he halved the angles involved. In modern language, he worked with sequences defined as follows:

$$a_0 = 2\sqrt{3},\ b_0 = 3;$$
$$a_{n+1} = \frac{2a_n b_n}{a_n + b_n},\quad b_{n+1} = \sqrt{a_{n+1} b_n} \text{ for } n \geq 0.$$

Then $a_n \to \pi$ and $b_n \to \pi$.

(a) Use a calculator or computer to evaluate these sequences for growing values of n. Start with small ones. When does the error become less than $0.00001 = 10^{-5}$?

(b) Look at a_n and b_n for larger n (writing a simple program is advisable). Can you determine π close to your machine's precision or do you start running into numerical errors? If you do obtain an almost precise π, what n is enough?

Remark: What does "almost precise" mean? If a machine works at N-digit precision, we cannot really expect results to be equally precise. However, with a bit of luck we may achieve errors of order $10^{-(N-1)}$. For instance, on my calculator working with 12-digit precision I cannot hope to get answers of calculations with errors of order 10^{-12}, but getting them down to 10^{-11} can happen. Try it and see what happens.

Hopefully things went well. This seems to be a very fast and robust algorithm. However, it has one disadvantage. The square root has to be calculated only approximately, which increases numerical error and takes too much time. In particular, it is unsuitable to hand calculations. We therefore try another approach.

(c) Find or create Taylor expansion for $\arctan(x)$. Modify it to an expansion for $4\arctan(x)$.
Then use this with suitable x to derive an expansion for π.

(d) Rewrite the resulting series into the form of a recurrent sequence. This sequence can be easily calculated by hand, but you can use a calculator or a computer to find values for $n = 5, 10, 15, \ldots$
Program a loop in your calculator or computer to find what n is needed to get the error under $0.001 = 10^{-3}$ and under $0.00001 = 10^{-5}$.

(e) Estimate the error that is made when the series is truncated as $\sum_{k=1}^{n} \frac{4 \cdot (-1)^{k+1}}{2k-1}$.
Hint: Comparison with a suitable integral looks hopeful. Another approach is to note that it is an alternating series.

Exercise 4.3.3.2. The number $\ln(2)$ is less popular than π, but it was not neglected. In fact, it appears in most calculus courses as the sum of the alternating harmonic series,

$$\ln(2) = \sum_{k=1}^{\infty} \frac{(-1)^{k+1}}{k} = 1 - \frac{1}{2} + \frac{1}{3} - \frac{1}{4} + \cdots.$$

(a) Find partial sums s_n for $n = 10, 20, 30$ and their errors as approximations for $\ln(2)$.

(b) Derive a theoretical estimate for the error $|\ln(2) - s_n|$.

(c) Less-known but still classical is the series representation

$$\ln(2) = \sum_{k=1}^{\infty} \frac{1}{2^k k}.$$

Find partial sums s_n for $n = 5, 10, 15, 20$ and determine their errors as approximations for $\ln(2)$. Derive a theoretical estimate for the error $|\ln(2) - s_n|$.
Hint: Comparison with a more tractable series may help.

4.3.4 Iterative approximation

Example. The square root is a very popular mathematical operation. However, as we explained in Chapter 2, when it comes to calculations with decimal numbers on computers, we have precise procedures only for addition, subtraction, and multiplication. Thus we can evaluate precisely only those expressions that can be reduced to these three basic operations. Unfortunately, the square root is not one of them, so we have to settle for a sufficiently good approximation. One popular iterative approach to approximating $\sqrt{A}$ for $A > 0$ is the recursive sequence

$$x_{n+1} = \frac{1}{2}\Big(x_n + \frac{A}{x_n}\Big).$$

Where does it come from? Does it really work? These are questions that we can answer without any deep theory.

We start with a bit of exploration. Already the ancient Babylonians and Greeks came up with the following idea. Take some positive number A. We start with a guess $x_0 > 0$ and observe (check for yourself), that if $x_0 > \sqrt{A}$, then $\frac{A}{x_0} < \sqrt{A}$, while if $x_0 < \sqrt{A}$, then $\frac{A}{x_0} > \sqrt{A}$. To put it together, whatever our guess is, the actual value $\sqrt{A}$ is bracketed between x_0 and $\frac{A}{x_0}$. Thus there is a hope that if we take the middle of this interval (average of the two bounds), we get a better estimate x_1.

Then we can repeat this process, until we arrive at a number that is good enough for our purposes. Mathematically, we construct a sequence x_n given by the recursive formula

$$x_{n+1} = \frac{1}{2}\Big(x_n + \frac{A}{x_n}\Big).$$

This is called the Babylonian method, or sometimes the Newton method since one can arrive at this formula also using methods from numerical mathematics (see Chapter 8).

To see how it performs, let us say that we want to find $\sqrt{4}$. For the initial guess we can take $x_0 = 1.3$ (13 is my favorite number) and start the procedure. In the chart we show a few iterates and, most importantly, the error $|\sqrt{4} - x_n|$.

$x_0 = 1.3$	$E_0 = 0.70000000000000000000$
$x_1 = 2.18846153...$	$E_1 = 0.188461538461538461538461538461538461538...$
$x_2 = 2.00811477...$	$E_2 = 0.008114776260646207922130593483844801947...$
$x_3 = 2.00001639...$	$E_3 = 0.000016395874015470171570530171918647782...$
$x_4 = 2.00000000...$	$E_4 = 0.000000000067205620235351236798379040512...$
$x_5 = 2.00000000...$	$E_5 = 0.000000000000000000001129148847766620333...$
$x_6 = 2.00000000...$	$E_6 = 0.000000000000000000000000000000000000000...$

Wow, that was wicked. Just six iterations and we have the root correct to 40 decimal digits. Of course I did not do it by hand, but in the days before calculators and computers, people called calculators did calculations like this with quill, ink, and paper.

Fast algorithms were in high demand. This algorithm is not just fast, but also flexible; for instance, it can also find imaginary roots like $\sqrt{3+4i} = 2+i$.

By the way, normally we would not know the error, since we would not know the exact value of $\sqrt{A}$. How do we tell when to stop? Actually, there is no reliable general answer to this question. However, there are some strategies that can work quite well; see Burden and Faires (2011) or another book on numerical analysis for stopping conditions. For this particular method the rule of thumb is that when iterations stop changing some leading digits, than these are most likely correct.

Now we put this algorithm on firmer mathematical footing. First note that if $A > 0$ and $x_0 > 0$ (we can always choose it like this), then $x_n > 0$ for all n.

Note also that if $x_n = \sqrt{A}$ for some n, then the same must be true for all successive iterations. Moreover, it is easy to show that in order to get $x_n = \sqrt{A}$ we need to have $x_{n-1} = \sqrt{A}$ and so on, so the only way to actually obtain the precise value for $\sqrt{A}$ is to start with it as x_0. Since there is zero likelihood of this happening by chance, we do not expect to see this.

(a) We claim that $x_{n+1} \geq \sqrt{A}$ for every $n \geq 0$.

Indeed,

$$\begin{aligned}
x_{n+1}^2 - A &= \left[\frac{1}{2}\left(x_n + \frac{A}{x_n}\right)\right]^2 - A = \frac{1}{4}\left(x_n^2 + 2A + \frac{A^2}{x_n^2}\right) - A \\
&= \frac{1}{4}\left(x_n^2 + 2A + \frac{A^2}{x_n^2} - 4A\right) = \frac{1}{4}\left(x_n^2 - 2A + \frac{A^2}{x_n^2}\right) \\
&= \frac{1}{4}\left(x_n - \frac{A}{x_n}\right)^2 \geq 0.
\end{aligned}$$

So this inequality is true for every x_n with a predecessor, that is, $x_n \geq \sqrt{A}$ is guaranteed for all but the initial guess x_0. Thus we can say that the sequence $\{x_n\}_{n=1}^{\infty}$ (note the starting index) is bounded from below by $\sqrt{A}$.

(b) We claim that the sequence $\{x_n\}_{n=1}^{\infty}$ is nonincreasing.

Indeed,

$$\begin{aligned}
x_{n+1} - x_n &= \frac{1}{2}\left(x_n + \frac{A}{x_n}\right) - x_n = \frac{1}{2}\left(x_n + \frac{A}{x_n} - 2x_n\right) \\
&= \frac{1}{2}\left(\frac{A}{x_n} - x_n\right) = \frac{1}{2x_n}(A - x_n^2) \leq 0.
\end{aligned}$$

The last inequality follows from our previous observation that $x_n \geq \sqrt{A}$ for $n \geq 1$. Note that if $\{x_n\}$ is nontrivial, that is, if it is not the constant sequence $x_n = \sqrt{A}$, then we get sharp inequality there, that is, the sequence is strictly decreasing.

Note that in our experiment $\{x_n\}$ behaved exactly as advertised, the first guess was too low but then the iteration jumped up to x_1 and decreased from then on.

(c) Because the sequence $\{x_n\}_{n=1}^{\infty}$ is nonincreasing and bounded from below, it must have a limit L. However, we do not know yet that $L = \sqrt{A}$. From the lower bound on x_n it follows that $L \geq \sqrt{A}$, but in general there is a possibility that such a sequence would level off before it reaches its lower bound.

However, once we know that the sequence has some limit L, we can pass to the limit in the recursive formula above. Then we have

$$\lim_{n\to\infty}(x_{n+1}) = \lim_{n\to\infty}\left(\frac{1}{2}\left(x_n + \frac{A}{x_n}\right)\right), \text{ and hence } L = \frac{1}{2}\left(L + \frac{A}{L}\right).$$

From this we deduce that $L^2 = A$, and since $L \geq 0$, we conclude that $L = \sqrt{A}$. We just proved that the sequence $\{x_n\}$ converges to $\sqrt{A}$.

(d) How fast does the sequence converge? In other words, we want to know how fast the error $|E_n| = |\sqrt{A} - x_n| = x_n - \sqrt{A}$ goes to zero. It is possible to show that $|E_{n+1}| \approx c \cdot E_n^2$. Roughly speaking, at every iteration the error gets squared. This means that once we get the error smaller than one, in the iterations that follow the number of zeros in the error after the decimal point should typically double after each iteration, which seems to fit with our experiment above. However, proving so is a bit tricky, so we start with something easier.

While error of approximation is the crucial parameter to investigate, another interesting parameter is the co-called residual, which is essentially a measure of how well our approximation does its job. In this example we want our x_n to do $x_n^2 = A$, so let us define $r_n = x_n^2 - A$ (we know now that this number is nonnegative for $n \geq 1$). We will investigate r_{n+1}; by a remarkable coincidence we already looked at it in part (a), so we can recycle the calculation. We write

$$r_{n+1} = x_{n+1}^2 - A = \frac{1}{4}\left(x_n - \frac{A}{x_n}\right)^2 = \frac{1}{4x_n^2}(x_n^2 - A)^2 = \frac{1}{4x_n^2}r_n^2.$$

If $A > 1$ (which is a typical case), then also $x_n \geq 1$ and we get $r_{n+1} \leq r_n^2$, which proves that the residual also decreases quadratically.

Remark: As a bonus, the following is a slightly trickier proof that E_n decreases quadratically:

$$\begin{aligned}
|E_{n+1}| &= x_{n+1} - \sqrt{A} = \frac{x_{n+1}^2 - A}{x_{n+1} + \sqrt{A}} = \frac{1}{4(x_{n+1} + \sqrt{A})}\left(x_n - \frac{A}{x_n}\right)^2 \\
&= \frac{1}{4x_n^2(x_{n+1} + \sqrt{A})}(x_n^2 - A)^2 = \frac{(x_n + \sqrt{A})^2}{4x_n^2(x_{n+1} + \sqrt{A})}(x_n - \sqrt{A})^2 \\
&= \frac{(1 + \frac{\sqrt{A}}{x_n})^2}{4(x_{n+1} + \sqrt{A})}E_n^2 \leq \frac{(1+1)^2}{4(\sqrt{A} + \sqrt{A})}E_n^2 = \frac{1}{2\sqrt{A}}E_n^2.
\end{aligned}$$

In the case of $A > 1$ we have $|E_{n+1}| \leq E_n^2$, which is great.

Exercise 4.3.4.1. While we do have a reasonable algorithm for division, it can be very time consuming when applied to numbers with many digits. People who need to divide fast came up with some interesting alternatives. First of all, we can always write $\frac{a}{d} = a \cdot \frac{1}{d}$, and since multiplication can be done quite efficiently (even for long numbers, but that is another fascinating story), the key problem is to approximate the reciprocal $\frac{1}{d}$.

The Newton–Raphson division works as follows.

(0) Choose some (positive) x_0.

(1) Iterate $x_{n+1} = x_n(2 - d \cdot x_n)$ until a sufficient number of decimal digits stays stable.

This algorithm is typically used with $d > 1$ and $x_0 > 0$. Since $d > 1$, we obviously want to choose $x_0 < 1$. There is a theory advising us on how to choose a good initial value x_0, but we will leave it to an interested reader.

(a) Investigate the expression $\frac{1}{d} - x_{n+1}$ to show that $x_n \leq \frac{1}{d}$ for all $n \geq 1$.
Hint: Complete a square, the expression $\sqrt{d}x_n$ will be helpful.

(b) Show that the sequence $\{x_n\}_{n=1}^{\infty}$ is nondecreasing.
Hint: The result of (a) will come handy.

(c) Argue that the sequence $\{x_n\}$ must have some limit $L > 0$, and determine its value.

(d) Show that the residual $r_n = 1 - dx_n$ satisfies the equality $r_{n+1} = r_n^2$.
Hint: If you do not see a way forward, try to substitute $1 - r_n$ for dx_n.

One can also prove a similar estimate for the error, $E_{n+1} \approx cE_n^2$. This shows that the algorithm converges very fast.

(e) Apply the algorithm to find $\frac{1}{2}$ with initial guess $x_0 = 0.13$. Prepare a table similar to the one above, with x_n and errors. You can do calculations by hand, with a calculator, or with a computer program.
Try it again, this time with initial guesses $x_0 = 0.87$ and $x_0 = 1.3$.
As a bonus, pick your favorite number and find its reciprocal using a reasonable initial guess.

(f) We know that $\frac{1}{1+i} = \frac{1}{2} - \frac{1}{2}i$. Apply the Newton–Raphson algorithm with $d = 1 + i$ and $x_0 = 0.5$. It may be a good idea to use some computer algebra system for the complex calculations. Try other initial values.

Exercise 4.3.4.2. In the previous exercise we tried to avoid division, as it is the least evaluation-friendly of the basic algebraic operations. This makes us worry about the Babylonian algorithm for the square root $\sqrt{A}$ in Example 4.3.4, because there we have to calculate $\frac{A}{x_n}$ at every step.

There is an interesting alternative: a division-free algorithm given by the formula

$$x_{n+1} = \frac{3}{2}x_n - \frac{1}{2A}x_n^3.$$

We actually do have some divisions there, but division by 2 is very simple, especially if we calculate in binary code, because then it is just a shift of digits to the right. We may also be lucky when it comes to dividing by A, or we just calculate the number $\frac{1}{A}$ before we run the algorithm and store it.

(a) Prove that if the resulting sequence x_n is positive with a limit L, then $L = \sqrt{A}$.

(b) Prove that if the resulting sequence x_n satisfies $0 < x_n < \sqrt{A}$, then it is increasing.

This algorithm is also very fast, namely, it has a quadratic error like the first algorithm, but it is more sensitive to the choice of the initial value x_0. Try some experiments.

Answers to exercises

4.3.1.1: $a_n \sim n^2 + \sqrt{n^6} = n^2 + n^3 \sim n^3$, so $a_n = \Theta(n^3)$. Confirmation:

$$\lim_{n\to\infty}\left(\frac{n^2+\sqrt{n^6+13n}}{n^3}\right) = \lim_{n\to\infty}\left(\frac{1}{n}+\sqrt{1+\frac{13}{n^5}}\right) = 0+\sqrt{1+0} = 1.$$

4.3.1.2: (a) $a_n = 2\ln(n) = \Theta(\ln(n))$.

(b) $a_n \approx \ln(n^2) = 2\ln(n)$ so $a_n = \Theta(\ln(n))$. Confirmation:

$$\lim_{n\to\infty}\left(\frac{\ln(n^2+1)}{\ln(n)}\right) = \lim_{x\to\infty}\left(\frac{\ln(x^2+1)}{\ln(x)}\right) \underset{\text{l'H}}{\overset{\frac{\infty}{\infty}}{=\!=}} \lim_{x\to\infty}\left(\frac{\frac{1}{x^2+1}2x}{\frac{1}{x}}\right) = \lim_{x\to\infty}\left(\frac{2x^2}{x^2+1}\right)$$
$= 2$.

4.3.1.3: $2^n = o(n2^n)$, but $n2^n = o(q^n)$ for any $q > 2$.

Confirmation:

$$\lim_{n\to\infty}\left(\frac{2^n}{n2^n}\right) = \lim_{n\to\infty}\left(\frac{1}{n}\right) = 0.$$

$\lim_{n\to\infty}\left(\frac{n2^n}{q^n}\right) = \lim_{n\to\infty}\left(\frac{n}{(\frac{q}{2})^n}\right) = 0$, since $\frac{q}{2} > 1$ and exponentials a^n dominate n for $a > 1$.

Thus $n2^n$ is a new type that fits into the gap between 2^n and q^n for any $q > 2$.

4.3.1.4: (a) $\lim_{x\to\infty}\left(\frac{\ln(x)}{x^a}\right) \underset{\text{l'H}}{\overset{\frac{\infty}{\infty}}{=\!=}} \lim_{x\to\infty}\left(\frac{\frac{1}{x}}{ax^{a-1}}\right) = \lim_{x\to\infty}\left(\frac{1}{ax^a}\right) = 0.$

(b) $\lim_{x\to\infty}\left(\frac{\ln^2(x)}{x^a}\right) \underset{\text{l'H}}{\overset{\frac{\infty}{\infty}}{=\!=}} \lim_{x\to\infty}\left(\frac{2\ln(x)\frac{1}{x}}{ax^{a-1}}\right) = \frac{2}{a}\lim_{x\to\infty}\left(\frac{\ln(x)}{x^a}\right) = 0$ by (a).

(c) Proof by induction. (0): For $m = 1$ proved above.

(1): Let $m \in \mathbb{N}$, assume that the assumption is true for m, so $\lim_{x\to\infty}\left(\frac{\ln^m(x)}{x^a}\right) = 0$.

Then $\lim_{x\to\infty}\left(\frac{\ln^{m+1}(x)}{x^a}\right) \underset{\text{l'H}}{\overset{\frac{\infty}{\infty}}{=\!=}} \lim_{x\to\infty}\left(\frac{(m+1)\ln^m(x)\frac{1}{x}}{ax^{a-1}}\right) = \frac{m+1}{a}\lim_{x\to\infty}\left(\frac{\ln^m(x)}{x^a}\right) =$ 0 by our assumption.

(d) If $b > 0$, then $B = \lceil b \rceil$ is a natural number, and hence by (c), $\lim_{x\to\infty}\left(\frac{\ln^B(x)}{x^a}\right) =$ 0. We also have $0 < \frac{\ln^b(x)}{x^a} \leq \frac{\ln^B(x)}{x^a}$ for $x > 1$, and hence by the squeeze theorem, $\lim_{x\to\infty}\left(\frac{\ln^b(x)}{x^a}\right) = 0.$

4.3.2.1: (a) $0.3333... = \sum_{n=1}^{\infty} 3\cdot 10^{-n} = 3\frac{1}{10}\frac{1}{1-\frac{1}{10}} = \frac{1}{3}$.

(b) $0.131313... = \sum_{n=1}^{\infty} 13 \cdot 100^{-n} = 13\frac{1}{100}\frac{1}{1-\frac{1}{100}} = \frac{13}{99}$.

4.3.2.2: (a) $a \le \sum_{n=1}^{\infty} 1 \cdot 2^{-n} = \frac{1}{2}\frac{1}{1-\frac{1}{2}} = 1$.

(b) $|E| \le \sum_{n=N+1}^{\infty} 1 \cdot 2^{-n} = \frac{1}{2^{N+1}}\frac{1}{1-\frac{1}{2}} = \frac{1}{2^N}$.

4.3.3.1: (a) b_n seems to have a smaller error.

n	a_n	E_n	b_n	E_n
3	3.146...	−0.004...	3.139...	0.002...
5	3.1419...	−0.0003...	3.1415...	0.0001...
7	3.14156...	−0.00002...	3.141584...	0.000009...
9	3.141594...	−0.000001...	3.1415921...	0.0000005...

Here is a plot of some values:

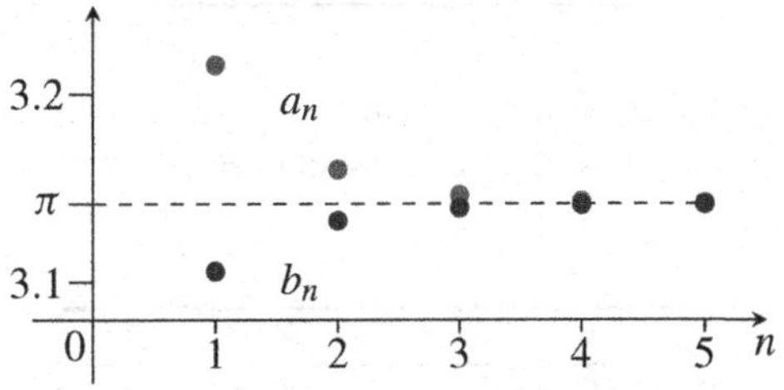

(b) I used a computer, but I kept the precision at 12 digits to have good comparison with my previous results done on my calculator. I got the error down to 10^{-11} for $n = 16$. This is a very fast convergence. I tried going up to $n = 13000$ and never had trouble with numerical errors.

(c) $4\arctan(x) = \frac{4x}{1} - \frac{4x^3}{3} + \frac{4x^5}{5} - \frac{4x^7}{7} + \cdots = \sum_{k=1}^{\infty} \frac{4\cdot(-1)^{k+1}x^k}{2k-1}$.

$x = 1 \Longrightarrow \pi = \frac{4}{1} - \frac{4}{3} + \frac{4}{5} - \frac{4}{7} + \frac{4}{5} - \frac{4}{9} + \cdots = \sum_{k=1}^{\infty} \frac{4\cdot(-1)^{k+1}}{2k-1}$.

(d) $s_1 = 4$, $s_n = s_{n-1} + (-1)^{n+1}\frac{4}{2n-1}$ for $n \ge 2$.

n	s_n	$\lvert E_n \rvert$
5	3.3...	0.2...
10	3.04...	0.099...
15	3.2...	0.067...
20	3.09...	0.05...

One needs $n = 1000$ and $n = 10000$. This is a very slow algorithm.

(e) If we use comparison with an integral to estimate the error, we obtain

$$|E_n| = \left| \sum_{k=n+1}^{\infty} \frac{4(-1)^{k+1}}{2k-1} \right| \le \sum_{k=n+1}^{\infty} \frac{4}{2k-1} \le \int_n^{\infty} \frac{4}{2x-1}\,dx = \infty.$$

Our estimate was obviously too rough to be any good. Another approach is needed. Our series is alternating, with terms (in absolute value) decreasing, and it is known that for such series, the sum is bounded by the first term. We obtain

$$|E| \le \frac{4}{2(n+1)-1} = \frac{4}{2n+1},$$

which goes to zero. However, it goes there relatively slowly, which is confirmed by our numerical experiments.

4.3.3.2: (a) Convergence is not really great:

n	s_n	$\lvert E_n \rvert$
10	0.64...	0.05...
20	0.669...	0.024...
30	0.677...	0.016...

(b) The series is alternating with decreasing terms (in absolute value), so $|E_n| \le \frac{1}{n}$. This goes to zero, but not too fast. Our experiments agree.

(c)

n	s_n	$\lvert E_n \rvert$
5	0.6885...	0.005...
10	0.69306...	0.00008...
15	0.693145...	0.000002...
20	0.69314714...	0.00000004...

Error: $|E_n| = \sum\limits_{k=n+1}^{\infty} \frac{1}{2^k k} \le \sum\limits_{k=n+1}^{\infty} \frac{1}{2^k} = \frac{1}{2^n}$.

This goes to zero pretty fast, which is borne out by our experiments.

A slightly better estimate goes as follows:

$|E_n| = \frac{1}{n} \sum\limits_{k=n+1}^{\infty} \frac{n}{2^k k} \le \frac{1}{n} \sum\limits_{k=n+1}^{\infty} \frac{1}{2^k} = \frac{1}{n2^n}$.

4.3.4.1: (a) $\frac{1}{d} - x_{n+1} = \frac{1}{d} - x_n(2 - dx_n) = \frac{1}{d} - 2x_n + dx_n^2 = \left(\frac{1}{\sqrt{d}} - \sqrt{d}x_n\right)^2 \ge 0$.

(b) $x_{n+1} - x_n = x_n(2 - dx_n) - x_n = x_n - dx_n^2 = dx_n\left(\frac{1}{d} - x_n\right) \ge 0$.

(c) Nondecreasing and bounded above sequence must have a proper limit L. Then also $L \ge x_0 > 0$, so $L > 0$.

$L = L(2 - dL) \implies 1 = 2 - dL \implies dL = 1 \implies L = \frac{1}{d}$.

(d) $r_{n+1} = 1 - dx_{n+1} = 1 - dx_n(2 - dx_n) = 1 - (1 + dx_n - 1)(1 + 1 - dx_n) = 1 - (1 - r_n)(1 + r_n) = 1 - 1 + r_n^2 = r_n^2$.

Bonus: $E_n = \frac{1}{d} - x_n$. Then $E_{n+1} = \frac{1}{d} - x_{n+1} = \frac{1}{d} - x_n(2 - dx_n) = \frac{1}{d} - dx_n\left(\frac{1}{d} + \frac{1}{d} - x_n\right) = \frac{1}{d} - d\left(\frac{1}{d} - E_n\right)\left(\frac{1}{d} + E_n\right) = \frac{1}{d} - d\left(\frac{1}{d^2} - E_n^2\right) = \frac{1}{d} - \frac{1}{d} + dE_n^2 = dE_n^2$.

(e, f) The algorithm converges for $x_0 < 1$, a real-valued initial value can find $\frac{1}{d}$ also for d complex. However, the algorithm diverges for $x_0 \geq 1$. I tried $x_0 = 1 + i$, $x_1 = 4$, $x_2 = -8 - 16i$, $x_3 = 432 - 96i$, which diverges.

4.3.4.2: (a) Passing to the limit in the recursive formula yields $L = \frac{3}{2}L - \frac{1}{2A}L^3$, that is, $L = \sqrt{A}$.

(b) $x_{n+1} - x_n = \frac{1}{2}x_n - \frac{1}{2A}x_n^3 = \frac{1}{2A}x_n(A - x_n^2) > 0$.

References

Stroud, K.A., 1970. Engineering Mathematics. Macmillan. Later editions with Dexter J. Booth, Industrial Press, Inc., New York.

Burden, R.L., Faires, J.D., 2011. Numerical Analysis, 9th Edition. Brooks/Cole, Cengage Learning.

Vibrations and harmonic analysis

5

Daniela Richtáriková, Miloš Musil
Slovak University of Technology in Bratislava, Bratislava, Slovakia

5.1 Basic theory background

Many processes in nature, sciences, and engineering can be modeled by linear combinations of functions, i.e.,

$$\sum_{i=1}^{\infty} a_i \xi_i = a_1\xi_1 + a_2\xi_2 + a_3\xi_3 + ..., \tag{5.1}$$

where ξ_i is a function of a real or complex variable and a_i is a number coefficient. Expression (5.1) stands for series (see Chapter 4), here a series of functions. For its practical usage, the property of convergence is very important.

Definition 5.1. Consider a series of functions $\sum_{n=n_0}^{\infty} f_n(x)$. We say that the series converges to a function $s(x)$ on a set M if

$$\lim_{N\to\infty} \left(\sum_{n=n_0}^{N} f_n(x) \right) = s(x) \tag{5.2}$$

for each $x \in M$. Otherwise we say that the series of functions $\sum_{n=n_0}^{\infty} f_n(x)$ diverges on M.

The function $s(x)$ is said to be the sum of the series $\sum_{n=n_0}^{\infty} f_n(x)$.

The function $s_N(x)$ is called the Nth partial sum of the series $\sum_{n=n_0}^{N} f_n(x) = s_N(x)$ and the function $s(x)$ can be approximated by the corresponding partial sum function $s_N(x)$.

The series convergence domain is represented by the set of all x for which the series $\sum_{n=n_0}^{\infty} f_n(x)$ is convergent.

Let us look closer at the definition and try to clarify some important relations in it. Functions f_n form a sequence $\{f_n\}_{n=n_0}^{N} = \{f_n(x)\}_{n=n_0}^{N}$. For each $x_0 \in M$, the functions f_n are defined, and their values $f_n(x_0)$ form a number sequence which is convergent: $\lim_{n\to\infty} f_n(x_0) = f(x_0)$. Then the sequence $\{f_n(x)\}_{n=n_0}^{N}$ converges on M (at all $x \in M$) to a new created function $f(x)$. Since the limit of continuous functions may not stay continuous, to preserve continuity, one has to consider a uniform convergence.

Definition 5.2. We say that the sequence of functions f_n uniformly converges to a limiting function f on M if for each $\varepsilon > 0$, a number n_0 exists such that for each $n > n_0$ and $x \in M$: $|f_n(x) - f(x)| < \varepsilon$ ($f_n(x)$ is arbitrarily close to f).

Calculus for Engineering Students. https://doi.org/10.1016/B978-0-12-817210-0.00012-6

Theorem 5.1. *If the sequence* $\{f_n\}_{n=n_0}^{N}$ *uniformly converges to the function* f *on the set* M *and functions* f_n *are continuous on* M *for each* n*, then the function* f *is also continuous.*

The ability to express the function s in the form of sum of functions f_n allows to apply the principle of superposition, by which a result of linear mapping $F(s)$ could be obtained as a sum of part by part responses:

$$F(s) = F(f_1) + F(f_2) + F(f_3) + \dots \tag{5.3}$$

This is a very valuable property, used in many engineering problems.

5.1.1 Taylor series

Theorem 5.2 (Taylor's theorem). *Let the function* f *be* $n+1$ *times differentiable on interval* (a,b)*, and* $x_0 \in (a,b)$*. Then for each* $x \in (a,b)$ *there exists such a number* ξ *between* x_0 *and* x *that* $f(x) = T_n(x) + R_n(x)$*, where*

$$\begin{aligned} T_n(x) &= \sum_{k=0}^{n} \frac{f^{(k)}(x_0)}{k!}(x-x_0)^k \\ &= f(x_0) + \frac{f'(x_0)}{1!}(x-x_0) + \frac{f''(x_0)}{2!}(x-x_0)^2 + \dots \\ &+ \frac{f^{(n)}(x_0)}{n!}(x-x_0)^n \end{aligned} \tag{5.4}$$

and

$$R_n(x) = \frac{f^{(n+1)}(\xi)}{(n+1)!}(x-x_0)^{n+1},$$

where $T_n(x)$ *is called the nth Taylor polynomial of a function* f *about* x_0 *and* $R_n(x)$ *is called a remainder term associated with* $T_n(x)$*. The series*

$$T(x) = \lim_{n\to\infty} T_n(x) = \sum_{k=0}^{\infty} \frac{f^{(k)}(x_0)}{k!}(x-x_0)^k$$

is called Taylor series. In the case $x_0 = 0$*, we speak about Maclaurin series.*

It is not difficult to show that $T_n(x_0) = f(x_0)$, and all kth derivatives $T_n^{(k)}(x_0) = f^{(k)}(x_0)$, $k = 1, 2, \dots, n$. The polynomial $T_n(x)$ can be considered to be the approximation of $f(x)$ with error $R_n(x)$ for any $x \in (a,b)$. With increasing distance from x_0, the accuracy of approximation decreases. If the value of remainder $R_n(x)$ decreases with increasing n: $\lim_{n\to\infty} R_n(x) = 0$, the function f can be represented by Taylor series $f(x) = T(x)$. Although the Taylor polynomial approximates the function f only in close vicinity of x_0, it has very important role in derivation of many numerical techniques and error bounding estimations, e.g., Newton's method in the root finding problem, interpolation, and numerical differentiation and integration (for more information consult for instance Burden and Faires, 2011). Taylor's theorem is one of the important building blocks and belongs to essential topics of elementary engineering mathematics (see e.g. Ivan, 1989).

5.1.2 Fourier series

Fourier series can stand for a function on a bounded interval $\langle a, b\rangle$ or for a periodic function with basic interval of periodicity $T = \langle a, b\rangle$. It is expressed by a series of sinusoids that can be stated in various forms. In essence, let us consider a pair of functions $\{\sin mt, \cos nt\}$, where t is a variable (usually time), and m, n are real multipliers of t, reflecting the length of interval $\langle a, b\rangle$. The pair belongs to the family of orthogonal functions which play the same role in the space of functions like orthogonal (perpendicular) vectors in Euclidean space (function space is a vector space with bilinear form).

Remark. Imagine three perpendicular vectors in $\boldsymbol{E}_3$ space: $\{e_1, e_2, e_3\}$. Each other vector u in $\boldsymbol{E}_3$ space can be expressed as their linear combination $u = a_1e_1 + a_2e_2 + a_3e_3$. The set $\{e_1, e_2, e_3\}$ is the orthogonal basis of $\boldsymbol{E}_3$ space.

Definition 5.3. The set of functions $\{\xi_i\}_{i=1}^N$ is said to be orthogonal on the bounded interval $\langle a, b\rangle$ if for each two functions from the set

$$\int_a^b \xi_m(t)\xi_n(t)dt = \begin{cases} 0, & \text{for } m \neq n, \\ K, & \text{for } m = n. \end{cases} \tag{5.5}$$

In the case $K = 1$, the set is said to be orthonormal.

Exercise 5.1. How can be find that the vectors $\{e_1, e_2, e_3\}$ are orthogonal in $\boldsymbol{E}_3$? Is there some similarity with the set $\{\xi_i\}_{i=1}^N$ of orthogonal functions?

Exercise 5.2. For each of the following pairs find out whether it is orthogonal (i is a complex unit, m, n are integers, and $a \neq 0$ is real):

(a) $\{\cos mt, \sin nt\},\ t \in \langle -a, a\rangle$
(b) $\{\cos mt, \cos nt\},\ t \in \langle 0, 2\pi\rangle$
(c) $\{t^4, t^2\},\ t \in \langle -a, a\rangle$
(d) $\{e^{mt}, e^{nt}\},\ t \in \langle -\pi, \pi\rangle$
(e) $\{e^{imt}, e^{-int}\},\ t \in \langle -\pi, \pi\rangle$.

Results: (a), (b), (e), yes; (c), (d) no.

Exercise 5.3. In the sense of the least square method, derive the generalized coefficient $a_i = \frac{1}{K}\int_a^b f(t)\xi_i(t)dt$ in the approximation $f(t) \approx \sum_{i=1}^N a_i\xi_i(t)$. (Hint: minimize $S = \int_a^b \left(f(t) - \sum_{i=1}^N a_i\xi_i(t)\right)^2 dt$, $\frac{\partial S}{\partial a_i} = 0$; and use the rules for orthogonal functions (5.5) or see for instance Ondráček (2008).)

Exercise 5.4. Approximate the function $f(t) = \begin{cases} 1 & \text{for } t \in \langle -1, 1 \rangle \\ -1 & \text{for } t \in (1, 3\rangle \end{cases}$, using the approximation from the previous exercise, if the set of functions $\left\{\mathrm{Cos}\left(m\frac{\pi}{2}t\right)\right\}_{m=1}^{6}$ is used. Results: $K_m = 2$, $a_m = \frac{4\,\mathrm{Sin}^3(m\pi/2)}{m\pi}$ for odd m, $a_m = 0$ for even m, $f(t) \approx \frac{4}{\pi}\mathrm{Cos}\left(\frac{1}{2}\pi t\right) - \frac{4}{3\pi}\mathrm{Cos}\left(\frac{3}{2}\pi t\right) + \frac{4}{5\pi}\mathrm{Cos}\left(\frac{5}{2}\pi t\right)$.

Definition 5.4 (Generalized Fourier series). The convergent linear combination of orthogonal functions $\sum_{i=0}^{\infty} a_i \xi_i$ is said to be a generalized Fourier series of a function $f(t)$ on $\langle a, b \rangle$ if $\{\xi_i\}_{i=1}^{\infty}$ is a complete set of orthogonal functions and a function $f(t)$ is square integrable, i.e., $\int_{-\infty}^{\infty} |f(t)|^2\,dt < \infty$ for $t \in \langle a, b \rangle$. We write $f(t) \sim \sum_{i=1}^{\infty} a_i \xi_i(t)$, where $a_i = \frac{1}{K}\int_a^b f(t)\xi_i(t)dt$, $K = \int_a^b \xi_i^2(t)dt$. If $\lim_{N\to\infty}\left(\sum_{i=1}^{N} a_i \xi_i(t)\right) = f(t)$, we write $f(t) = \sum_{i=1}^{\infty} a_i \xi_i(t)$ and say that the series is a generalized Fourier series expansion of a function $f(t)$, or a function $f(t)$ is decomposed into a generalized Fourier series $\sum_{i=0}^{\infty} a_i \xi_i$.

There are many sets of orthogonal functions which can be used for generalized Fourier decomposition of a function. The most known follow trigonometric functions, exponential functions, and special polynomials such as Legendre, Chebyshev, Hermite, Laguerre, etc.

Definition 5.5 (Fourier series). The series

$$f(t) \sim \frac{a_0}{2} + \sum_{n=1}^{\infty} a_n \cos(n\Omega t) + b_n \sin(n\Omega t) \tag{5.6}$$

is said to be a Fourier series associated with a function $f(t)$ on bounded interval $\langle a, b \rangle$ or a Fourier series associated with a periodic function $f(t)$ with basic interval of periodicity $\langle a, b \rangle$ and a fundamental period T, where the basic angular frequency $\Omega = \frac{2\pi}{T}$, T is the length of interval $\langle a, b \rangle$, $n = 1, 2, 3 \ldots$. Fourier coefficients a_0, a_n, b_n are calculated by the following formulas:

$$a_n = \frac{2}{T}\int_a^b f(t)\cos(n\Omega t)\,dt, \qquad b_n = \frac{2}{T}\int_a^b f(t)\sin(n\Omega t)\,dt \tag{5.7}$$

if $n = 0$ $a_0 = \frac{2}{T}\int_a^b f(t)dt$, $b_0 = 0$.

Denote

$$s(t) = \frac{a_0}{2} + \sum_{n=1}^{\infty} a_n \cos(n\Omega t) + b_n \sin(n\Omega t). \tag{5.8}$$

Definition 5.5 does not provide any information about convergence. To be expressed by Fourier series, the function $f(t)$ has to satisfy so-called Dirichlet conditions.

Dirichlet conditions. Function $f(t)$ has the following properties:

1. $\int_a^b |f(t)|dt < \infty$; $f(t)$ is absolutely integrable on $\langle a, b\rangle$, i.e., the integral $\int_a^b f(t)dt$ is finite;
2. $f(t)$ is bounded on $\langle a, b\rangle$; $f(t)$ has a finite number of finite extrema;
3. $f(t)$ is piecewise continuous on $\langle a, b\rangle$; $f(t)$ is continuous on $\langle a, b\rangle$, or it has a finite number of points d with jump discontinuity: $\lim_{t\to d^-} |f(t)| < \infty$ and $\lim_{t\to d^+} |f(t)| < \infty$.

Theorem 5.3 (Convergence of Fourier series). *If $f(t)$ satisfies Dirichlet conditions, then for each $t_0 \in \langle a, b\rangle$*

$$s(t_0) = \frac{1}{2}\left(\lim_{t\to t_0^-} f(t) + \lim_{t\to t_0^+} f(t)\right). \tag{5.9}$$

That means:

$s(t_0) = f(t_0)$ in points t_0, $a < t_0 < b$, where $f(t)$ is continuous,
$s(d) = \frac{1}{2}\left(\lim_{t\to d^-} f(t) + \lim_{t\to d^+} f(t)\right)$ in points of jump discontinuity d, $a < d < b$,
$s(a) = s(b) = \frac{1}{2}\left(\lim_{t\to a^+} f(t) + \lim_{t\to b^-} f(t)\right)$ in endpoints of interval $\langle a, b\rangle$.

And we write

$$f(t) = \frac{a_0}{2} + \sum_{n=1}^{\infty} a_n \cos(n\Omega t) + b_n \sin(n\Omega t) \tag{5.10}$$

for all points t, where $s(t) = f(t)$.

Remark. Satisfying Dirichlet conditions, the Fourier series $s(t)$ can differ from the function $f(t)$ only in a finite number of jump discontinuity points on $\langle a, b\rangle$.

Example 5.1. Explore the convergence of the Fourier series $s(t)$ associated with $f(t)$ and decompose $f(t)$ into Fourier series.

$$f(t) = \begin{cases} 2, & 0 \le t \le 1, \\ -t + 2, & 1 < t \le 2. \end{cases}$$

Solution.
Convergence. The function $f(t)$ satisfies Dirichlet's conditions. It is absolutely integrable and bounded on interval $\langle 0, 2\rangle$ and it is continuous on interval $\langle 0, 2\rangle$ except the point $d = 1$, where the point of jump discontinuity is placed (Fig. 5.1). The Fourier series $s(t)$ associated with $f(t)$ converges (see Fig. 5.3) and

$s(t_0) = f(t_0)$ in points t_0, $0 < t_0 < 1$ or $1 < t_0 < 2$, $f(t)$ is continuous here;
$s(1) = \frac{1}{2}\left(\lim_{t\to 1^-} f(t) + \lim_{t\to 1^+} f(t)\right) = \frac{1}{2}(2 + 1) = \frac{3}{2}$ in the point of jump discontinuity $d = 1$;
$s(0) = s(2) = \frac{1}{2}\left(\lim_{t\to 0^+} f(t) + \lim_{t\to 2^-} f(t)\right) = \frac{1}{2}(2 + 0) = 1$ in endpoints of interval $a = 0$, $b = 2$.

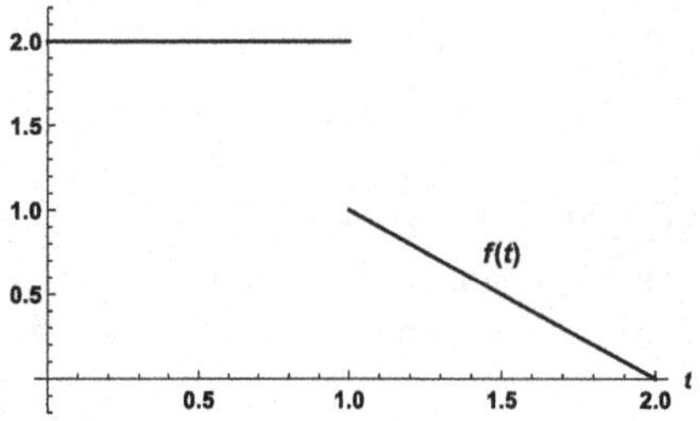

Figure 5.1 Function $f(t)$.

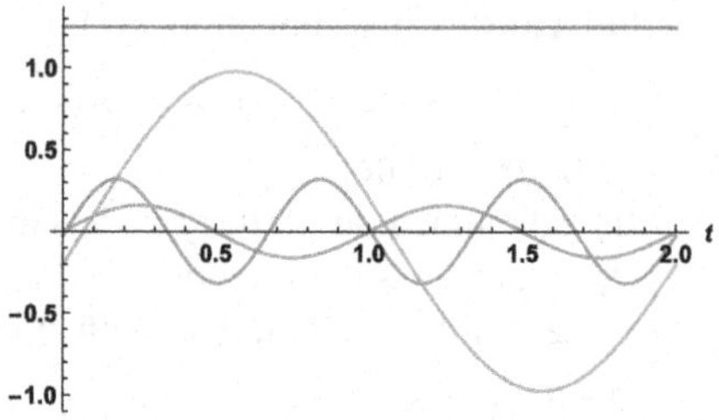

Figure 5.2 The first 4 terms of $s(t)$.

Decomposition.

The length of the interval $\langle 0, 2\rangle$ is $T = 2$, and $\Omega = \pi$. Fourier coefficients (5.7) are

$$a_0 = \left(\int_0^1 2dt + \int_1^2 (-t+2)\,dt\right) = \frac{5}{2},$$

$$a_n = \left(\int_0^1 2\cos(n\pi t)\,dt + \int_1^2 (-t+2)\cos(n\pi t)\,dt\right) = \frac{-1+(-1)^n}{n^2\pi^2},$$

$$b_n = \left(\int_0^1 2\sin(n\pi t)\,dt + \int_1^2 (-t+2)\sin(n\pi t)\,dt\right) = \frac{2-(-1)^k}{k\pi}.$$

Fourier series (5.8) is

$$s(t) = \frac{5}{4} + \sum_{n=1}^{\infty} \frac{-1+(-1)^n}{n^2\pi^2}\cos(n\pi t) + \frac{2-(-1)^n}{n\pi}\sin(n\pi t) =$$

$$= \frac{5}{4} + \left(-\frac{2\cos(\pi t)}{\pi^2} + \frac{3\sin(\pi t)}{\pi}\right) + \frac{\sin(2\pi t)}{2\pi}$$

$$+ \left(-\frac{2\cos(3\pi t)}{9\pi^2} + \frac{\sin(3\pi t)}{\pi}\right) + \ldots$$

The constant term and the first three sinusoids of the Fourier series are shown in Fig. 5.2; and periodic $f(t)$ with basic interval of periodicity $\langle 0, 2\rangle$ is shown with its Fourier expansion in Fig. 5.4.

Despite the fact that the Fourier series $s(t)$ converges to the function $f(t)$ in all points where $f(t)$ is continuous and to the middle value in points of jump discontinuity, the partial sum $s_N(t)$ does not hold this property and it overshoots in the neighborhoods of jump discontinuity (Figs. 5.3 and 5.5). Increasing N, the point of maximal overshoot moves closer to the point of discontinuity, but the value of overshoot reaches only its limit, about 9% of the jump (Fig. 5.6). This is known as Gibb's phenomenon, named after J. Willard Gibbs (1899). Nevertheless, it was already described 50 years earlier by Henry Wilbraham (1848) (Carslaw, 1925).

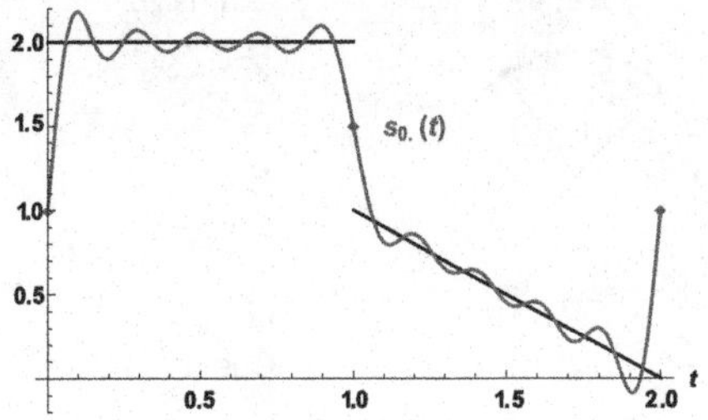

Figure 5.3 $f(t)$ and $s_{10}(t)$, $t \in \langle 0, 2 \rangle$.

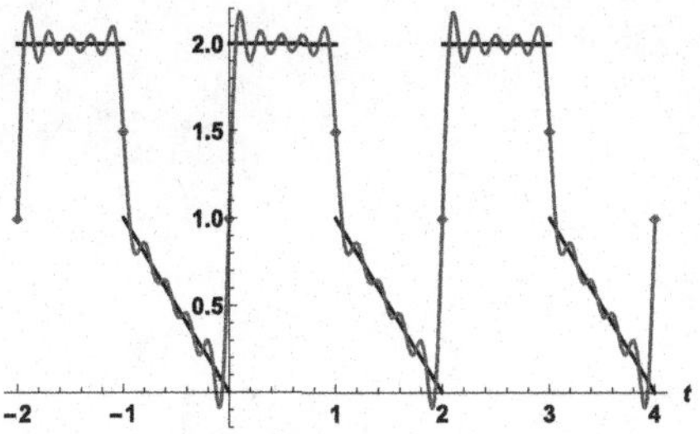

Figure 5.4 Periodic $f(t)$ and $s_{10}(t)$.

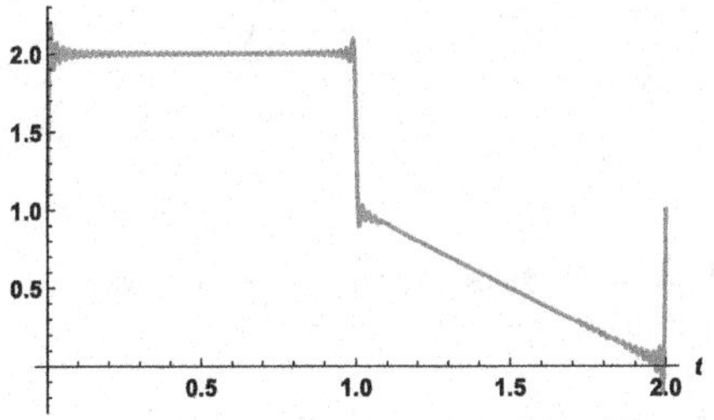

Figure 5.5 $s_{100}(t)$.

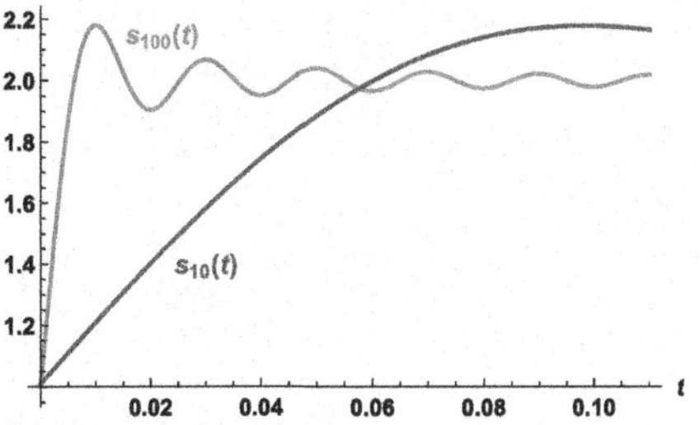

Figure 5.6 Detail of $s_{100}(t)$ and $s_{10}(t)$.

5.1.2.1 Real forms of Fourier series

The term

$$f_n(t) = a_n \cos(n\Omega t) + b_n \sin(n\Omega t) \tag{5.11}$$

represents a sinusoid, real function $f_n(t)$, where t is a real variable, $n\Omega$, a_n, b_n, are real constants, and n is a natural number. As a sinusoid oscillates around the zero value, the constant term $a_0/2$ in (5.10) stands for the average value of the function $f(t)$. Expression (5.11) is said to be a trigonometric or goniometric form of a sinusoid and the Fourier series is said to be written in trigonometric or goniometric form. Applying sine and cosine addition or subtraction formulas we obtain four possible expressions, i.e.,

$$\begin{aligned} f_n(t) &= A_n \cos(n\Omega t + \varphi_n), \qquad & f_n(t) &= A_n \cos(n\Omega t - \varphi_n), \\ f_n(t) &= A_n \sin(n\Omega t + \varphi_n), \qquad & f_n(t) &= A_n \sin(n\Omega t - \varphi_n), \end{aligned} \tag{5.12}$$

said to be amplitude-phase forms, where $A_n > 0$ is the amplitude of the sinusoid, the maximal deviation of the curve from the axis t, calculated by the formula

$$A_n = \sqrt{a_n^2 + b_n^2}, \tag{5.13}$$

and an angle $\varphi_n \in (-\pi, \pi\rangle$ is the initial phase (or only phase) of the sinusoid.

Describing one particular wave $f_n(t)$, the amplitude A_n is determined uniquely, whereas the value of phase φ_n differs with respect to the used expression. For instance,

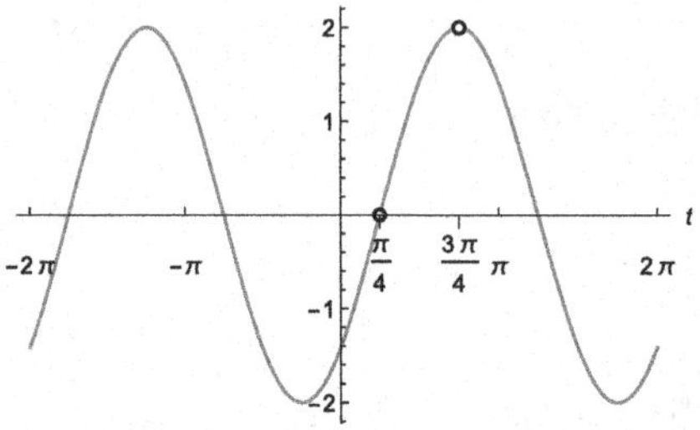

Figure 5.7 $f_1(t) = \sqrt{2}\sin t - \sqrt{2}\cos t$.

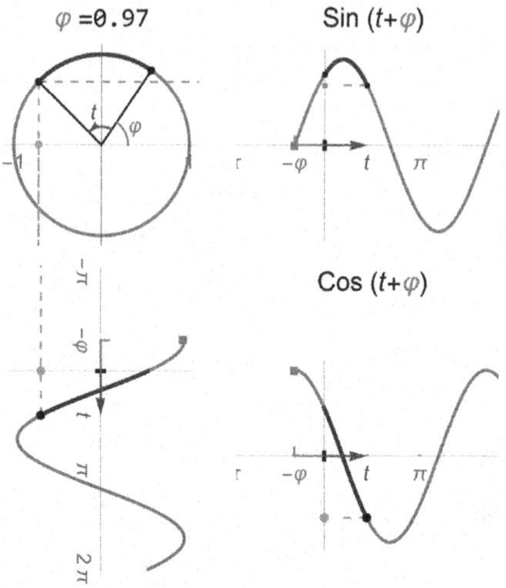

Figure 5.8 Sinusoids with ambiguous phase and unit circle.

the sinusoid $f_1(t) = \sqrt{2}\sin t - \sqrt{2}\cos t$ (Fig. 5.7) can be expressed in amplitude-phase form as $2\cos(t + (-\frac{3\pi}{4})) = 2\cos(t - \frac{3\pi}{4}) = 2\sin(t + (-\frac{\pi}{4})) = 2\sin(t - \frac{\pi}{4})$. The amplitude is 2 and under the phase we can see four values, reflecting the sine/cosine reference (initial) point shift $\frac{\pm\varphi_n}{n\Omega}$.

In technical applications, the amplitudes and phases are very important characteristics of sinusoids. Solving the engineering problem, the most suitable formula (5.12) is picked up ambiguously using the same term phase notation φ_n, so one has to be very careful which of the expressions (5.12) to consider, and not to mix phase values coming from different representations (Fig. 5.8).

Let us consider two basic amplitude-phase forms of the same sinusoid wave,

$$f_n(t) = A_n \cos(n\Omega t + \varphi_n), \qquad f_n(t) = A_n \sin(n\Omega t + \psi_n), \tag{5.14}$$

where φ_n is the phase of the sine wave and ψ_n is the phase of the cosine wave, $\varphi_n = \psi_n - \frac{\pi}{2}$, keeping $\varphi_n \in (-\pi, \pi\rangle$ and $\psi_n \in (-\pi, \pi\rangle$ by adding or subtracting the basic period 2π. Applying (5.11) and trigonometric addition formulas, the formulas for phase calculation are derived:

$$\psi_n \sim \arctan\frac{a_n}{b_n}, \qquad \varphi_n \sim \arctan\frac{-b_n}{a_n}. \tag{5.15}$$

Nonetheless, arctangent function values lie only in interval $(-\pi/2, \pi/2)$. To cover the phase range interval $(-\pi, \pi\rangle$, the period of tangent function π has to be added for the phase from $(\pi/2, \pi\rangle$, and subtracted for the phase from $(-\pi, -\pi/2)$. In points where the arctan fraction denominator is equal to zero, the phase is $\pm\pi/2$, or 0/undefined. For details of computation see the table in Fig. 15.1 in Chapter 15.

Remark. For number arctan calculations of fractions with domain of results $(-\pi, \pi\rangle$, CAS software offers a special command considering two arguments, for instance

ArcTan [denominator, numerator] in Wolfram Mathematica, or atan2 (denominator, numerator) in MATLAB®.

5.1.2.2 Complex form of Fourier series

Applying the Euler formula $e^{\mathrm{i}(n\Omega t)} = \cos(n\Omega t) + \mathrm{i}\sin(n\Omega t)$ and its complex conjugate, we substitute $\cos(n\Omega\, t)$ and $\sin(n\Omega\, t)$ in (5.10), and get the Fourier series in complex form,

$$f(t) = \sum_{n=-\infty}^{\infty} f_{n,c}(t) = \sum_{n=-\infty}^{\infty} c_n e^{\mathrm{i}n\Omega t}, \tag{5.16}$$

where c_n is a complex Fourier coefficient

$$c_n = \frac{1}{T}\int_a^b f(t)e^{-\mathrm{i}n\Omega t}dt, \quad \text{if } n = 0,\ c_0 = \frac{1}{T}\int_a^b f(t)dt. \tag{5.17}$$

The form (5.16) is said to be the exponential (also complex, or polar) form of the Fourier series. The amplitude of the complex Fourier term $f_{n,c}(t)$ is the absolute value of complex coefficient $|c_n|$, and the phase φ_n is the argument of complex coefficient $\arg(c_n)$, which is calculated under the same rules as seen in the table in Fig. 15.1 in Chapter 15 considering arctan ($\mathrm{Im}(c_n)/\mathrm{Re}(c_n)$). Then the amplitude-phase form is

$$c_n = |c_n| e^{\mathrm{i}\varphi_n}. \tag{5.18}$$

Complex conjugate $\bar{c}_n = c_{-n}$, and the real wave (5.11)

$$f_{n,\mathrm{real}}(t) = f_{n,c}(t) + f_{-n,c}(t) = c_n e^{\mathrm{i}n\Omega t} + \bar{c}_n e^{-\mathrm{i}n\Omega t} = 2\,\mathrm{Re}\left(c_n e^{\mathrm{i}n\Omega t}\right).$$

In comparison of real and complex Fourier coefficients, we get transformation formulas between real and complex parameters,

$$c_n = \frac{a_n - \mathrm{i}b_n}{2}, \quad a_n = c_n + c_{-n} = 2\,\mathrm{Re}(c_n), \quad b_n = \mathrm{i}(c_n - c_{-n}) = -2\,\mathrm{Im}(c_n), \tag{5.19}$$

$$|c_n| + |c_{-n}| = 2|c_n| = A_n, \ \arg(c_n) \sim \arctan\frac{\mathrm{Im}(c_n)}{\mathrm{Re}(c_n)} = \arctan\frac{-b_n}{a_n} \sim \varphi_n.$$

The complex form is short and apparent in expression; it usually simplifies computations (the differential equations are modified into algebraic ones), and moreover, the complex coefficient c_n does not depend on variable t, but provides information on the amplitude and *unique* phase of $f_{n,c}(t)$.

5.2 Fourier series in applications

The sinusoid (5.11) can be revealed in many technical and natural phenomena. It mathematically depicts oscillating waves, circular uniform movement, or any periodic movement or phenomenon which can be approximated by uniform movement along a circle (Fig. 5.8). Besides sinusoids, two other terms are common. We can speak about monochromatic waves while depicting electromagnetic waves of one particular frequency, or about harmonics, the name of which comes from description of ideal uniform oscillating movement of a mass hanging on a spring, called simple harmonic motion. The latter term is wildly used across technical branches, in physics, mechanics, acoustics, electric energy transmission (current and voltage), magnetic field description, or information propagation by signals, and others. In the following, we will use the term harmonic or sinusoid. Composing harmonic waves, periodic nonharmonic signals arise, which occur in steady behavior, and are explored and controlled by means of Fourier series.

Each term of Fourier series, in addition to the time domain formula and graph, is uniquely determined by its amplitude (A_n or $|c_n|$), angular frequency ($\Omega_n = n\Omega$), and phase (φ_n or ψ_n), providing a triple (*frequency, amplitude, phase*) or two ordered pairs (*frequency, amplitude*) and (*frequency, phase*) in the frequency domain. Geometrically, we get two frequency graphs of discrete points for Fourier series, an "amplitude-frequency spectrum" – harmonics points [*frequency, amplitude*], and a "phase-frequency spectrum" – harmonics points [*frequency, phase*]. Sinusoids are called nth harmonic ($n \geq 1$). The constant term is known also as the DC term, referring to direct current in electrotechnics, for which $\Omega_0 = 0$, $A_0 = |a_0/2| = |c_0|$ and phase is not considered (sometimes for $c_0 \geq 0$, $\varphi_0 = 0$). The first harmonic is also called the fundamental harmonic; sinusoids with $n > 1$ are higher harmonics. The expansion of $f(t)$ into Fourier series providing its frequency spectra is called harmonic analysis (or Fourier analysis); and harmonic synthesis (or Fourier synthesis) stands for a function composition from Fourier series terms or from corresponding frequency spectra. Depending on real or complex forms, we speak about real or complex harmonics, one-sided frequency spectra for real Fourier series (Fig. 5.27), and two-sided frequency spectra for complex Fourier series (Fig. 5.10). The complex coefficient is called also phasor. Taking into account that $|c_n| = |c_{-n}|$ and $\arg(c_{-n}) = -\arg(c_n)$, the complex amplitude spectra are even and complex phase spectra are odd (Fig. 5.10, Fig. 5.22); and frequently, only the right-handed half of two-sided graphs is used (Fig. 5.12). Despite the fact that time domain Fourier series are not of very high importance in professional applications (calculations are complicated), the frequency spectra are widely used for property description and regulations of physical quantities. The amplitude spectrum reflects the power properties of signal (see (5.21)), and phase spectrum reflects the initial angle shift of the harmonics, important in their synthesis. Two sinusoidal signals are said to be in phase when they have the same phase, otherwise they are said to be out of phase. In the case the phases differ by π they are said to be in antiphase.

Example 5.2. Inverters convert direct current (DC) power to alternating current (AC) power. AC voltage is generated by alternating the load connection to the positive and

(A)

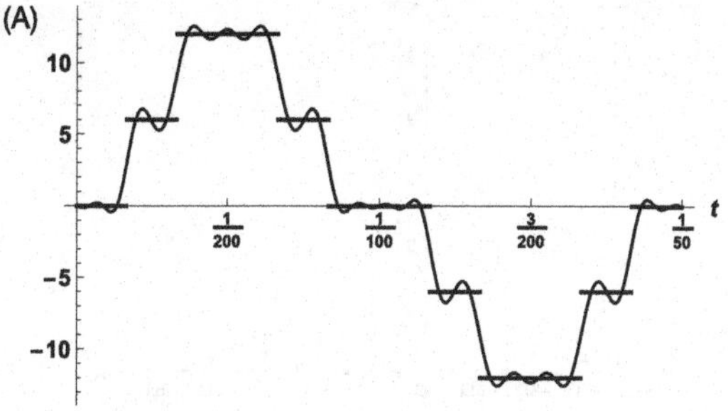

(B)

$n\Omega$	c_n	$amp_n = \vert c_n \vert$	$\varphi_n = \arg(c_n)$
$-700\,\pi$	0. − 0.199731 i	0.199731	−1.5708
$-500\,\pi$	0. − 0.279623 i	0.279623	−1.5708
$-300\,\pi$	0. − 1.27324 i	1.27324	−1.5708
$-100\,\pi$	0. + 5.21783 i	5.21783	1.5708
$100\,\pi$	0. − 5.21783 i	5.21783	−1.5708
$300\,\pi$	0. + 1.27324 i	1.27324	1.5708
$500\,\pi$	0. + 0.279623 i	0.279623	1.5708
$700\,\pi$	0. + 0.199731 i	0.199731	1.5708

Figure 5.9 (A) Stair-shaped signal $U(t)$ and its approximation $U(t) \approx \sum_{n=-10}^{10} c_n e^{i100\pi nt}$. (B) Frequency table for eight odd complex harmonics.

negative poles of the DC power supply. The generated alternating voltage is thus periodic, but naturally nonharmonic. Its frequency depends on the switching speed of the inverter switches.

Let the electrical circuits of the power inverter generate a stair-shaped periodic output signal from the input DC voltage (Fig. 5.9). Have a look at the complex Fourier series form of given signal, the table of harmonics amplitudes and phases, and the frequency graphs.

Solution.
Fourier series. The stair-shaped output signal satisfies Dirichlet conditions, so it may be expanded into Fourier series. Usually, a piecewise function (see Example 5.1) is used for its description, or it can be defined using the Heaviside step function θ, which is often accepted in software for further calculations. Let the signal be

$$\begin{aligned} U(t) = v\,\theta(t-11s) + v\,\theta(t-10s) - v\,\theta(t-8s) - v\,\theta(t-7s) - \\ - v\,\theta(t-5s) - v\,\theta(t-4s) + v\,\theta(t-2s) + v\,\theta(t-s), \end{aligned} \tag{5.20}$$

where maximal pitch is 12 V and frequency $f = 50$ Hz. Then partial interval length is $s = 1/600$ s; partial pitch value $v = 6$ V, basic period is $T = 1/50$ s; and angular frequency $\Omega = 2\pi/T = 2\pi f = 100\pi$ rad/s.

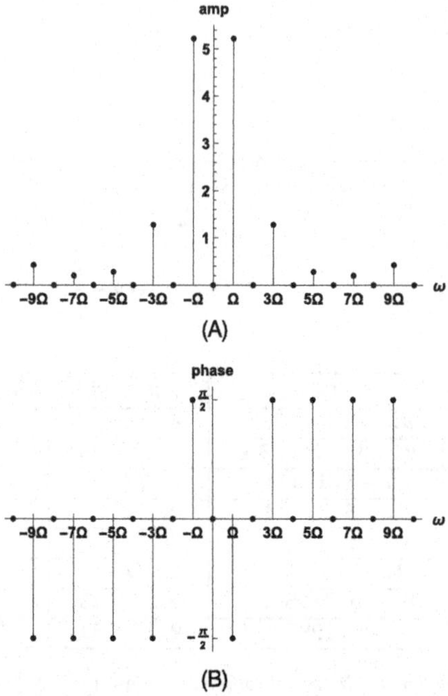

Figure 5.10 Two-sided frequency spectra of complex Fourier series. (A) Amplitude spectrum. (B) Phase spectrum.

Complex Fourier coefficients are calculated by formulas (5.17) and the Fourier expansion of the output signal by (5.16),

$$U(t) = \sum_{n=-\infty}^{\infty} c_n e^{in\Omega t} = \ldots - \frac{4\mathrm{i}}{\pi}\mathrm{e}^{-300\pi it} + \frac{6\mathrm{i}\left(1+\sqrt{3}\right)}{\pi}\mathrm{e}^{-100\pi it} - \frac{6\mathrm{i}\left(1+\sqrt{3}\right)}{\pi}\mathrm{e}^{100\pi it} + \frac{4\mathrm{i}}{\pi}\mathrm{e}^{300\pi it} + \ldots$$

The signal approximation by partial sum of Fourier series where $N = 10$ is shown in Fig. 5.9A.

Frequency table and graphs. Amplitudes and phases are determined by amplitudes and arguments of complex coefficients. Since $c_n = 0$ for even n, in the frequency table (Fig. 5.9B), there are only values for odd frequencies present; c_n are imaginary numbers, that is why the phase is $\pm\pi/2$. Providing sets of points $\{[n\Omega, |c_n|]\}_{n=-N}^{N}$ and $\{[n\Omega, \arg(c_n)]\}_{n=-N}^{N}$, the two-sided frequency spectra are easily drawn (Fig. 5.10).

The inverters of DC to AC represent one group of semiconductor converters used in power and electrical engineering that generally change parameters of the power supply system according to the load (appliance) needs. Their use is very extensive. In the field of consumer electrical equipment, converters serve as controllable power sources for light and heat appliances, for induction heating and hardening, for charging accumulator batteries, or as components of standby or emergency power sources, and for many other purposes. However, their main use is mostly in regulating electric drives in industry and electric traction.

Regarding control methods, we distinguish phase control converters, where the switch-on angle of the semiconductor device varies, and amplitude control converters, where the amplitude of the output voltage or current varies. With respect to their particular role, converters are able to follow nonharmonic behavior of current or voltage and generate a compensating image of filtered nondesired content.

The polygonal shape of signal includes always higher harmonics (Fig. 5.10), which can cause problems in powering electric engines, or it can reduce active power supply. As an example, consider phase control of converter due to which specific conditions arise in the supply network – the mains voltage is sinusoidal (within the electricity mains guarantee) and the current (phase current) is periodic, but not harmonic.

Let us inspect the total power factor λ defined as the ratio of the active power to the total apparent power. The total apparent power input from the m-phase system is given as $S = m\ U\ I$, where U is the effective value of the harmonic supply voltage, and I is the effective value of the nonharmonic phase supply current. The effective values are calculated as (root mean square) r.m.s. values, for current

$$I = \sqrt{\frac{1}{T}\int_T I^2} = \sqrt{\sum_{n=1}^{\infty} I^2_{n\,\mathrm{real}}}, \tag{5.21}$$

where T is the basic period of the current signal, I_n is amplitude of its nth real harmonic. Active power P represents mean instantaneous power over one period. In the case the voltage is harmonic, only the first harmonic of current can be involved and $P = m\ U\ I_1\ \cos\varphi_1$, where φ_1 is the phase difference (the angle) between voltage and current. The higher harmonics participate on so-called deforming power, and the total power factor λ is then calculated as

$$\lambda = \frac{P}{S} = \frac{mUI_1\cos\varphi_1}{mUI} = \frac{I_1}{I}\cos\varphi_1.$$

For more detailed information on electrical principles (Borba, 2013) or another fundamentals of electrical engineering.

Example 5.3. Two music instruments playing one tone of 233 Hz ($B_{\flat 3}$) were recorded together by one microphone. Their joined sound complex half-spectra data are shown in Table 5.1. Can we separate the sounds of instruments if we know the spectra data (Table 5.2) of one of them?

Table 5.1 Joined sound complex half-spectra data.

f_n	233	466	699	932	1165	1398	1631	1864
$\lvert c_n \rvert$	1.	0.295	0.397	0.04	0.158	0.172	0.03	0.05
φ_n	-0.698	-1.115	1.438	0.824	-1.157	-1.233	-0.1	-1.571

Table 5.2 Trombone complex half-spectra data.

f_n	233	466	699	932	1165	1398
$\|c_n\|$	0.5	0.26	0.055	0.005	0.02	0.01
φ_n	-0.698	-1.047	-0.3	1.1	-0.35	-3.1

f_n	233	466	699	932	1165	1398
$n\Omega$	466 π	932 π	1398 π	1864 π	2330 π	2796 π
$\|c_n\|$	0.5	0.26	0.055	0.005	0.02	0.01
φ_n	-0.698	-1.047	-0.3	1.1	-0.35	-3.1
$\varphi_{n,\text{ anti}}$	2.443	2.094	2.842	-2.042	2.792	0.042

(A)

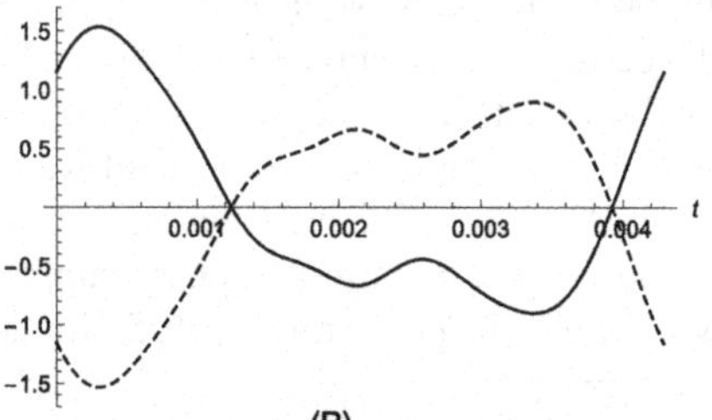

(B)

Figure 5.11 (A) Frequency table. (B) Trombone and its antiphase (dashed) waves.

Remark. The data were approximated by the author from publicly available waves characteristics of trombone, and clarinet and they can differ from true data.

Solution. Although both instruments play one tone at given frequency, they differ in higher harmonics that make their sound specific. To separate them, let us look at the principle active noise reduction headphones work on. Simply said, a small microphone receives the outside noise sound, sends it to electronic circuit, which generates the sound in antiphase, and plays it together with listened content into the loudspeaker (see e.g. Woodford, 2008). We will proceed in a similar way. We generate the trombone antiphase signal in the frequency domain: $|c_{n,\text{anti}}| = |c_n|$, $\varphi_{n,\text{anti}} = \varphi_n + \pi$, $\varphi_{n,\text{anti}} \in (-\pi, \pi)$, $\Omega = 2\pi f_1 = 2\pi \cdot 233$ (Fig. 5.11A). Now the frequency half- or two-sided complex spectra can be drawn (try as exercise). To represent signals in the time domain, we carry out the Fourier synthesis (5.16), where $|c_n| = |c_{-n}|$ and $\arg(c_{-n}) = -\arg(c_n)$. Applying (5.18) and the partial sum of Fourier series (5.16) for $N = 6$, where $c_0 = 0$, we obtain approximations of trombone sound and trombone antiphase sound in the time domain with $T = \frac{2\pi}{\Omega} = \frac{1}{f} \cong 0.004$ s (Fig. 5.11B).

The sum of the trombone signal and its antiphase counterpart makes the zero signal (verify), and so the trombone signal can be eliminated by its antiphase counterpart from the mixed signal. To compose the recorded mixed signal with the antiphase trombone signal, it will be worth to work in the frequency domain with phasors (5.18). The complex coefficients of the final wave $c_{n,\text{fin}} = c_{n,\text{mix}} + c_{n,\text{anti}} = |c_{n,\text{mix}}|e^{i\varphi_{n,\text{mix}}} + |c_{n,\text{anti}}|e^{i\varphi_{n,anti}}$ provide amplitudes and phases of the final signal (Table 5.3), the sec-

Table 5.3 Phasors of mixed, antiphase, and final sound with amplitude and phase of final sound.

$n\Omega$	c_nMIX	c_nANTI	c_nFIN	$\lvert c_n\text{FIN}\rvert$	φ_nFIN
466 π	0.76604 - 0.64279 i	-0.38302 + 0.32139 i	0.38302 - 0.32139 i	0.5	-0.69813
932 π	0.13 - 0.26517 i	-0.13 + 0.22517 i	0. - 0.04 i	0.04	-1.5708
1398 π	0.05254 + 0.39375 i	-0.05254 + 0.01625 i	0. + 0.41 i	0.41	1.5708
1864 π	0.02702 + 0.0292 i	-0.00227 - 0.00446 i	0.02475 + 0.02475 i	0.035	0.7854
2330 π	0.06359 - 0.14476 i	-0.01879 + 0.00686 i	0.04481 - 0.1379 i	0.145	-1.25664
2796 π	0.05698 - 0.16209 i	0.00999 + 0.00042 i	0.06697 - 0.16168 i	0.175	-1.1781
3262 π	0.02985 - 0.003 i	0	0.02985 - 0.003 i	0.03	-0.1
3728 π	0. - 0.05 i	0	0. - 0.05 i	0.05	-1.5708

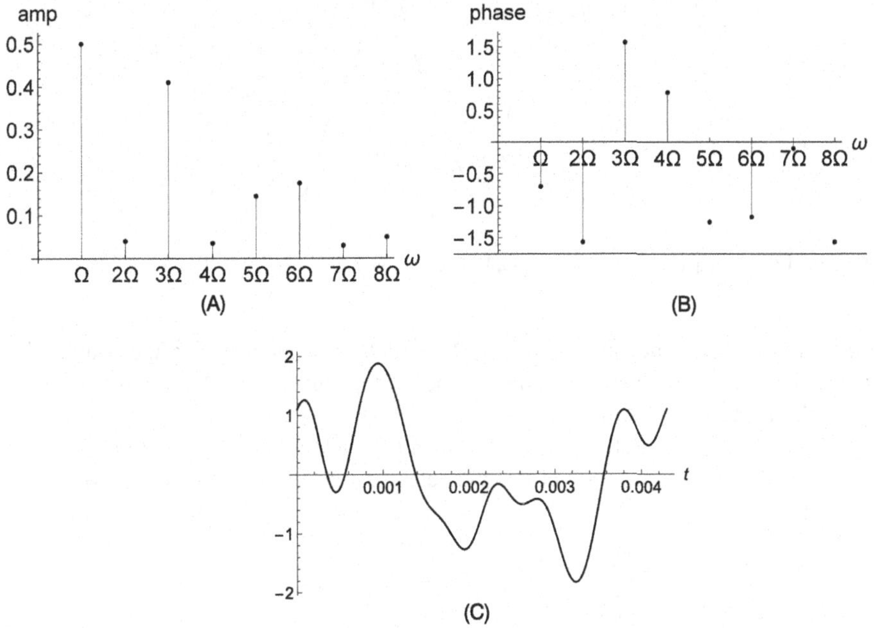

Figure 5.12 Half-frequency spectra and time domain graph of the final (clarinet) wave.

ond instrument sound (clarinet) (Fig. 5.12). In the time domain,

$$f_{\text{fin}}(t) \cong \sum_{n=-8}^{8} c_{n,\text{fin}} e^{in\Omega t}.$$

Sound synthesizers imitating various musical instruments are based on Fourier synthesis of harmonics. Setting modified higher harmonics amplitudes and phases allows to mix one's distinctive sound.

Example 5.4. Current I, voltage U, and impedance load Z are bounded by the formula $I = \frac{U}{Z}$. Consider a stable voltage source from an inverter with properties from

nΩ	$cI_n = c_n/Z_n$	$\lvert cI_n \rvert$	$\arg(cI_n)$
-500 π	0.00323272 - 0.00102901 i	0.003	-0.308
-300 π	0.0210847 - 0.0111858 i	0.024	-0.488
-100 π	-0.0940206 + 0.149638 i	0.177	2.132
100 π	-0.0940206 - 0.149638 i	0.177	-2.132
300 π	0.0210847 + 0.0111858 i	0.024	0.488
500 π	0.00323272 + 0.00102901 i	0.003	0.308

(A)

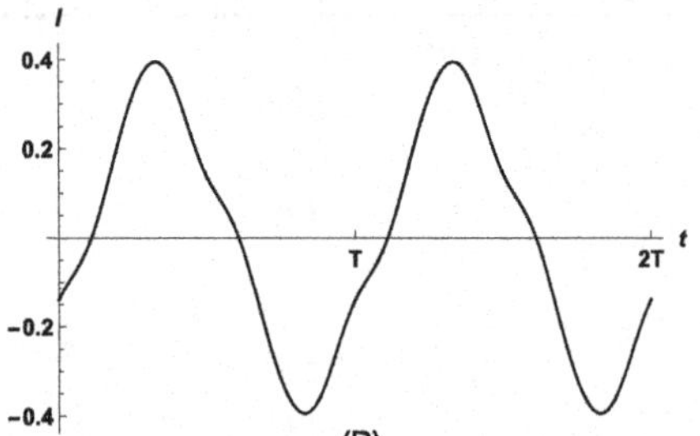

(B)

Figure 5.13 (A) Frequency table and (B) graph of the current approximation in the time domain.

Example 5.3. The impedance load $Z = 25 + 0.05i\omega$ is connected to the source. We would like to know the Fourier characteristics of the current signal under the load and its graph in the time domain.

Solution. In the case all items are linear in the system, the superposition property can be used, and the current can be calculated as the sum of voltage harmonics responses,

$$I = \sum_{n=-\infty}^{\infty} \frac{U_n}{Z_n} = \sum_{n=-\infty}^{\infty} \frac{c_n}{Z_n} e^{in\Omega t} \cong \sum_{n=-5}^{5} \frac{c_n}{Z_n} e^{in100\pi t},$$

where U_n is the nth voltage harmonic, Z_n is the value of impedance for $\omega = n\ \Omega$, and c_n is the complex coefficient of U_n from Example 5.3. The result is here approximated by the finite partial sum of 10 complex harmonics and the constant term c_0. Characteristics of six odd complex current coefficients ($c_n = 0$ for even n) and the current approximation graph are shown in Fig. 5.13. Note the irregular current shape. It is caused by higher harmonics. Considering the impedance load of the engine, the engine operation will be also irregular, leading to gradual destruction of a device.

Exercise. What will be the effective value of the current? (Hint: look at (5.21).)

Exercise 5.5. Compare the properties of the magnetic field by means of Fourier analysis in magnetic bearing with magnetic pole arrangement SSNN and SNSN (where S means South and N means North) in a stator (Vlnka et al., 1995). Hint: Construct the rectangle function with values 1, −1, and 0 within the interval of periodicity with respect to the layout of magnets along the stator circle.

Today, phased array systems are widely used in radars, sonars, ultrasound imaging scanners, or acoustics. The radar antenna series consists of a set of several primary emitters. By adjusting amplitudes and phases of their radio waves, it is possible to significantly influence the antenna performance. Even the radar beam could be deflected only electronically, without the need of mechanical antenna component movements, when only phases are changed, keeping the source amplitudes. In computations, modified general Fourier series or Fourier series are used.

Amplitude-frequency graphs are of great importance. They reveal the resonance regions where the amplitudes reach their maximal values (Figs. 5.18, 5.23, and 5.28). They uncover the range of operational frequency, within which participating signals, although being small, have similar frequency and phase, and incredibly increase the gain. In radiotechnics, in communication appliances, or in pendular motion, the resonance contributes to effective performance (e.g., the antenna gain increase); on the other hand, in energetics, building constructions, or in circular motion of engine components, the resonance could be very dangerous, causing dangerously high voltage, currents, or vibrations.

Extension of Fourier series. In the case the interval of periodicity is prolonged to infinity, a Fourier series is transformed into Fourier transform which enables us to follow amplitude and phase properties, as well as to do manipulations in the frequency domain (which are usually less complicated than in the time or space domain) also for nonperiodic signals, extending the use of Fourier analysis and synthesis in many branches from industry through medical services to social sciences, as for instance in nuclear magnetic resonance, chemical spectroscopy, detection of defective products, sound technique, radio communication, or space research. The fast Fourier transform (FFT), an algorithm processing sampled signals, is built-in in many software packages and represents one of the basic exploration methods.

The Fourier series and Fourier transform were named after the French mathematician Jean-Baptiste Joseph Fourier (1768–1830), who first introduced the trigonometric series in solving the heat equation (see Chapter 13), the partial differential equation (PDE), which had not been generally solved before. Fourier series and Fourier transform simplify the process of PDE solution.

5.3 Mechanical vibration forced by periodic force with viscous damping: harmonic analysis of a force, stabilized output movement obtained by principle of superposition

In machinery, vibration diagnosis, fault detection, or equipment service, lifetime is closely related to vibration analysis of particular machine parts. In most cases, oscillation occurs as an undesirable phenomenon. The specific type of vibration is the so-called torsional vibration. The most common causes of torsional vibration include

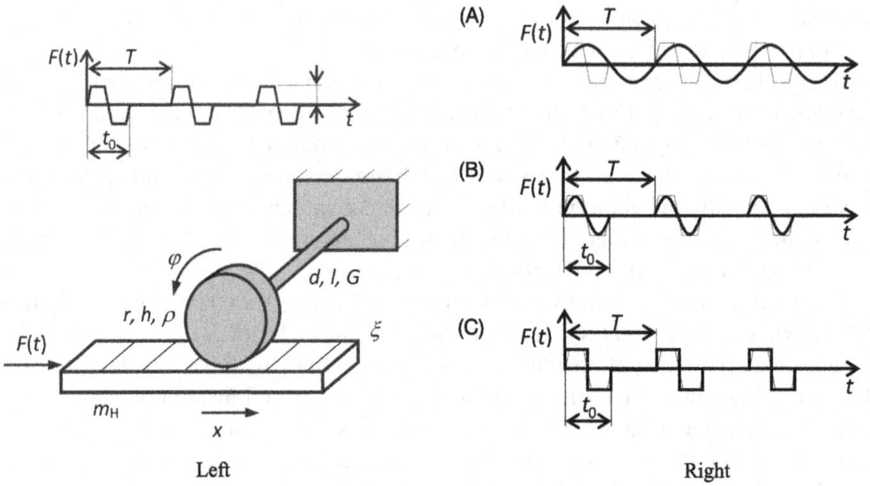

Figure 5.14 Manipulator torsion bar strain. Credit: Musil (2013).

rotary devices, e.g., drives from electric motors, cylinder engines, working technologies, changing friction in bearings, or, e.g., uneven running due to structural imperfections. All mentioned mechanisms represent sources of excitation in the form of a periodic function and can cause forced periodic torsional vibration. A typical example of torsional vibration excitation is also a sudden (impulse) change in the turbine load, either due to the sudden loss of drive, or due to the contact of rotating parts with the stator. Torsional vibrations reduce power output and can result even in unexpected fatal damage of structural parts. Comparing with other forms of vibration, their danger lies mostly in problematic detection. Therefore it is necessary to pay attention to their careful analysis already at the design stage.

A motion of a rack-pinion gear (Fig. 5.14) is a typical case of torsional vibration. The given periodic force can represent uneven running of the gearing. The scheme can also describe the roll of the profiling line, where $F(t)$ represents the periodic forming of the resistance of the profiled material, or drive of a production line with uneven motion. One possible application of said description is a pellet press where the source of periodic excitation is the mechanical resistance of the molded material in a closed or open press chamber. A wood crusher, a tamping machine, or a briquetting press can work on a similar principle.

Problem

Consider a simplified manipulator with one degree of freedom. The device consists of a rotating member (cylinder) interconnected with a translational member (plate) through a flexible shaft (Fig. 5.14). It is driven by a motor with constant angular speed within an operating range defining the angular frequencies of excitation. The excitation causes oscillations of the system and its possible resonances. The task is to examine the vibration behavior of the device.

Remark. The other mechanical systems, also with more degrees of freedom, which can be studied by Fourier analysis, can be found in Musil et al. (2015).

Mathematically, the motion is described by the second-order linear differential equation with constant coefficients,

$$m\ddot{x}(t) + b\dot{x}(t) + kx(t) = F(t), \tag{5.22}$$

maybe the most frequently used equation in mechanics, which allows to follow the behavior and consequently analyze also the fundamental properties of the system.

The expected solution of Eq. (5.22), $x(t) = x_G(t) + x_P(t)$, consists of a general solution $x_G(t)$ related to $e^{\lambda t} = e^{(-\delta+\omega_D i)t} = e^{-\delta t}(\cos\omega_D t + i\sin\omega_D t)$, $x_G(t) = e^{-\delta t}(c_1\cos\omega_D t + c_2\sin(\omega_D t))$, which is a solution of homogeneous equation $m\ddot{x}(t) + b\dot{x}(t) + kx(t) = 0$ (not forced system), and one particular solution $x_P(t)$ of Eq. (5.22) which correspondents to the special type of the right-hand side function (for details see Chapter 12). Under the influence of damping ($e^{-\delta t}$), the general solution decreases to zero, and stabilizes to the steady solution $x(t) = x_P(t)$, which we are going to deal with in the following.

Let the periodic excitation be defined by experimental data (Fig. 5.14). We will approximate it by three types of periodic functions: (a) harmonic (Fig. 5.14A), (b) harmonic with working hold-ups (Fig. 5.14B), and (c) polygonal periodic chain (Fig. 5.14(c)). We will show the Fourier analysis for each type. Moreover, we will focus also on the construction of desired output amplitude dependence on operational frequency. The solution of (c) can be compared with the solution presented in Chapter 12 (see Problem 2 and Fig. 12.6 in Chapter 12), where numerical methods for ordinary differential equations are used.

(a) The system is forced by harmonic function $F(t)$.

The force at the right-hand side $F(t)$ (Fig. 5.14A) is harmonic. With respect to the special type of the right-hand side function (see Chapter 12), the expected solution wave $x(t)$ is also harmonic. The functions can be expressed in real goniometric forms as

$$F(t) = a_F\cos(\omega_p t) + b_F\sin(\omega_p t) = A_F\cos(\omega_p t + \varphi_F),$$
$$x(t) = a_x\cos(\omega_p t) + b_x\sin(\omega_p t) = A_x\cos(\omega_p t + \varphi_x)$$

or they can be expressed in the complex exponential form

$$F(t) = c_F e^{i\omega_p t},$$
$$x(t) = c_x e^{i\omega_p t}.$$

Let us show the solution in both forms. We will start with the exponential one.

a1. Solution in exponential form.

Substituting $F(t)$ and $x(t)$ in Eq. (5.22), the differential equation transforms into an algebraic one, and we solve

$$m\left(-c_x\omega_p^2 e^{i\omega_p t}\right) + bi\omega_p c_x e^{i\omega_p t} + kc_x e^{i\omega_p t} = c_F e^{i\omega_p t},$$

from which

$$c_x = \frac{c_F}{k + \mathrm{i}b\omega_p - m\omega_p^2}.$$

The amplitude of solution is $|c_x| = \sqrt{\mathrm{Re}(c_x)^2 + \mathrm{Im}(c_x)^2}$, and phase $\varphi_x = \arg(c_x)$, which can be calculated as $\arctan\left(\frac{\mathrm{Im}(c_x)}{\mathrm{Re}(c_x)}\right)$ or $\arctan\left(\frac{\mathrm{Im}(c_x)}{\mathrm{Re}(c_x)}\right) \pm \pi$ or it can equal 0 (see rules in the table in Fig. 15.1 in Chapter 15). Since the complex coefficient c_x can be considered to be a function of ω_p, we can easily model the dependence of amplitude volume or phase on the excitation operational frequency ω_p.

Let us demonstrate the solution with the following physical parameters of the system: mass of a plate $m_{\mathrm{H}} = 4$ kg; disk radius $r = 0.09$ m; disk thickness $h = 0.03$ m; disk density $\rho = 7800$ kg/m^3; shear modulus $G = 0.8\times10^{11}$ MPa; damping ratio $\xi = 0.1$; shaft length $l = 0.5$ m; shaft diameter $d = 0.008$ m, operational angular frequency $\omega_p = 10$ rad/s, driving force $F_M = 40$ N, period $T = \frac{2\pi}{\omega_p}$, $t_0 = 0.55$ T, and the excitation $F(t) = F_M \sin(\omega_p t)$. Translation $x(t)$ satisfies Eq. (5.22). Inertial properties of the system are described through simple relations, where reduced mass of the whole mechanism $m = m_H + \frac{I_D}{r^2} = 6.97729$, mass inertia of disk $I_D = \frac{1}{2} m_D r^2$, mass of disk $m_{\mathrm{D}} = \pi\, r^2 h\rho$; damping $b = 2m\xi = 47.0837$, natural frequency of undamped system $\omega_0 = \sqrt{\frac{k}{m}} = 33.7407$, natural frequency of damped system $\omega_D = \sqrt{\omega_0^2 - \delta^2} = 33.5716$, $\delta = \xi\omega_0$, spring coefficient $k = \frac{k_t}{r^2} = 7943.19$, $k_t = \frac{GJ_P}{l}$, cross-sectional moment of inertia $J_P = \frac{\pi d^4}{32}$. Then Eq. (5.22) will have the form

$$6.97729\ddot{x}(t) + 47.0837\dot{x}(t) + 7943.19x(t) = 40\sin(\omega_p t).$$

After the calculation of the harmonic force complex coefficient (5.19)

$$c_F = \frac{a_x - \mathrm{i}b_x}{2} = \frac{0 - 40\mathrm{i}}{2} = -20i,$$

we obtain the complex coefficient of the result wave

$$c_x = \frac{c_F}{k + \mathrm{i}b\omega_p - m\omega_p^2} = -0.000179 - 0.002749i,$$

for $\omega_p = 10$, and the resulting complex wave

$$x(t) = c_x e^{\mathrm{i}\omega_p t} = (-0.000179 - 0.002749\mathrm{i})\, e^{10\mathrm{i}t}.$$

Amplitude $|c_x| = \sqrt{\mathrm{Re}(c_x)^2 + \mathrm{Im}(c_x)^2} = \sqrt{(-0.000179)^2 + (-0.002749)^2} = 0.00275454$; and phase $\varphi_x = \arg(c_x) = -1.63569$. To obtain the real wave (Fig. 5.15), we calculate

$$x_{\mathrm{real}}(t) = c_x e^{\mathrm{i}\omega_p t} + \bar{c}_x e^{-\mathrm{i}\omega_p t} =$$
$$= (-0.000179 - 0.002749\mathrm{i})\, e^{10\mathrm{i}t} + (-0.000179 + 0.002749\mathrm{i})\, e^{-10\mathrm{i}t}.$$

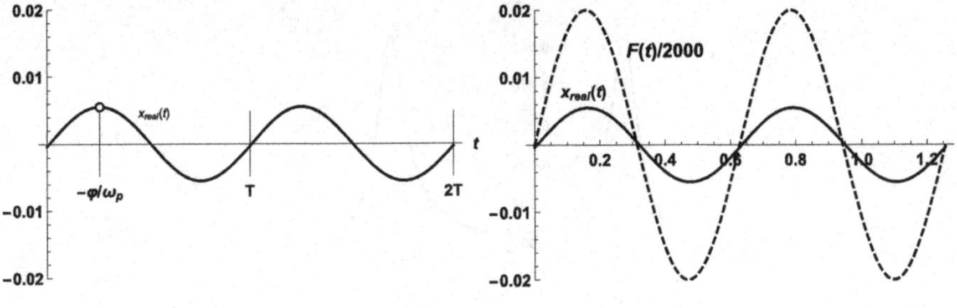

(A) (B)

Figure 5.15 (A) Real wave $x_{real}(t)$. (B) Comparison of $F(t)/2000$ (dashed) and $x_{real}(t)$.

The amplitude of the real wave is $2|c_x| = |c_x| + |\bar{c}_x| = 0.00550908$, and the phase of the cosine wave stays $\varphi_x = -1.63569$, so we get the following amplitude-phase form of the real wave:

$$x_{\mathrm{real}}(t) = 0.005509 \cos(10t - 1.63569).$$

Remark. Since the amplitude of result wave is ca. 10^{-3} smaller than the amplitude of the force wave, in comparison we display $F(t)/2000$ (Figs. 5.15, 5.21, and 5.26).

In the frequency domain, besides spectra of the particular result wave ($\omega_p = 10$), we are interested mainly in amplitude and phase dependence on operational frequency.

The complex coefficient as a function of ω_p is

$$c_x(\omega_p) = \frac{-20\mathrm{i}}{7943.19 + 47.0837\mathrm{i}\omega_p - 6.97729\omega_p^2} = \frac{-2.86644\mathrm{i}}{1138.43 + 6.74814\mathrm{i}\omega_p - \omega_p^2},$$

where real and imaginary parts are

$$\mathrm{Re}(c_x(\omega_p)) = \frac{19.3431\omega_p}{1296033.99 - 2231.33\omega_p^2 + \omega_p^4},$$

$$\mathrm{Im}(c_x(\omega_p)) = \frac{-3263.26 + 2.86644\omega_p^2}{1296034 - 2231.33\omega_p^2 + \omega_p^4}.$$

Then the amplitude

$$\mathrm{amp}(\omega_p) = |c_x| = \sqrt{\mathrm{Re}(c_x)^2 + \mathrm{Im}(c_x)^2}$$
$$= \sqrt{\frac{1.0648865 \times 10^7 - 18333.7\omega_p^2 + 8.21648\omega_p^4}{\left(1296034 - 2231.33\omega_p^2 + \omega_p^4\right)^2}}$$

and the phase

$$\varphi_x(\omega_p) = \arg(c_x(\omega_p)) = \mathrm{Arctan}(\mathrm{Re}(\omega_p), \mathrm{Im}(\omega_p)).$$

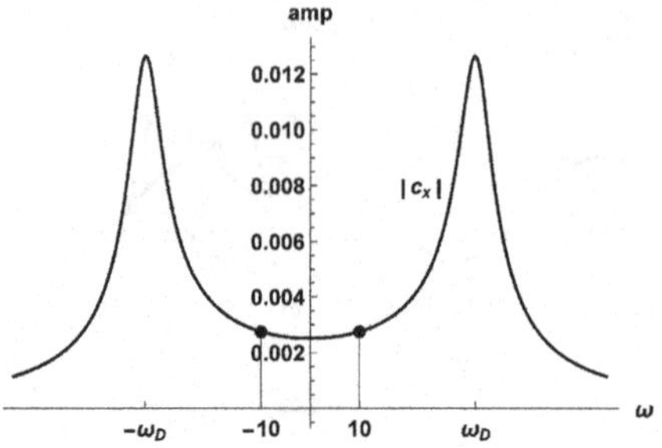

Figure 5.16 Two-sided amplitude spectrum. The system forced by the sine function $F(t)$, exponential form. The impact of the force operational frequency on the amplitude of the resulting vibration $x(t)$.

In basic arctan form (for details see the table in Fig. 15.1 in Chapter 15),

$$\varphi_x(\omega_p) = \arctan\left(\frac{\operatorname{Im}(\omega_p)}{\operatorname{Re}(\omega_p)}\right) + \pi \quad \text{for } \omega_p > \omega_0, (\operatorname{Re}(\omega_p) < 0, \operatorname{Im}(\omega_p) > 0),$$

$$\varphi_x(\omega_p) = \arctan\left(\frac{\operatorname{Im}(\omega_p)}{\operatorname{Re}(\omega_p)}\right) - \pi \quad \text{for } 0 \le \omega_p \le \omega_0, (\operatorname{Re}(\omega_p) < 0, \operatorname{Im}(\omega_p) <= 0),$$

$$\varphi_x(\omega_p) = \arctan\left(\frac{\operatorname{Im}(\omega_p)}{\operatorname{Re}(\omega_p)}\right) \quad \text{for } \omega_p < 0, (\operatorname{Re}(\omega_p) > 0),$$

$$\varphi_x(\omega_p) = -\frac{\pi}{2} \quad \text{for } \omega_p = 0, (\operatorname{Re}(\omega_p) = 0, \operatorname{Im}(\omega_p) < 0).$$

Fig. 5.16 shows the amplitude of the output sinusoidal vibration change with respect to operational frequency, and Fig. 5.17A shows its phase dependence on ω_p.

Remark. The point ω_0 in Fig. 5.17A cannot be considered a point of discontinuity; it appears only as a result of keeping the basic interval of periodicity.

To know how the phase of output sinusoid differs from the phase of input force sinusoid with respect to changing operational frequency, we express

$$\varphi_{\text{dif}}(\omega_p) = \arg(c_x(\omega_p)) - \arg(c_F(\omega_p)) = \arg(c_x(\omega_p)) + \frac{\pi}{2}, \quad \varphi_{\text{dif}} \in (-\pi, \pi\rangle$$
$$(\text{for } \omega_p > \omega_0,\ \varphi_{\text{dif}}(\omega_p) = \varphi_x(\omega_p) + \frac{\pi}{2} - 2\pi).$$

The amplitude and phase expressions as functions of ω_p and corresponding spectra (Figs. 5.16 and 5.17; see also Figs. 5.23 and 5.28) are the very valuable output for a designer to tune the system and avoid the frequency range where the output vibrations are dangerous, or to control the motor running.

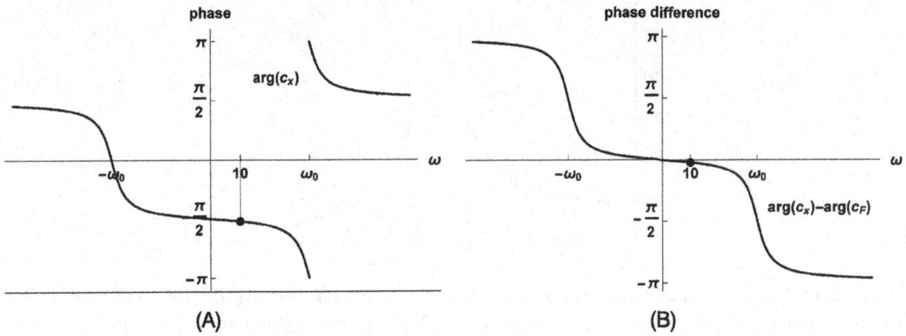

Figure 5.17 Two-sided phase spectra. The system is forced by the sine function $F(t)$, exponential form. The impact of the force operational frequency on (A) the phase of the result vibration $x(t)$ and (B) the difference of phases between output $x(t)$ and input $F(t)$.

a2. Solution in real goniometric form.

Substituting $F(t)$ and $x(t)$ in Eq. (5.22) by their goniometric form we obtain

$$\cos\left(\omega_p t\right)\left(a_x\left(k-m\omega_p^2\right)+bb_x\omega_p\right)+\sin\left(\omega_p t\right)\left(b_x\left(k-m\omega_p^2\right)-ba_x\omega_p\right)$$
$$=a_F\cos\left(\omega_p t\right)+b_F\sin\left(\omega_p t\right),$$

in particular

$$\sin\left(\omega_p t\right)\left(b_x\left(7943.19-6.97729\omega_p^2\right)-47.0837a_x\omega_p\right)+$$
$$+\cos\left(\omega_p t\right)\left(a_x\left(7943.19-6.97729\omega_p^2\right)+47.0837b_x\omega_p\right)=40\sin\left(\omega_p t\right).$$

Comparing the items along $\cos\left(\omega_p t\right)$ and $\sin\left(\omega_p t\right)$, we get

$$a_x=\frac{ka_F-m\omega_p^2a_F-b\omega_pb_F}{b^2\omega_p^2+k^2-2km\omega_p^2+m^2\omega_p^4}$$
$$=\frac{1883.35\omega_p}{6.3094225\times10^7-108626.9\omega_p^2+48.6825\omega_p^4},$$
$$b_x=\frac{b\omega_pa_F+kb_F-m\omega_p^2b_F}{b^2\omega_p^2+k^2-2km\omega_p^2+m^2\omega_p^4}$$
$$=\frac{317727.5-279.091\omega_p^2}{6.3094225\times10^7-108626.9\omega_p^2+48.6825\omega_p^4}.$$

For $\omega_p=10$

$$a_x=-0.00035724 \text{ and } b_x=0.00549748,$$
$$x(t)=-0.000357\cos(10t)+0.005497\sin(10t).$$

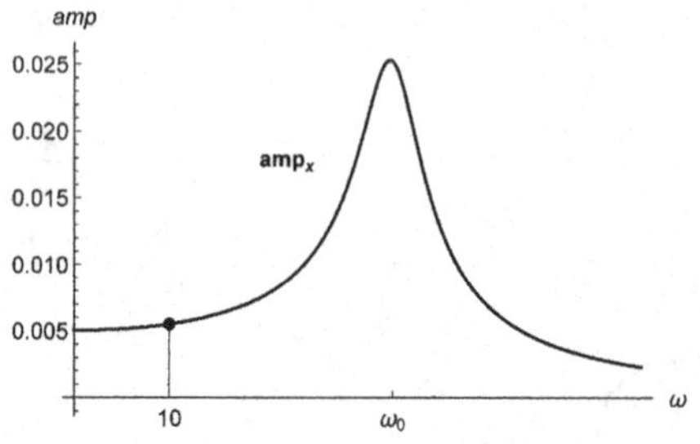

Figure 5.18 One-sided amplitude spectrum. The system is forced by the sine function $F(t)$, real form. The impact of the force operational frequency on the amplitude of the resulting vibration $x(t)$.

The amplitude of solution sinusoid is $A_x = \sqrt{a_x^2 + b_x^2} = 0.005509$, and the phase of the cosine wave is $\varphi_x = \arctan(a_x, -b_x) = \arctan\left(\frac{-b_x}{a_x}\right) - \pi = -1.63569$. In amplitude-phase form we obtain

$$x(t) = 0.005509 \cos(10t - 1.63569).$$

Coefficients a_x and b_x are also functions of ω_p, so we can directly express the dependence of solution amplitude volume on operational frequency ω_p, i.e.,

$$A_x(\omega_p) = \sqrt{\frac{a_F^2 + b_F^2}{k^2 + b^2\omega_p^2 - 2km\omega_p^2 + m^2\omega_p^4}}$$
$$= 40\sqrt{\frac{1}{63094225 - 108626.9\omega_p^2 + 48.6825\omega_p^4}}.$$

The phase (Fig. 5.19A; for computation details see the table in Fig. 15.1 in Chapter 15)

$$\varphi_x(\omega_p) = \arctan\left(\frac{-b_x}{a_x}\right) - \pi = \arctan\left(\frac{168.704 - 0.148189\omega_p^2}{\omega_p}\right) - \pi$$

for $0 \le \omega_p \le \omega_0$ $(a_x < 0, b_x <= 0)$

$$\varphi_x(\omega_p) = \arctan\left(\frac{-b_x}{a_x}\right) + \pi = \arctan\left(\frac{168.704 - 0.148189\omega_p^2}{\omega_p}\right) + \pi$$

for $\omega_p > \omega_0$ $(a_x < 0, b_x > 0)$.

The difference between phases of output sinusoid and input force sinusoid (Fig. 5.19B) is calculated as

$$\varphi_{\text{dif}}(\omega_p) = \varphi_x(\omega_p) - \varphi_F(\omega_p) = \varphi_x(\omega_p) + \frac{\pi}{2}, \quad \varphi_{\text{dif}} \in (-\pi, \pi\rangle$$

$$(\text{for } \omega_p > \omega_0,\ \varphi_{\text{dif}}(\omega_p) = \varphi_x(\omega_p) + \frac{\pi}{2} - 2\pi).$$

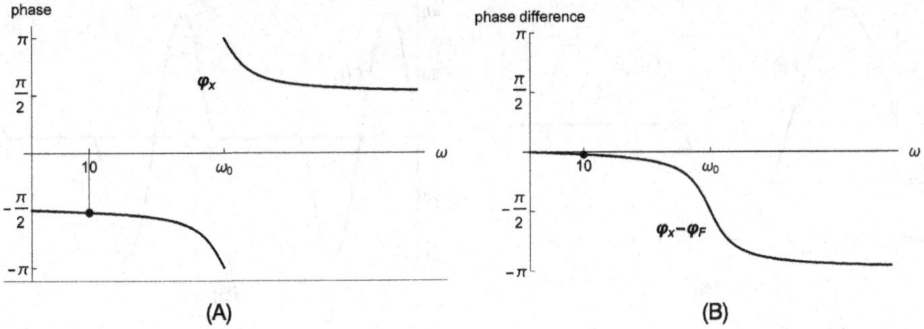

Figure 5.19 One-sided phase spectra. The system is forced by the sine function $F(t)$, real form. The impact of the force operational frequency on (A) the phase of the result vibration $x(t)$ and (B) the difference of phases between output $x(t)$ and input $F(t)$.

(b) The system is forced by a harmonic function with working hold-ups.

The harmonic force function with working hold-ups (Fig. 5.14B) on the right side can be expressed as the piecewise determined function

$$F(t)=\begin{cases} F_M \sin\frac{2\pi t}{t_0} & \text{for } 0\le t\le t_0, \\ 0 & \text{for } t_0<t\le T \end{cases} \tag{5.23}$$

or as

$$F(t)=F_M\cdot\sin\left(\frac{2\pi t}{t_0}\right)\cdot\theta\left(-t+t_0\right), \tag{5.24}$$

where θ stands for Heaviside step function. Taking into account the given particular parameters of the system and $\Omega=\omega_p=10$,

$$F(t)=40\sin(18.1818t)\ \text{for}\ 0\le t\le t_0,\quad \text{and}\quad F(t)=0\ \text{for}\ t_0<t\le T. \tag{5.25}$$

The harmonic force function with working hold-ups $F(t)$ (Fig. 5.20A) satisfies the Dirichlet conditions and it can be expressed by the Fourier series of harmonics (5.10) (Fig. 5.20B). The result function $x(t)$ will have the same form and it will be calculated as the sum of harmonics responses using the principle of superposition. In real goniometric forms,

$$\begin{aligned} F(t) &= \frac{a_{F0}}{2}+\sum_{n=1}^{\infty}F_n(t) \\ &= \frac{a_{F0}}{2}+\sum_{n=1}^{\infty}a_{Fn}\cos(n\Omega t)+b_{Fn}\sin(n\Omega t) \\ &= \frac{a_{F0}}{2}+\sum_{n=1}^{\infty}A_{Fn}\cos(n\Omega t+\varphi_{Fn}), \end{aligned} \tag{5.26}$$

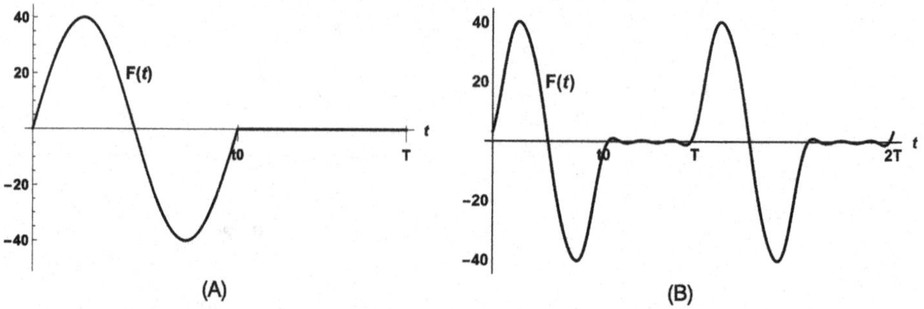

Figure 5.20 The system is forced by a harmonic function with working hold-ups. Excitation function $F(t)$. (A) Harmonic function with working hold-ups $F(t)$, $t \in \langle 0, T \rangle$. (B) $F(t) \cong s_7(t)$, $\Omega = \omega_p = 10$, $t \in \langle 0, 2T \rangle$.

$$\begin{aligned} x(t) &= \frac{a_{x0}}{2} + \sum_{n=1}^{\infty} x_n(t) \\ &= \frac{a_{x0}}{2} + \sum_{n=1}^{\infty} a_{xn} \cos(n\Omega t) + b_{xn} \sin(n\Omega t) \\ &= \frac{a_{x0}}{2} + \sum_{n=1}^{\infty} A_{xn} \cos(n\Omega t + \varphi_{xn}) \end{aligned} \tag{5.27}$$

and in the exponential form

$$F(t) = \sum_{n=-\infty}^{\infty} F_n(t) = \sum_{n=-\infty}^{\infty} c_{Fn} e^{in\Omega t}, \tag{5.28}$$

$$x(t) = \sum_{n=-\infty}^{\infty} x_n(t) = \sum_{n=-\infty}^{\infty} c_{xn} e^{in\Omega t}. \tag{5.29}$$

We will show the solution in exponential form.

Solution in exponential form.

First we generally express the solution for the nth harmonic. Substituting $F(t)$ and $x(t)$ in Eq. (5.22) by harmonics $F_n(t)$ and $x_n(t)$ in exponential form, we obtain

$$m\left(-n^2\Omega^2 c_{xn} e^{in\Omega t}\right) + bin\Omega c_{xn} e^{in\Omega t} + kc_{xn} e^{in\Omega t} = c_{Fn} e^{in\Omega t}, \tag{5.30}$$

from which

$$c_{xn} = \frac{c_{Fn}}{k + ibn\Omega - mn^2\Omega^2}. \tag{5.31}$$

The amplitude of the solved harmonic is $|c_{xn}| = \sqrt{\mathrm{Re}(c_{xn})^2 + \mathrm{Im}(c_{xn})^2}$, and the phase $\varphi_{xn} = \arg(c_{xn})$.

In (5.28)

$$c_{Fn} = \frac{1}{T}\int_0^T F(t)\mathrm{e}^{-in\Omega t}\mathrm{d}t = \frac{1}{T}\int_0^{t_0} F_M \cdot \sin\left(\frac{2\pi t}{t_0}\right) \cdot \mathrm{e}^{-in\Omega t}\mathrm{d}t.$$

In particular,

$$c_{Fn} = \frac{1157.49 - 1157.49\mathrm{e}^{(-0.345575\mathrm{i})n\Omega}}{330.579 - n^2\Omega^2},$$

and $c_{F0} = 0$. Then we get the right-hand side force function expressed as the Fourier series (Fig. 5.20), i.e.,

$$F(t) = \sum_{n=-\infty}^{\infty} F_n(t) = \sum_{n=-\infty}^{\infty} \frac{1157.49 - 1157.49\mathrm{e}^{(-0.345575\mathrm{i})n\Omega}}{330.579 - n^2\Omega^2} e^{in\Omega t}.$$

The first seven real harmonics of $F(t)$, where $\Omega = \omega_p = 10$, with their graphs and forms of expressions, are presented in the table in Fig. 15.2 in Chapter 15.

From (5.31), where $\Omega = \omega_p = 10$,

$$c_{xn} = \frac{11.5749 - 11.5749\mathrm{e}^{(0.-3.45575\mathrm{i})n}}{\left(470.837\mathrm{i}n - 697.729n^2 + 7943.19\right)\left(3.30579 - n^2\right)}.$$

Time domain

We have

$$x(t) = \sum_{n=-\infty}^{\infty} x_n(t) = \sum_{n=-\infty}^{\infty} c_{xn}e^{in\Omega t}$$

$$= \sum_{n=-\infty}^{\infty} \frac{11.5749 - 11.5749\mathrm{e}^{(0.-3.45575\mathrm{i})n}}{\left(470.837\mathrm{i}n - 697.729n^2 + 7943.19\right)\left(3.30579 - n^2\right)} e^{in10t}.$$

For visualization we used the approximation by $x(t) \cong sx_7(t) = \sum_{n=-7}^{7} c_{xn}e^{in\Omega t}$, $\Omega = \omega_p = 10$, and its comparison with the force wave (Fig. 5.21). The first seven real harmonics of $x(t)$, with their graphs and forms of expressions, are presented in the table in Fig. 15.3 in Chapter 15.

Frequency domain

We are interested in complex harmonics amplitudes and phases with respect to their frequency $n\Omega$. Considering the first seven real harmonics, we calculate the amplitudes and phases for the sequence of complex coefficients $\{c_{xn}\}_{n=-7}^{7}$ (Table 5.4) and construct corresponding spectra (Fig. 5.22).

As we can see in Fig. 5.22, the most dominating are amplitudes of the first five harmonics, which decisively take part in the maximal value of $x(t)$.

In order to analyze the impact of the force operational frequency on the resulting vibration, it is necessary to represent somehow the highest value (also amplitude) of

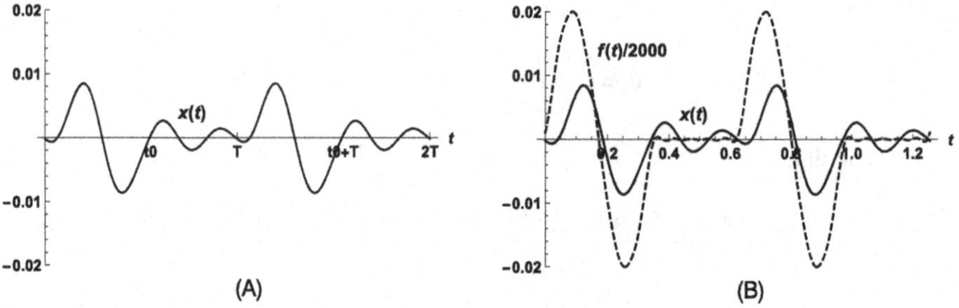

Figure 5.21 (A) $x(t) \cong sx_7(t)$. (B) comparison of $F(t)/2000$ (dashed) and $x(t)$.

Table 5.4 Frequency table of amplitudes and phases of complex harmonics.

$n\Omega$	c_n	Abs (c_n)	Arg (c_n)
-70	$4.88211 \times 10^{-6} + 7.19525 \times 10^{-6}$ i	8.69521×10^{-6}	0.974633
-60	0.0000231327 - 0.0000234094 i	0.0000329109	-0.791343
-50	0.0000660258 + 0.000039801 i	0.0000770943	0.542483
-40	0.0000284401 - 0.000285908 i	0.000287319	-1.47165
-30	-0.000639651 - 0.00153162 i	0.00165982	-1.96641
-20	-0.000934481 + 0.00173135 i	0.00196744	2.06573
-10	0.00133223 + 0.000300673 i	0.00136574	0.221972
0	0. + 0. i	0.	0.
10	0.00133223 - 0.000300673 i	0.00136574	-0.221972
20	-0.000934481 - 0.00173135 i	0.00196744	-2.06573
30	-0.000639651 + 0.00153162 i	0.00165982	1.96641
40	0.0000284401 + 0.000285908 i	0.000287319	1.47165
50	0.0000660258 - 0.000039801 i	0.0000770943	-0.542483
60	0.0000231327 + 0.0000234094 i	0.0000329109	0.791343
70	$4.88211 \times 10^{-6} - 7.19525 \times 10^{-6}$ i	8.69521×10^{-6}	-0.974633

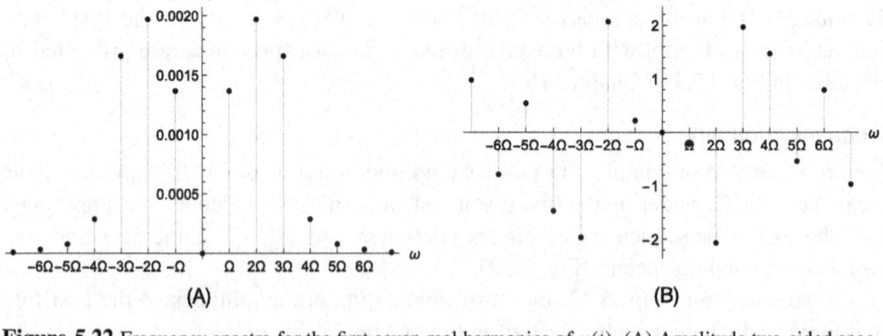

Figure 5.22 Frequency spectra for the first seven real harmonics of $x(t)$. (A) Amplitude two-sided spectrum. (B) Phase two-sided spectrum.

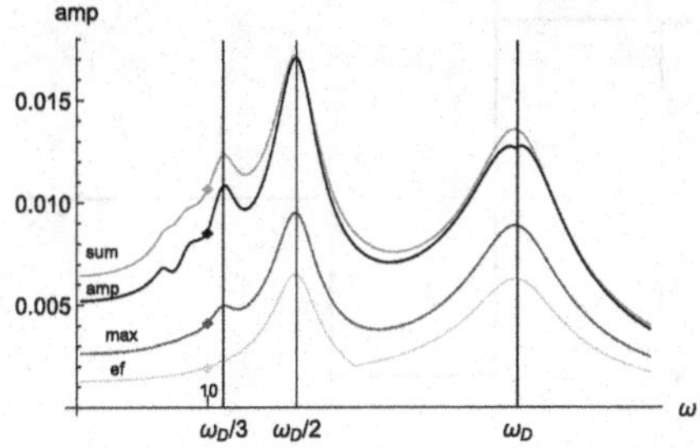

Figure 5.23 The system forced by a harmonic function with working hold-ups, exponential form. The impact of the force operational frequency on the amplitude of the resulting vibration $x(t)$.

$x(t)$ (5.29) for each $\Omega = \omega_p$ within an operational range of ω. To find the maximal value of $x(t)$ analytically is not a simple task. In the case we are really interested in its particular value we can use some of the numerical tools built-in in software (e.g., FindMaximum in Mathematica). As numerical methods work on specific values of arguments, they are suitable mainly for individual calculations and individually composed algorithms. Much more easier with much less computational and time effort is to consider the maximal value substitution by some reference value, which can serve for comparison. Let us look at three possible representations:

(a) the sum of determining harmonic amplitudes (usually the first five real harmonics, $sum = \sum_{n=-5}^{5} |c_n|$),

(b) the maximal amplitude of determining harmonics $max = \max\{|c_n|\}_{n=-5}^{5}$,

(c) the effective value calculated under the formula: $ef = \sqrt{\sum_{n=-5}^{5} |c_n|^2}$.

For $\omega_p = 10$, we receive $sum = \sum_{n=-5}^{5} |c_n| = 0.0107148$, $max = \max\{|c_n|\}_{n=-5}^{5} = 0.00196744$, and $ef = 0.00414236$, compared to an approximate maximal value found by software of $amp = 0.00847577$ and its one half related to complex amplitude volume 0.00423788.

Remark. Due to the fact that the two-sided amplitude spectrum is even, it is enough to consider only its positive right-sided half ($\omega \geq 0$).

As seen in Fig. 5.23, all three representations (*sum*, *max*, *ef*) have very similar behavior like the actual amplitude *amp*, so one of them can be efficiently used for the explored purpose. The dangerous resonance regions lie in the vicinity of the system's natural frequency ω_D (ω_0) and its fractions ω_D/n (ω_0/n).

(c) The system is forced by a polygonal periodic chain $F(t)$.

The polygonal force function $F(t)$ (Fig. 5.14(c)) on the right side can be expressed as the piecewise determined function

$$F(t) = \begin{cases} F_M & \text{for } 0 \leq t < t_0/2, \\ -F_M & \text{for } t_0/2 \leq t < t_0, \\ 0 & \text{for } t_0 \leq t \leq T \end{cases} \tag{5.32}$$

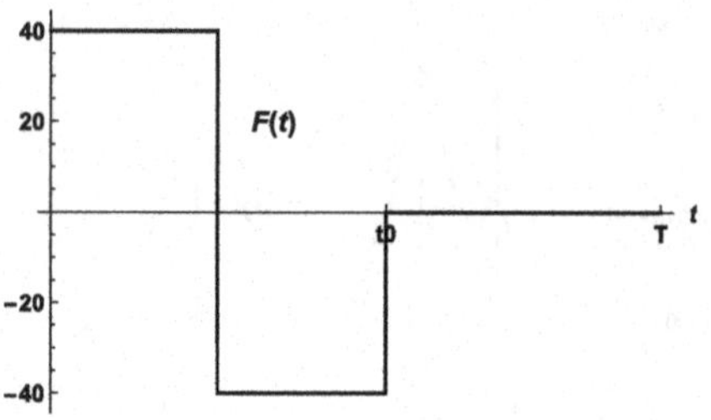

Figure 5.24 $F(t)$ represented by (5.34).

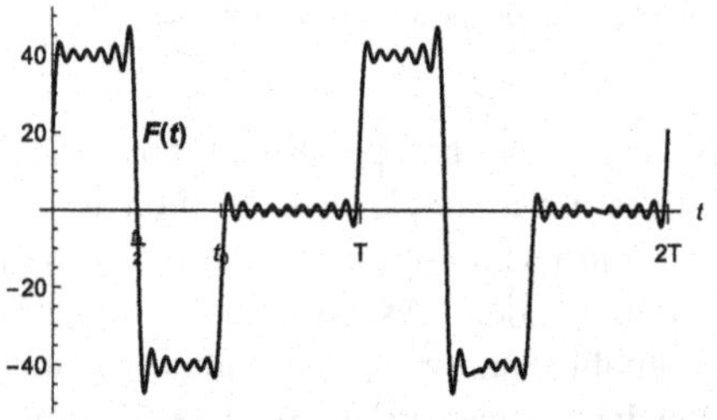

Figure 5.25 $F(t) \cong s_{20}(t)$ and $\Omega = \omega_p = 10$.

or as

$$F(t) = F_M \cdot \theta t - 2F_M \cdot \theta\left(t - \frac{t_0}{2}\right) + F_M \cdot \theta(t - t_0), \tag{5.33}$$

where θ stands for the Heaviside step function. Taking into account the given particular parameters of the system

$$F(t) = \begin{cases} 40 & \text{for } 0 \le t < t_0/2, \\ -40 & \text{for } t_0/2 \le t < t_0, \\ 0 & \text{for } t_0 \le t \le T \end{cases} \tag{5.34}$$

or

$$F(t) = 40 \cdot \theta t - 80 \cdot \theta\left(t - \frac{t_0}{2}\right) + 40 \cdot \theta(t - t_0), \tag{5.35}$$

the polygonal chain (Fig. 5.24) satisfies the Dirichlet conditions (give the reason) and it can be expressed by a Fourier series of harmonics (5.10). The solution function will have the same general polyharmonic form as in (b) and it will be calculated as the sum of harmonic responses using the principle of superposition in a similar way.

We will show the solution in goniometric form.

Solution in real goniometric form.

The formulas for calculation of Fourier coefficients a_{xn} and b_{xn} in (5.27) will be derived in a similar way as in (a). Comparing the items along $\cos(\omega_p t)$ and $\sin(\omega_p t)$ in

(5.22) where goniometric forms are involved, we get

$$a_{xn} = \frac{a_{Fn}\left(k - mn^2\Omega^2\right) - bn\Omega b_{Fn}}{b^2n^2\Omega^2 + k^2 - 2kmn^2\Omega^2 + m^2n^4\Omega^4}$$
$$= \frac{a_{Fn}\left(7943.19 - 6.97729n^2\Omega^2\right) - 47.0837n\Omega b_{Fn}}{48.6825n^4\Omega^4 - 108626.9n^2\Omega^2 + 63094225},$$
$$b_{xn} = \frac{bn\Omega a_{Fn} + b_{Fn}\left(k - mn^2\Omega^2\right)}{b^2n^2\Omega^2 + k^2 - 2kmn^2\Omega^2 + m^2n^4\Omega^4}$$
$$= \frac{47.0837n\Omega a_{Fn} + b_{Fn}\left(7943.19 - 6.97729n^2\Omega^2\right)}{48.6825n^4\Omega^4 - 108626.9n^2\Omega^2 + 6.3094225},$$

where

$$a_{Fn} = \frac{2}{T}\int_0^T F(t)\cos(n\Omega t)\,\mathrm{d}t$$
$$= \frac{2}{T}\left(\int_0^{t_0/2} F_M\cos(n\Omega t)\,\mathrm{d}t - \int_{t_0/2}^{t_0} F_M\cos(n\Omega t)\,\mathrm{d}t\right),$$
$$b_{Fn} = \frac{2}{T}\int_0^T F(t)\sin(n\Omega t)\,\mathrm{d}t$$
$$= \frac{2}{T}\left(\int_0^{t_0/2} F_M\sin(n\Omega t)\,\mathrm{d}t - \int_{t_0/2}^{t_0} F_M\sin(n\Omega t)\,\mathrm{d}t\right).$$

In particular,

$$a_{Fn} = \frac{10\,(80.\sin(0.173n\Omega) - 40.\sin(0.346n\Omega) - 40.\sin(0.628n\Omega) + 40.\sin(\pi n\Omega/5))}{\pi n\Omega},$$
$$b_{Fn} = \frac{10\left(80.\sin^2(\pi n\Omega/10) - 80.\cos(0.173n\Omega) + 40.\cos(0.346n\Omega) + 40.\cos(0.628n\Omega)\right)}{\pi n\Omega},$$

$a_{F0} = 0$, and the time domain formula of the force expansion to Fourier series will hold (5.10). We can see its approximation by 20 harmonics in Fig. 5.25.

Computing coefficients a_{xn} and b_{xn}, we get the particular values in the time domain formula for $x(t)$ (5.27), $a_{x0} = 0$. Its approximation and the comparison with $F(t)/2000$ are shown in Fig. 5.26.

Remark. The particular Fourier series formulas of $F(t)$ and $x(t)$ are not presented here due to their long expressions.

In the frequency domain, considering the first seven real harmonics of $x(t)$, we calculate their amplitudes and phases (Table 5.5) and construct the frequency spectra (Fig. 5.27). The impact of the force operational frequency on the amplitude of the resulting vibration $x(t)$ is carried out in a similar way like in (b) (see Fig. 5.28).

Comparing the graphs of the force operational frequency impact on the amplitude of the resulting vibration $x(t)$ (compare Figs. 5.16, 5.23, and 5.28) we can see that the approximation of the exiting force by a pure sinusoid is the worse one, as it is not able

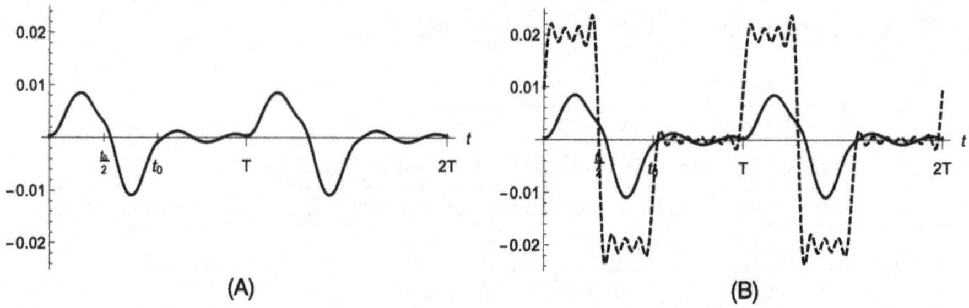

Figure 5.26 (A) $x(t) \cong sx_7(t)$. (B) Comparison of $F(t)/2000$ (dashed) and $x(t)$ approximations.

Table 5.5 The system is forced by a polygonal periodic chain, $\Omega = \omega_p = 10$. Amplitudes and phases of the first seven result wave harmonics.

$n\Omega$	10	20	30	40	50	60	70
amp_n	0.004056	0.004743	0.002124	0.000326	0.000888	0.000387	0.000015
φ_n	-0.22197	-2.06573	1.96641	-1.66994	2.59911	0.79134	-0.97463

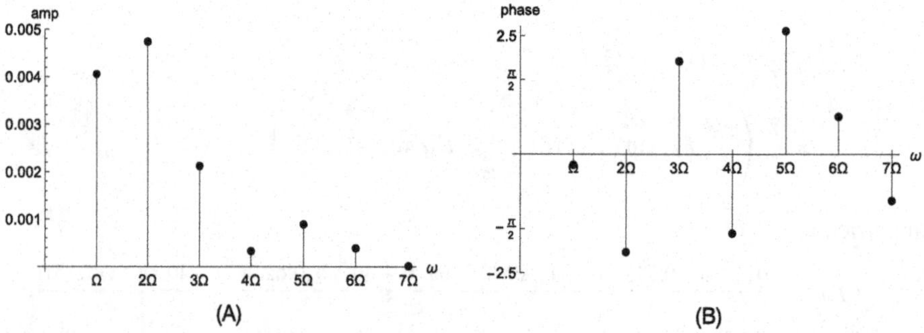

Figure 5.27 The system is forced by a polygonal periodic chain $\Omega = \omega_p = 10$. Frequency spectra for the first seven real harmonics of $x(t)$. (A) Amplitude one-sided spectrum. (B) Phase one-sided spectrum.

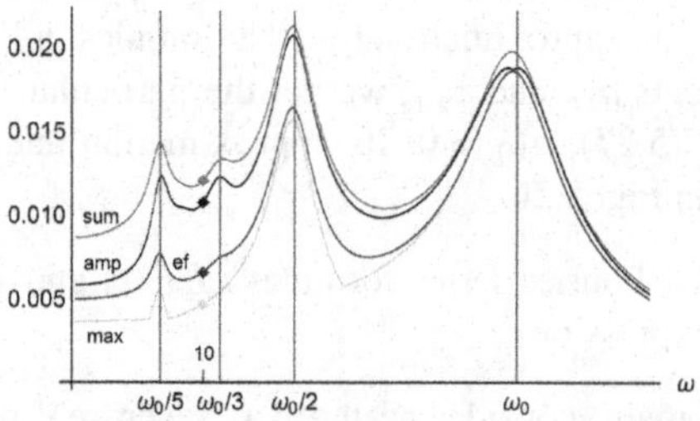

Figure 5.28 The system forced by a polygonal periodic chain, goniometric form. The impact of the force operational frequency on the amplitude of the resulting vibration $x(t)$.

to reflect the influence of any higher harmonics. It omits the corresponding amplitude peaks and so cannot provide an appropriate model for a designer, especially in the range where the frequency is less than the natural frequency of the system.

Exercise 5.6. Solve the problem for $F(t) = F_M \cos(\omega_p t)$, $F_M = 30$.

Exercise 5.7. Find the solution and perform the harmonic analysis for the system forced by a polygonal chain in exponential form.

Exercise 5.8. Approximate the experimental data (Fig. 5.14) by

(a) a zig-zag function with hold-ups,
(b) a polygonal function of a similar shape to experimental data,

and perform the harmonic analysis of the ODE solution.

All computations and figures in this chapter were generated by Wolfram Mathematica (Wolfram, 2018).

References

Borba, L., 2013. Elektrické Pohony a Výkonová Elektronika (Electric Drives and Power Electronics). Vyd. Nakladatel'stvo STU, Bratislava. ISBN 978-80-227-3858-3, pp. 54–126. 227 p. (in Slovak).

Burden, Richard L., Faires, J. Douglas, 2011. Numerical Analysis, 9th Edition. Brooks/Cole, Cengage Learning.

Carslaw, H.S., 1925. A historical note on Gibbs' phenomenon in Fourier's series and integrals. Bulletin of the American Mathematical Society 31 (8), 420–424.

Ivan, J., 1989. Matematika 2 (Mathematics 2). Vyd. Alfa, Bratislava. ISBN 80-05-00114-2. p. 631 (in Slovak).

Musil, M., 2013. Základy dynamiky strojov s Matlabom (Fundamentals of machine dynamics with Matlab). 1. Vyd. Nakladatel'stvo STU, Bratislava. ISBN 978-80-227-3938-2. 98 p. (in Slovak).

Musil, M., Úradníček, J., Havelka, F., 2015. Torzné Kmitanie Rotačných Mechanických Sústav (Torsional Vibration of Rotary Mechanical Systems). 1. Vyd. Nakladatel'stvo STU, Bratislava. ISBN 978-80-227-4481-2. p. 100 (in Slovak).

Ondráček, O., 2008. Signály a sústavy (Signals and Systems). 3. Vyd. STU v Bratislave, Bratislava. ISBN 978-80-227-2956-7, pp. 21–77. p. 341 (in Slovak).

Vlnka, J., Ševčovič, D., Lababneh, W., 1995. The influence of pole distribution magnets on magnetic bearing. In: Zbornik Vedeckych Prac Strojnickej Fakulty STU, Bratislava. ISBN 80-227-0753-8, pp. 203–214 (in Slovak).

Wolfram Research, Inc., 2018. Mathematica, Version 11.3. Champaign, IL.

Woodford, Ch., 2008. Noise-cancelling headphones. https://www.explainthatstuff.com/noisecancellingheadphones.html. (Accessed 1 August 2019).

Applications of integral calculus

6

Ion Mierluş-Mazilu[a], Ştefania Constantinescu[a], Lucian Niţă[a], César de Santos-Berbel[b]
[a]Technical University of Civil Engineering, Bucharest, Romania, [b]Universidad Politécnica de Madrid, Madrid, Spain

6.1 Key ideas on the calculus of primitive integrals

We know that in many scientific domains we need to compute the areas of figures described by the graphs bounded by straight lines of equations $x = a$, $x = b$, $a < b$ and $y = f(x)$, $x \in [a, b]$ (Fig. 6.1).

To compute such areas, mathematicians and scientists as Newton, Leibniz, and many others created an integral calculus system relying on some results, which are presented below.

Definition 6.1. Let $A \subseteq \mathbb{R}$ be an interval of real numbers set and $f : A \to \mathbb{R}$ defined on A. Any differentiable function $F : A \to \mathbb{R}$ such that $(\forall)x \in A$, $F'(x) = f(x)$ and it is called a *primitive* of f on A. **This is also often known as the *antiderivative* or the *integral* of f on A.**

A primitive F of f is denoted by the notation $\int f(x)dx$ and it is called *an integral of f with respect to variable x.*

Theorem 6.1. *Any function $f : A \to \mathbb{R}$ continuous on A admits primitives on A.*

Remark 6.1. If a function $f : A \to \mathbb{R}$ is not continuous on A, this does not mean that f does not admit primitives on A, but if a function $f : A \to \mathbb{R}$ does not admit primitives on A, it certainly is not continuous on A.

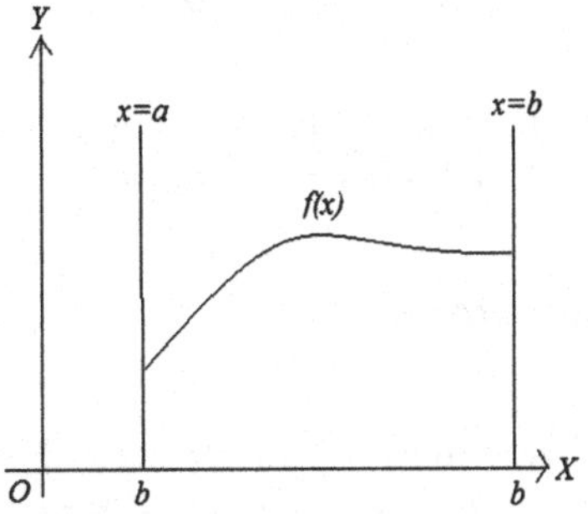

Figure 6.1 The graphs bounded by straight lines.

Calculus for Engineering Students. https://doi.org/10.1016/B978-0-12-817210-0.00013-8

Theorem 6.2. *If a function $f : A \to \mathbb{R}$ admits primitive F on A, then f admits an infinity of primitives of the form $F + C$, $(\forall)\, C \in \mathbb{R}$, so if F_1, F_2 are two primitives of function f, then they differ by a constant, namely, $(\exists)\, c \in \mathbb{R}$ such that $F_2(x) = F_1(x) + c$, $(\forall)\, x \in A$.*

Remark 6.2. The set of all primitives of a function f is called the *indefinite integral of function* f and it is denoted by the notation $\int f(x)dx = F(x) + C$.

The main problem remains to "guess" the form of the primitives of a function. We will give in Table 6.1 the basic elementary functions, mostly used in exercises of integral calculus.

Table 6.1 Primitives of the basic elementary functions.

Function	Primitive	Domain
$x^n, n \in \mathbb{N}$	$\int x^n dx = \frac{x^{n+1}}{n+1} + C$	$\mathbb{R}$
$x^{r+1}, r \in \mathbb{R}$	$\int x^r dx = \frac{x^{r+1}}{r+1} + C$	$(0, \infty)$
e^x	$\int e^x dx = e^x + C$	$\mathbb{R}$
$a^x, a > 0, a \neq 1$	$\int a^x dx = \frac{a^x}{\ln a} + C$	$\mathbb{R}$
$\frac{1}{x}$	$\int \frac{1}{x} dx = \ln\|x\| + C$	$\mathbb{R}\backslash\{0\}$
$\sin x$	$\int \sin x dx = -\cos x + C$	$\mathbb{R}$
$\cos x$	$\int \cos x dx = \sin x + C$	$\mathbb{R}$
$\frac{1}{\cos^2 x}\ (1 + \tan^2 x)$	$\int \frac{1}{\cos^2 x} dx = \tan x + C$	$\mathbb{R} \setminus \left\{\frac{\pi}{2} + k\pi \mid k \in \mathbb{Z}\right\}$
$-\frac{1}{\sin^2 x}(-(1 + \cot^2 x))$	$\int -\frac{1}{\sin^2 x} dx = \cot x + C$	$\mathbb{R}\backslash \{k\pi \mid k \in \mathbb{Z}\}$
$\frac{1}{\sqrt{a^2-x^2}}, a > 0$	$\int \frac{1}{\sqrt{a^2-x^2}} dx = \arcsin \frac{x}{a} + C$	$(-a, a)$
$\frac{1}{\sqrt{x^2+a^2}}$	$\int \frac{1}{\sqrt{x^2+a^2}} d$ = $\ln\left(+\sqrt{x^2+a^2}\right)+C$	$\mathbb{R}$

Operations with functions which admit primitives.

Theorem 6.3. *Let $f, g : I \to \mathbb{R}$, f, g admit primitives on I and $a \in \mathbb{R}$. Then the following properties are valid:*

1. *$f + g$, $f - g$ admit primitives and $\int (f(x) \pm g(x))\, dx = \int f(x)dx \pm \int g(x)dx$,*
2. *af admits primitives and $\int af(x)dx = a \int f(x)dx$.*

6.1.1 Methods of integration

1. *Integration by parts*
   Let $f, g : I \to \mathbb{R}$ be two functions of class $C^1(I)$. If we have to compute the primitive function f, which seems to be of form $f(x) = g(x) \cdot h'(x)$, then $\int f(x)dx = \int g(x)h'(x)dx = g(x)h(x) - \int g'(x)h(x)dx$.

2. *Integration by substitution*
Let I, J be two intervals of $\mathbb{R}$, $f : J \to \mathbb{R}$, $\varphi : I \to \mathbb{R}$, $\text{Im}\varphi \subset J$ such that φ is differentiable on I and f admits primitives on J (let F be a primitive of f). Then function $(f \circ \varphi) \cdot \varphi'$ admits primitives and $F \circ \varphi$ is a primitive of it.
3. *Integration of rational functions*
One can prove that any rational function can be expressed as a finite sum of simple rational function and thus the calculus of primitives reduces to the calculus of primitives of simple rational functions.
4. *Integration of trigonometric functions*
The general method is represented by the convert of variable $x \to \tan\frac{x}{2}$. Further, we use the trigonometric formulas $\cos x = \frac{1-\tan^2\frac{x}{2}}{1+\tan^2\frac{x}{2}}$, $\sin x = \frac{2\tan^2\frac{x}{2}}{1+\tan^2\frac{x}{2}}$. Then

$$\tan\frac{x}{2} = t \Rightarrow \frac{x}{2} = \arctan t \to dx$$
$$\Rightarrow \int R(\cos x, \sin x)\,dx = \int R\left(\frac{1-t^2}{1+t^2}, \frac{2t}{1+t^2}\right)\frac{2}{1+t^2}dx.$$

Now, since we have enunciated the main methods of computing integrals, we will lay down the most important formulas in finding lengths, areas, and volumes.

6.1.2 The construction of the Riemann integral

Definition 6.2. Let $[a, b] \subset R$ a closed interval. The set of points

$$\Delta = \{a = x_0 < x_1 < \cdots < x_{n-1} < x_n = b\}$$

is called a division of $[a, b]$.

Definition 6.3. Let $\Delta =$ division of interval $[a, b]$. The number $\|\Delta\| \overset{\text{def}}{=} \max_{1\le k\le n}(x_k - x_{k-1})$ is called *the norm of the division.*

Now let $f : [a, b] \to R$ be a function and Δ a division of $[a, b]$. In each interval $[x_{k-1}, x_k]$ we choose an intermediate point c_k.

Definition 6.4. The number $\sigma_\Delta(f, c_k) \overset{\text{def}}{=} \sum_{k=1}^{n} f(c_k)(x_k - x_{k-1})$ is called the *Riemann sum* associated to f, Δ and to points $(c_k)_{1\le k\le n}$.

Definition 6.5. We will say that function $f : [a, b] \to R$ is *Riemann integrable* on $[a, b]$ if $\exists I \in R$ such that $\forall \varepsilon > 0$, $\exists \delta_\varepsilon > 0$ such that $\forall \Delta$ – division of $[a, b]$ with $\|\Delta\| < \delta_\varepsilon$ and $\forall c_k \in [x_{k-1}, x_k]$ $(1 \le k \le n)$, we have $|\sigma_\Delta(f, c_k) - I| < \varepsilon$. In brief, $\lim_{\|\Delta\|\to 0}\sigma_\Delta(f, c_k) = I$. The number I is called the *integral from a to b of f* and is denoted by $\int_a^b f(x)dx$.

6.1.2.1 Operations with integrable functions

Theorem 6.4. *Let $f, g : [a, b] \to \mathbb{R}$ be two functions integrable on $[a, b]$ and $a \in \mathbb{R}$. Then the following properties are valid:*

1. $f + g, f - g$ *are integrable and* $\int_a^b (f(x) \pm g(x))\,dx = \int_a^b f(x)dx \pm \int_a^b g(x)dx$,
2. af *is integrable and* $\int_a^b af(x)dx = a \int_a^b f(x)dx$.

Theorem 6.5 (The Leibniz–Newton formula). *Let $f : [a, b] \to \mathbb{R}$ an integrable function which admits primitives on $[a, b]$. Then for any primitive F of f, $\int_a^b f(x)dx = F(b) - F(a)$.*

Theorem 6.6. *Let $f : [a, b] \to \mathbb{R}$ be a positive, continuous function. Let $\Gamma_f = \{(x, y) \in \mathbb{R}^2 \mid a \le x \le b, 0 \le y \le f(x)\}$ have area and area$(\Gamma_f) = \int_a^b f(x)dx$.*

The area of the set bounded by the graphs of two functions (the area delimited by the intersection of two graphs) can be computed using the following result:

$$\Gamma_{f,g} = \{(x, y) \mid a < x < b, f(x) \le y \le g(x)\,(g(x) \le y \le f(x))\}.$$

Theorem 6.7. *Let $f, g : [a, b] \to \mathbb{R}$ be integrable functions. Let $\Gamma_f = \{(x, y) \in \mathbb{R}^2 \mid a \le x \le b, 0 \le y \le f(x)\}$. Then $\Gamma_{f,g}$ has area and area$(\Gamma_{f,g}) = \int_a^b |g(x) - f(x)|\,dx$.*

Theorem 6.8. *Let $f : [a, b] \to \mathbb{R}$ be a differentiable function with f' continuous. Then its graph has the length equal to $l(G_f) = \int_a^b \sqrt{1 + |f'(x)|^2}dx$.*

Theorem 6.9. *If $f : [a, b] \to (0, \infty)$ is a continuous function, then the rotational solid determined by f has volume and the volume of it is equal to vol $\left(C_f\right) = \pi \int_a^b f^2(x)dx$.*

Theorem 6.10. *Let $f : [a, b] \to \mathbb{R}$ be two continuous functions on $[a, b]$ such that $f(x) \le g(x)$, $\forall x \in [a, b]$. Then the rotational solid obtained by rotating $\Gamma_{f,g}$ around the Ox axis has volume and vol $\left(C_{f,g}\right) = \pi \int_a^b \left(g^2(x) - f^2(x)\right)dx$.*

Theorem 6.11. *Let $f : [a, b] \to (0, \infty)$ a continuous function and nonidentically null. Then the coordinates of the mass center of Γ_f, seen as a homogenous, plane bar of negligible thickness, are $x_G = \frac{\int_a^b xf(x)dx}{\int_a^b f(x)dx}$, $y_G = \frac{\int_a^b \frac{1}{2}f^2(x)dx}{\int_a^b f(x)dx}$.*

Theorem 6.12. *Let $F : [a, b] \to \mathbb{R}$ be a force (variable) which actions along the closed interval $[a, b]$. The work effectuated by F is $L = \int_a^b F(x)\,dx$.*

6.2 Description of general problems and areas where they are very common

The number of areas where an engineer (or any scientist in general) can find a situation that requires calculating an integral is huge.

The calculation of integrals is very useful in many physical problems, such as the calculation of distances based on certain data where speeds, accelerations, forces, or powers of any vehicle are provided. There are also many problems where it is necessary to calculate the mass, the center of gravity, or moments of inertia, which require the calculation of integrals. Such problems are very common for industrial engineers.

In addition integrals appear in many problems related to civil engineering or topography. In many of these cases, the most common type of problem is the calculation of area, volume, or a line integral.

Also, some concepts require the idea of derivative and (therefore) antiderivative intuitively: Since instant velocity is calculated as the derivative of the position function, and acceleration as the derivative of the velocity, many problems for mechanical and civil engineers, for example, can be solved by calculating integrals.

It is almost impossible to give an exhaustive list of situations where real problems require integral calculus.. In this section we outline some basic examples, whilst in the next section we will show more complex exercises which will be modeled by integrals.

6.2.1 Introductory problems

1. Let $v(t)$ be the velocity of the vehicle and $v(0) = 72$ km/h. When the brakes are applied, the acceleration is $a = -5$ m/s^2. How far will it travel after the brakes have been applied?

   Solution:

$$v_1 - v_0 = \int_0^t a(x)\,dx,$$

$$\int_0^t -5dx = -5x\big|_0^t = -5t,$$

$$\Rightarrow v(t) = v_0 - 5t,$$

$$v_0 = 72 \text{ km/h} = \frac{72000 \text{ m}}{3600 \text{ s}} = 20 \text{ m/s}.$$

   The car will come to a stop when $v(t) = 0$; $v(t) = 0 \Rightarrow 20 - 5t = 0 \Rightarrow t = 4$ s $\Rightarrow$ $\int_0^4 v(t)dt = \int_0^4 (20 - 5t)\,dt = 20t\big|_0^4 - 5\frac{t^2}{2}\big|_0^4 = 80 - 40 = 40$ m.

2. The floor of a ship is like the intersection of two parabolas described by the functions $f(x) = -\frac{1}{110}x^2 + 10$, $g(x) = \frac{1}{110}x^2 - 10$. How much paint is needed if the length of the ship is 80 m and 0.25 kg of paint is required for each m^2 ?

   Solution:

$$0.25 \int_{-40}^{40} \left| -\frac{1}{110}x^2 + 10 - \frac{1}{110}x^2 + 10 \right| dx$$

$$= 0.5 \int_0^{40} \left(-\frac{2}{110}x^2 + 20 \right) = \int_0^{40} \left(-\frac{1}{110}x^2 + 10 \right)$$

$$= -\frac{1}{110}\frac{40^3}{3} + 400 \approx 206 \text{ kg}.$$

3. Determine the quantity of electrical energy which passes through a conductor in $t \in [5, 6]$ if the intensity of an electrical current is given by the formula $I(t) = 6t^2 - 4t$.

Solution: $\int_5^6 (6t^2 - 4t)\, dt = 160J$.

4. A material point moves after $v(t) = \frac{1}{\sqrt{6t+2}}$. Determine the distance covered after 5 s.

Solution: $\int_0^5 \frac{1}{\sqrt{6t+2}} dt$. We note $\sqrt{6t+2} = u6t + 2 = u^2 \Rightarrow 6dt = 2udu \Rightarrow \frac{1}{6}\int_{\sqrt{2}}^{\sqrt{32}} \frac{2udu}{u} = \frac{1}{3}\sqrt{2}(4-1) = \sqrt{2}$.

5. A meteorite has the form of a rotational solid obtained by rotating about the Ox-axis of the asteroid: $x^{\frac{2}{3}} + y^{\frac{2}{3}} = a^{\frac{2}{3}}$. Determine the volume of the meteorite if $a = 10$ km.

Solution:
Since $y = \left(a^{\frac{2}{3}} - x^{\frac{2}{3}}\right)^{\frac{3}{2}}$, according to Theorem 6.9 we have

$$V = \pi \int_{-a}^{a} (a^{\frac{2}{3}} - x^{\frac{2}{3}})^3 dx = 2\pi \int_{-a}^{a} (a^2 - 3a^{\frac{4}{3}} \cdot x^{\frac{2}{3}} + 3a^{\frac{2}{3}} \cdot x^{\frac{4}{3}} - x^2) dx = \frac{32\pi a^3}{105}.$$

Particularly, $V = \frac{32 \cdot 3.14 \cdot 1000}{105} \cong 6.7 \text{ km}^3$.

6. Calculate the mass M and the coordinate x_c of the center of gravity of a bar with density $\rho(x) = x$ and length $l = 1$.

Solution:
We have $M = \int_0^1 (x+1)dx = \frac{x^2}{2}\Big|_0^1 + x\Big|_0^1 = \frac{3}{2}$ and $x_c = \frac{\int_0^1 x(x+1)dx}{\int_0^1 (x+1)dx} = \frac{\frac{x^3}{3}\Big|_0^1 + \frac{x^2}{2}\Big|_0^1}{\frac{x^2}{2}\Big|_0^1 + x\Big|_0^1} = \frac{\frac{1}{3}+\frac{1}{2}}{\frac{3}{2}} = \frac{5}{3}$.

7. Find the mass M and the coordinate x_c of the center of gravity of a bar with density $\rho(x) = \frac{1}{x^2}$, $x \in [1, 2]$.

Solution:
We have

$$M = \int_1^2 \frac{1}{x^2} dx = -\frac{1}{x}\Big|_1^2 = -\frac{1}{2} + 1 = \frac{1}{2};$$

$$x_c = \frac{\int_1^2 x \cdot \frac{1}{x^2} dx}{M} = \frac{\int_1^2 \frac{1}{x} dx}{\frac{1}{2}} = \frac{\ln x\Big|_1^2}{\frac{1}{2}} = 2\ln 2 = \ln 4.$$

6.3 Challenging problems

6.3.1 Approximate the mass/position of the gravity center of a bar

Consider a unidimensional bar of length $l = 3$, having as density the function (supposed as continuous) $\rho : [0, 3] \to (0, \infty)$. From measurements, we know that $\rho(0) = 1$, $\rho(1) = \frac{3}{2}$, $\rho(2) = 2$, $\rho(3) = 3$ (we remember that for an unidimensional body we measure in mass unities/length unities).

a) Approximate the mass of the bar.

b) Approximate the position of the gravity center of the bar.

Solution:

a) 1^0) For the bar, as it has no constant density, we cannot apply the formula $M = \rho l$ (M is the mass of the bar). We will deduce a formula for the calculation of the mass, when the density is variable. We will suppose that the bar occupies a part of an interval $[a, b]\mathbb{R}$. Let $\Delta = \{a = x_0 < x_1 < \cdots < x_{n-1} < x_n = b\}$ be a division of $[a, b]$, with $x_k - x_{k-1} = \frac{b-a}{n}$, $\forall k \in \{1, 2, \ldots, n\}$. We suppose that n is big enough such that on the interval $[x_{k-1}, x_k]$ we can approximate the density with its value in an intermediate point $c_k \in [x_{k-1}, x_k]$. Then we have the approximation $M \cong \sum_{k=1}^{n} \rho(c_k)(x_{k-1}, x_k)$. The right member of this relation is a Riemann sum associated to ρ, to divisions Δ and to points $(c_k)_{1 \le k < n}$. As $\|\Delta\| \to 0$ (for $n \to \infty$) and ρ is continuous, for the Riemann integrable for $n \to \infty$ we find the following formula:

$$M = \int_a^b \rho(x)dx.$$

2^0) The theorem deduced at 1^0) would solve the problem if we knew the analytical expression of the density function. Only, we know the values of the density just in four points. From this reason we will try to approximate the density function using the Lagrange's polynomial of interpolation P_L:

$$\begin{aligned}\rho(x) \cong P_L(x) \overset{\text{def}}{=} & \frac{(x - x_1)(x - x_2)(x - x_3)}{(x_0 - x_1)(x_0 - x_2)(x_0 - x_3)} y_0 \\ & + \frac{(x - x_1)(x - x_2)(x - x_3)}{(x_1 - x_0)(x_1 - x_2)(x_1 - x_3)} y_1 \\ & + \frac{(x - x_0)(x - x_1)(x - x_3)}{(x_2 - x_1)(x_2 - x_2)(x_2 - x_3)} y_2 \\ & + \frac{(x - x_0)(x - x_1)(x - x_2)}{(x_3 - x_0)(x_3 - x_1)(x_3 - x_2)} y_3,\end{aligned}$$

where $x_0 = 0$, $x_1 = 1$, $x_2 = 2$, $x_3 = 3$, $y_0 = \rho(x_0) = 1$, $y_1 = \rho(x_1) = \frac{3}{2}$, $y_2 = \rho(x_2) = 2$, $y_3 = \rho(x_3) = 3$. By replacing these values in Lagrange's polynomial, we obtain

$$\rho(x) \cong \frac{(x-1)(x-2)(x-3)}{-1\cdot(-2)\cdot(-3)}\cdot 1 + \frac{x(x-2)(x-3)}{1\cdot(-1)\cdot(-2)}\cdot\frac{3}{2}$$
$$+\frac{(x-1)(x-3)}{2\cdot 1\cdot(-1)}\cdot 2 + \frac{x(x-1)(x-2)}{3\cdot 2\cdot 1}\cdot 3.$$

After making the calculations, we find

$$\rho(x) \cong \frac{1}{12}x^3 - \frac{1}{4}x^2 + \frac{2}{3}x + 1.$$

3^0) Now we will use the formula deduced at 1^0). We have

$$M = \int_0^3 \rho(x)\,dx \cong \int_0^3 \left(\frac{1}{12}x^3 - \frac{1}{4}x^2 + \frac{2}{3}x + 1\right) dx$$
$$= \frac{x^4}{48}\Big|_0^3 - \frac{x^3}{12}\Big|_0^3 + \frac{x^2}{3}\Big|_0^3 + x\Big|_0^3$$
$$\cong \frac{87}{16} = 5.4375 \text{ (mass unities)}.$$

b) The position of the gravity center:

$$x_G = \frac{\int_0^3 x\rho(x)dx}{\int_0^3 \rho(x)dx} = \frac{\int_0^3 \left(\frac{1}{12}x^4 - \frac{1}{4}x^3 + \frac{2}{3}x^2 + x\right) dx}{\frac{87}{16}}$$
$$= \frac{\frac{x^5}{60}\Big|_0^3 - \frac{x^4}{16}\Big|_0^3 + \frac{2x^3}{9}\Big|_0^3 + \frac{x^2}{2}\Big|_0^3}{\frac{87}{16}} = \frac{759}{435} \cong 1.7448.$$

Remark. The position on the bar of the gravity center, obtained using the approximation of the density, corresponds to intuition: $x_G \cong 1.7448 > 1.5$, hence x_G is on the right side of the middle of the bar. This thing is normal, because function ρ (approximated by P_1) has ever bigger values as we get closer to the right margin of the bar.

6.3.2 Determine the moment of inertia of a wire

We consider a wire, with a negligible cross section, dispensed in the plane after the curve $y = x^2, x \in [1, 2]$, having as density function $\rho(x) = x$. We propose to determine the moment of inertia of the wire with respect to the Ox- and Oy-axis.

1^0) We will determine a formula for the calculus of the moments of inertia of a wire that occupies in plan the position $y = f(x), x \in [a, b]$, having as density the continuous function $\rho : [a, b] \to (0, \infty)$.

We remember that for a system of material points $M_1, M_2, \ldots, M_n$ the moment of inertia of this system of points with respect to an axis or a point is $I = \sum_{k=1}^{n} m_k d_k^2$, where, for $k = \{1, 2, \ldots, n\}$, m_k is the mass of M_k and d_k is the distance of M_k to the axis or the point with respect to which we calculate the moment of inertia.

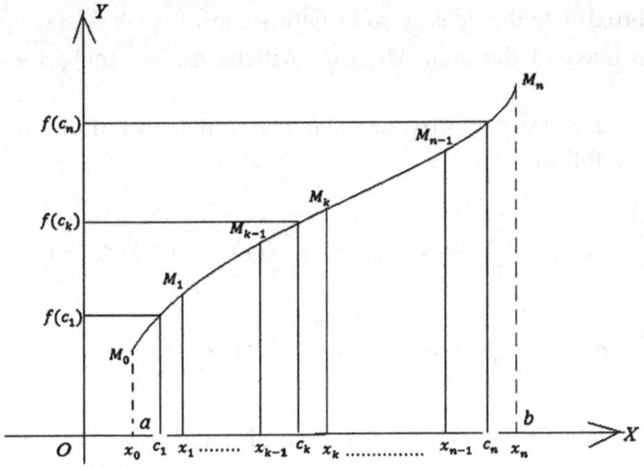

Figure 6.2 Division of interval.

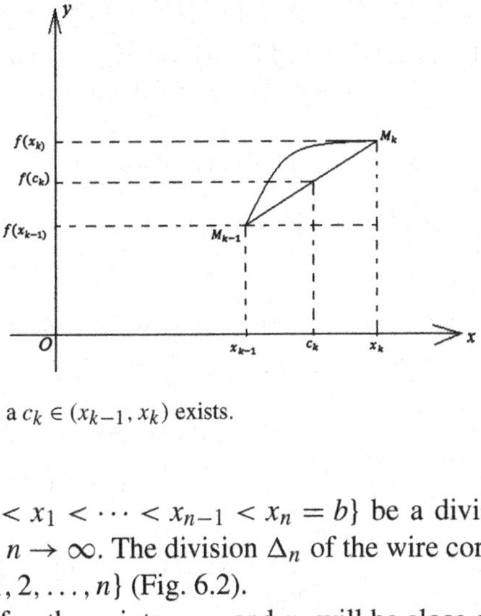

Figure 6.3 Mean theorem, a $c_k \in (x_{k-1}, x_k)$ exists.

Let $\Delta_r = \{a = x_0 < x_1 < \cdots < x_{n-1} < x_n = b\}$ be a division of interval $[a, b]$ with $\|\Delta_r\| \to 0$, when $n \to \infty$. The division Δ_n of the wire corresponds to the points $M_k(x_k, f(x_k))$, $k = \{1, 2, \ldots, n\}$ (Fig. 6.2).

For bigger values of n, the points x_{k-1} and x_k will be close enough for making the following approximations.

i) the arch $M_{k-1}M_k$ will be associated to a material point;

ii) the length of arch $M_{k-1}M_k$ will be approximated;

iii) the density of the wire is constant on arch $M_{k-1}M_k$, the length of the segment $M_{k-1}M_k$, i.e.,

$$\sqrt{(x_k - x_{k-1})^2 + (f(x_k) - f(x_{k-1}))^2}.$$

According to the mean theorem, a $c_k \in (x_{k-1}, x_k)$ exists (Fig. 6.3) such that

$$f(x_k) - f(x_{k-1}) = f'(c_k)(x_k - x_{k-1}).$$

We will approximate the density $\rho(x)$ with value $\rho(c_k)$, $\forall x \in [x_{k-1}, x_k]$. In these conditions, the mass of the arch $M_{k-1}M_k$ will be $m_k = \rho(c_k)\sqrt{1+f'^2(c_k)}(x_k - x_{k-1})$.

In consequence, one can approximate the inertia moments of the wire with respect to Ox and Oy as follows:

$$I_{OX} \cong \sum_{k=1}^{n} m_k d_{k_x}^2 \cong \sum_{k=1}^{n} \rho(c_k)\sqrt{1+f'^2(c_k)} \cdot f^2(c_k)(x_k - x_{k-1}),$$

$$I_{OY} \cong \sum_{k=1}^{n} m_k d_{k_y}^2 \cong \sum_{k=1}^{n} \rho(c_k)\sqrt{1+f'^2(c_k)} \cdot c_k^2(x_k - x_{k-1}),$$

where we have denoted with d_{k_x} and d_{k_y} the distances from arch $M_{k-1}M_k$ to Ox and to Oy, approximated by $f(c_k)$ and c_k, respectively. The sums from the right sides in those two relations are Riemann sums. For $n \to \infty$, we obtain $I_{OX} = \int_a^b \rho(x)\sqrt{1+f^2(x)} \cdot f^2(x)dx$, $I_{OX} = \int_a^b \rho(x)\sqrt{1+f^2(x)} \cdot x^2 dx$.

2^0) We will use the formulas deduced from the calculation of the moment of inertia with respect to Ox and Oy for the wire given in the problem:

$$f(x) = x^2 \Rightarrow f'(x) = 2x, x \in [1,2], \rho(x) = x$$
$$\Rightarrow I_{OX} = \int_1^2 x\left(x^2\right)^2\sqrt{1+(2x)^2}dx.$$

We make the change of variable $t = \sqrt{4x^2+1}$. We deduce successively: $x^2 = \frac{1}{4}t^2 - \frac{1}{4}$, $xdx = \frac{1}{4}tdt$, $x = 1 \Rightarrow t = \sqrt{5}$, $x = 2 \Rightarrow t = \sqrt{17}$, and

$$I_{OX} = \frac{1}{64}\int_{\sqrt{5}}^{\sqrt{17}} \left(t^2-1\right)^2 t^2 dt = \frac{1}{64}\left(\frac{t^7}{7} - \frac{2t^5}{5} + \frac{t^3}{3}\right)\Bigg|_{\sqrt{5}}^{\sqrt{17}} \cong 37.801,$$

$I_{OY} = \int_1^2 xx^2\sqrt{1+(2x)^2}dx = \int_1^2 x^3\sqrt{1+4x^2}dx$. With the same change of variable, we obtain, after calculation, $I_{OY} \cong 12.968$.

Observation. The concept of moment of inertia (as the one of gravity center) is very important for many disciplines from theoretical mechanics to resistance of materials. Hence, this concept is very important for any future engineer.

6.3.3 Rolling motion in mechanics, remarkable curve

Now, we will introduce a remarkable curve, coming also from mechanics. In an $\boldsymbol{xOy}$ system of coordinates, we will consider a circle of radius $\boldsymbol{a > 0}$, situated in the superior semiplane, tangent in O to $\boldsymbol{Ox}$. The circle is endowed with a rolling motion (without friction) on the positive semiaxis $\boldsymbol{Ox}$. We note with $\boldsymbol{M}$ the point on the circle. The set of the points from the plan covered by $\boldsymbol{M}$ as the circle rolls is called a *cycloid*.

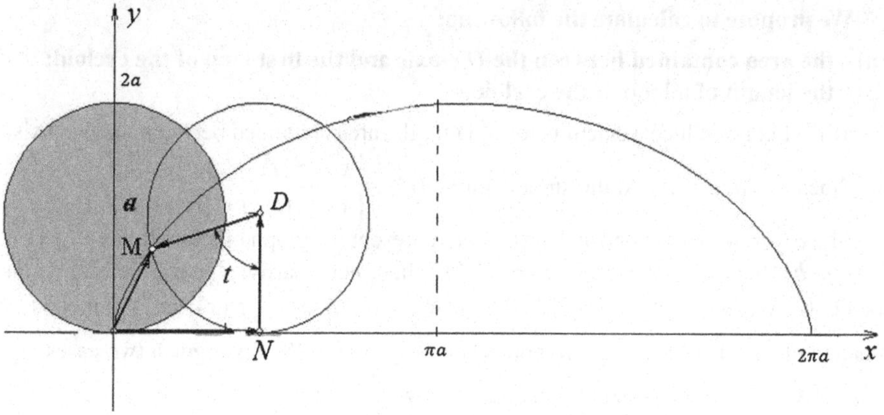

Figure 6.4 D is the center of the circle corresponding to the position of M.

The point M will touch Ox infinitely many times, the distance between two consecutive positions on Ox of M being equal to the length of the circle.

For describing the points on the chord we will use a parameter: for a certain position of M, we will take as parameter t, the measure in radians of angle $\angle MDN$, where N is the tangent point of Ox and D is the center of the circle corresponding to the position of M (see Fig. 6.4).

Obviously, ON is equal to the length of $\overline{MN}$, i.e., at, and hence $x_M = at - \cos\left(t - \frac{\pi}{2}\right)$, $y_M = a + a\sin\left(t - \frac{\pi}{2}\right)$. Hence, the cycloid is described by the following equations:

$$\begin{cases} x = a\,(t - \sin t)\,, \\ y = a(1 - \cos t), t \geq 0. \end{cases}$$

Observations. 1) When a curve is described by equations $\begin{cases} x = f\,(t)\,, \\ y = g\,(t)\,, \end{cases}$ we say that we have *a parametrical representation of the curve*, and those two functions are called *parametrical equations of the curve*.

2) Sometimes, from parametrical equations one can deduce the Cartesian equation of the curve. For example, if $\begin{cases} x = r\cos t, \\ y = r\sin t, \end{cases}$ it results that $x^2 + y^2 = r^2$ (the Cartesian equation of the circle of center O, of radius r).

In other situations (as in the case of a cycloid) it is difficult to obtain the Cartesian equation of the curve. In such cases, if we want to calculate the area contained under the curve or the length of it, we need the formula which utilizes the parametrical equations.

We will need a *cycloid loop* that contains part of the curve between $t = 2k\pi$ and $t = 2\ (k+1)\,\pi$, $k \in N$. For $k = 0$, we obtain the first loop of cycloid ($t \in [0, 2\pi]$).

We propose to calculate the following:

a) the area contained between the Ox-axis and the first loop of the cycloid;

b) the length of a loop of the cycloid.

a) 1^0) Let us deduce a calculus formula for the area contained between the Ox-axis, the lines $x = a$, $x = b$ and the smooth curve $(C) : \begin{cases} x = x(t), \\ y = y(t), \ t \in [t_1, t_2]. \end{cases}$

Moreover, we will suppose that (C) can represent the graph of a function $y = f(x)$, $x \in [a, b]$; hence, for $t \neq t^*$, $x(t) \neq x(t^*)$. Then, necessarily, x' has a constant sign on $[a, b]$. We note with A the asked area. We have $A = \int_a^b |f(x)|\, dx$. We make the change of variable $x = x(t)$. We note $y(t) \overset{\text{def}}{\longrightarrow} f(x(t))$. We distinguish two cases:

i) If $x'(t) < 0$, then $x(t_1) = b$, $x(t_2) = a \Rightarrow$

$$A = \int_{t_1}^{t_2} |y(t)|\, x'(t)\, dt = \int_{t_1}^{t_2} |y(t)|\, (-x'(t))dt.$$

ii) If $x'(t) > 0$, then $x(t_1) = a$, $x(t_2) = b \Rightarrow$

$$A = \int_{t_1}^{t_2} |y(t)|\, x'(t)\, dt.$$

Reuniting those two cases, we can write $\int_{t_1}^{t_2} \left|y(t)\, x'(t)\right| dt$.

2^0) We have $\begin{cases} x = a(t - \sin t), \\ y = a(1 - \cos t), \quad t \in [0, 2\pi]. \end{cases}$

Using the deduced formula, we will have

$$\begin{aligned} A &= \int_0^{2\pi} |\, a(1 - \cos t) \cdot a(1 - \cos t)\, |\, dt \\ &= a^2 \int_0^{2\pi} (1 - 2\cos t + \cos^2 t)dt \\ &= a^2 \left[(t - 2\sin t)\big|_0^{2\pi} + \int_0^{2\pi} \frac{1 + \cos 2t}{2} dt \right] \\ &= a^2 \left[2\pi + \frac{1}{2}\left(t + \frac{1}{2}\sin 2t\right)\Big|_0^{2\pi} \right] = 3\pi a^2. \end{aligned}$$

b) 1^0) We will give, without demonstration, the length of the curve $(C) : \begin{cases} x = x(t), \\ y = y(t), \quad t \in [t_1, t_2]. \end{cases}$

We denote with L_C the length of the curve. We have

$$L = \int_{t_1}^{t_2} \sqrt{x'^2(t) + y'^2(t)} dt.$$

2^0) In the case of the first loop of the cycloid, we obtain

$$\begin{aligned} L &= \int_0^{2\pi} \sqrt{a^2(1-\cos t)^2 + a^2 \sin^2 t}\, dt \\ &= a \int_0^{2\pi} \sqrt{2 - 2\cos t}\, dt = 2a \int_0^{2\pi} \sqrt{\frac{1-\cos t}{2}}\, dt \\ &= 2a \int_0^{2\pi} \left| \sin \frac{t}{2} \right| dt \overset{\frac{t}{2} \in [0,\pi]}{=} 2a \int_0^{2\pi} \sin \frac{t}{2} dt = -\, 4a \cos \frac{t}{2} \Big|_0^{2\pi} = 8a. \end{aligned}$$

6.3.4 Speed prediction models

Speed is an essential factor in road geometric design and operation. In this sense, the operating speed is the distribution of the speed values chosen by the drivers at a particular location and driving direction. To characterize these distributions of speeds, percentile speeds are used. In particular, the 85th percentile is set in road design. It is the value of speed that is assumed not to be exceeded by 85% of the drivers at a particular location.

Departments of transportation need to check that adequate sight distance is provided to road users so that they adapt their speed to the road features, ensuring safe design, for example, when entering a curve, approaching an intersection or a toll plaza, or initiating a passing maneuver.

Where naturalistic operating speed data are lacking, speed prediction models are devised. Particularly, the transition profiles between alignments that require to be traveled at significantly different speeds are to be schemed. These transitions have been assumed to be performed at constant acceleration and deceleration rates (Castro et al., 2008; Ottesen and Krammes, 2000). However, this assumption does not seem to fit to the naturalistic data when the difference between the initial speed and the final speed is great. In addition, it runs against the principles of dynamics, because, as speed increases, so does aerodynamic drag. Hence, a number of authors proposed variable acceleration and deceleration rates (Fitzpatrick et al., 2000; Perco et al., 2012; Pérez-Zuriaga et al., 2013).

According to the TWOPAS model, vehicle acceleration and deceleration rates on level terrain vary linearly, but inversely, with speed (Fitzpatrick et al., 2000; St John and Kobett, 1978). To account for grade effects, another term must be added to the performance equation, which reads

$$a(v) = a_0 \left(1 - \frac{v}{v_{max}}\right) + 9.81G,$$

where

a = vehicle acceleration/deceleration at speed v (m/s^2);

a_0 = highest acceleration/deceleration achieved by the vehicle at zero speed (m/s^2);

v = vehicle speed (m/s);

v_{max} = maximum speed attainable by the vehicle on level terrain (m/s);

G = grade (positive for upgrades and negative for downgrades) (m/m).

It should be noted that these models do not need to be based on the maximum acceleration performed by the engine of a vehicle, but need to outline the general scheme of naturalistic data. As mentioned above, drivers do not usually fully utilize the acceleration capabilities, so reduction coefficients are used in the TWOPAS model. It is applied sequentially over a 1-s interval.

6.3.4.1 The problem of estimating the distance traveled

A highway segment ends at an intersection that is visible from a distance of 360 m. The highway profile is featured by a 4% downgrade. To ensure a safe design, this distance must be greater than that needed by both a light vehicle and a heavy vehicle approaching that might need to stop completely at comfortable deceleration rates. Whereas the 85th percentile speed of light vehicles is assumed to be 118 km/h, the corresponding heavy vehicle speed is 95 km/h at the location where the intersection is first seen. Both types of vehicles initiate the deceleration at −0.5 m/s^2. During the stopping maneuver, the light vehicle and the heavy vehicle should not exceed a deceleration of −4.8 m/s^2 and −3.6 m/s^2, respectively. Is 360 m enough to guarantee the safety conditions in the deceleration of both vehicles?

Solution:

To deduce the distance traveled by either of the vehicles described during the deceleration, time needs to be expressed as a function of speed. To this end, we need to clear t and obtain a primitive(integral) of the equation of speed of the TWOPAS model.

$$a(v) = \frac{dv}{dt} \Rightarrow dt = \frac{dv}{a(v)}.$$

For the sake of simplicity, let us reformulate the equation of the TWOPAS model as follows:

$$a(v) = a_0\left(1 - \frac{v}{v_{max}}\right) + 9.81G = A + Bv,$$

where

$$A = a_0 + 9.81G,$$
$$B = -\frac{a_0}{v_{max}}.$$

Therefore,

$$\int_0^t dt = \int_{v_0}^{v} \frac{dv}{A + Bv} = \left[\frac{\ln(A + Bv)}{B}\right]_{v_0}^{v} + C.$$

We can consider the origin of time at the beginning of the deceleration ($C = 0$) without loss of generality, resulting in

$$t = \frac{ln(A + Bv)}{B} - \frac{\ln(A + Bv_0)}{B} = \frac{1}{B}\ln\frac{A + Bv}{A + Bv_0} \Rightarrow e^{Bt} = \frac{A + Bv}{A + Bv_0}.$$

Next, v must be found, expressed as a function of t, i.e.,

$$v = v(t) = \left(\frac{A}{B} + v_0\right) e^{Bt} - \frac{A}{B}.$$

Then the distance traveled can be deduced as follows:

$$v(t) = \frac{dx}{dt} \Rightarrow dx = v(t)\,dt \Rightarrow \int_{x_0}^{x} dx = \int_0^t \left[\left(\frac{A}{B} + v_0\right) e^{Bt} - \frac{A}{B}\right] dt$$

$$\Rightarrow x - x_0 = \left[\frac{A + Bv_0}{B^2} e^{Bt} - \frac{A}{B}t\right]_0^t = \frac{A + Bv_0}{B^2}\left(e^{Bt} - 1\right) - \frac{A}{B}t.$$

Substituting the expression of t in the previous equation,

$$x - x_0 = \frac{A + Bv_0}{B^2}\left(\frac{A + Bv}{A + Bv_0} - 1\right) - \frac{A}{B^2}\ln\frac{A + Bv}{A + Bv_0}$$

$$= \frac{A + Bv_0}{B}\frac{v - v_0}{A + Bv_0} - \frac{A}{B^2}\ln\frac{A + Bv}{A + Bv_0} = \frac{v - v_0}{B} - \frac{A}{B^2}\ln\frac{A + Bv}{A + Bv_0}.$$

Once the expression that relates the distance traveled to the initial and final speeds is known, the parameters A and B must be obtained. They can be deduced using the deceleration equation of the TWOPAS model. The maximum deceleration would correspond to level terrain, which will be produced at the moment the vehicle stops completely, i.e.,

$$a_{max} = a(0) = a_0\left(1 - \frac{0}{v_{max}}\right) + 9.81G = a_0 - 9.81 \cdot (-0.04).$$

In the case of the light vehicle, we will have $a_{0,l} = -5.192$ m/s^2, and in the case of the heavy vehicle $a_{0,h} = -3.992$ m/s^2.

Next, the equation is evaluated for the contour conditions at the beginning of deceleration to obtain the maximum speeds associated with the respective equations, i.e.,

$$a(v) = a_{max}\left(1 - \frac{v_0}{v_{max}}\right) - 9.81 \cdot (-0.04) = -0.5 \text{ m/s}^2 \Rightarrow$$

$$v_{max} = \frac{v_0 a_{max}}{a_{max} + 1 + 9.81G}.$$

Replacing the respective values in the equation, we will have $v_{max,l} = 120.497$ km/h in the case of the light vehicle and $v_{max,l} = 97.631$ km/h in the case of the heavy vehicle. Noting that the units have been converted to the International System, the counterpart values of the parameters are

$$\begin{cases} A_l = -5.585 \text{ m/s}^2, \\ B_l = 0.155 \text{ s}^{-1}, \end{cases}$$

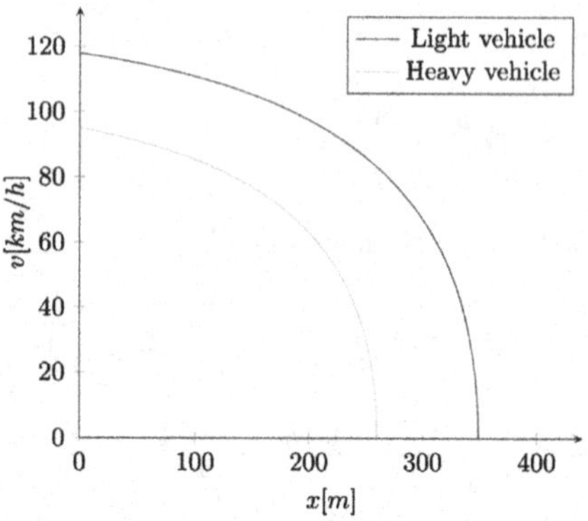

Figure 6.5 Deceleration curves for light and heavy vehicles as per the distance traveled.

$$\begin{cases} A_h = -4.385 \text{ m/s}^2, \\ B_h = 0.147 \text{ s}^{-1}. \end{cases}$$

After replacing the corresponding values in the equation of the distance traveled, and assuming $x_0 = 0$,

$$x_l = \frac{0-118}{3.6B_l} - \frac{A_l}{B_l^2} \ln \frac{A_l + B_l \frac{0}{3.6}}{A_l + B_l \frac{118}{3.6}} = 348.737 \text{ m},$$

$$x_h = \frac{0-95}{3.6B_h} - \frac{A_h}{B_h^2} \ln \frac{A_h + B_h \frac{0}{3.6}}{A_h + B_h \frac{95}{3.6}} = 260.056 \text{ m}.$$

Therefore, there is sufficient sight distance for road users to anticipate the intersection and decelerate within the specified limits.

Fig. 6.5 plots the theoretical deceleration curves obtained for light and heavy vehicles as a function of the distance traveled.

References

Castro, M., Sánchez, J.A., Vaquero, C.M., Iglesias, L., Rodríguez-Solano, R., 2008. Automated GIS-based system for speed estimation and highway safety evaluation. Journal of Computing in Civil Engineering 22, 325–331.

Fitzpatrick, K., Elefteriadou, L., Harwood, D.W., Collins, J.M., Mcfadden, J., Anderson, I.B., Krammes, R.A., Irizarry, N., Parma, K.D., Bauer, K.M., Passetti, K., 2000. Speed Prediction for Two-Lane Rural Highways. Federal Highway Administration, Washington, D.C.

Ottesen, J.L., Krammes, R., 2000. Speed-profile model for a design-consistency evaluation procedure in the United States. Transportation Research Record 1701, 76–85.

Perco, P., Marchionna, A., Falconetti, N., 2012. Prediction of the operating speed profile approaching and departing intersections. Journal of Transportation Engineering 138, 1476–1483.

Pérez-Zuriaga, A.M., Camacho-Torregrosa, F.J., García, A., 2013. Tangent-to-curve transition on two-lane rural roads based on continuous speed profiles. Journal of Transportation Engineering 139, 1048–1057.

St John, A.D., Kobett, D.R., 1978. Grade effects on traffic flow stability and capacity. NCHRP Report 185. Transportation Research Board, Washington, D.C.

Multiple integrals in mechanical engineering

7

Deolinda M.L. Dias Rasteiro[a], *Manuel Rodríguez-Martín*[b],
Pablo Rodríguez-González[c]
[a]Coimbra Polytechnic – ISEC, Coimbra, Portugal, [b]University of Salamanca, Salamanca, Spain, [c]University of León, León, Spain

7.1 Background

The aim of this section is to provide the reader with essential concepts that will be needed to better understand what multiple integrals are and their possible calculus procedures.

Double integrals

We will start with the determination of the volume V of the solid that is formed between all the points of the region R and below the surface $z = f(x, y)$, where R is a limited region of the XOY-plane and f is a continuous and nonnegative function in R. To best understand the concepts and other related theorems and definitions we recommend Stewart (2016) or Jerrold and Marsden (2003). See Fig. 7.1.

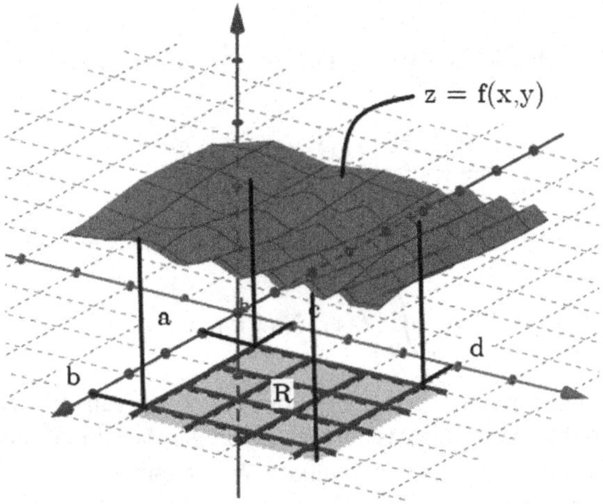

Figure 7.1 Assuming $R = [a, b] \times [c, d]$.

Calculus for Engineering Students. https://doi.org/10.1016/B978-0-12-817210-0.00014-X

Consider a rectangle in the plane containing the region R. Assume that the rectangle is subdivided into subrectangles, not necessarily equal (*grid*). Suppose that the number of rectangles all contained in R are n and that the area of the kth subrectangle is ΔA_k. The subrectangles not all contained in R will not be considered.

In each subrectangle consider a point. Let $(x_k^\star, y_k^\star)$ be the point that corresponds to the kth subrectangle. The product

$$f(x_k^\star, y_k^\star)\Delta A_k$$

is the volume of the parallelepiped whose base has an area equal to ΔA_k and hight $f(x_k^\star, y_k^\star)$. Thus, the sum

$$\sum_{k=1}^{n} f(x_k^\star, y_k^\star)\Delta A_k$$

is an approximate value of the volume V, which is the objective that we intend to accomplish.

By approximating V with the above value we perform two types of errors:

- the upper face of the parallelepiped is plane and, in general, the surface $z = f(x, y)$ is not;
- the subrectangles in which R is subdivided may not all be completely contained in that region.

Although, if the referred process is repeated for a number of subdivisions of R such that the rectangle's length and width tend to zero, we may assume that the error will also tend to zero. Thus, the **exact volume** is

$$V = \lim_{n\to+\infty} \sum_{k=1}^{n} f(x_k^\star, y_k^\star)\Delta A_k.$$

Definition. We call **the double integral** of $f(x, y)$ under the region R the value

$$\iint_R f(x, y)\, dA = \lim_{n\to+\infty} \sum_{k=1}^{n} f(x_k^\star, y_k^\star)\Delta A_k.$$

Therefore, the volume V that we intended to determine is given by

$$V = \iint_R f(x, y)\, dA = \iint_R f(x, y)\, dx\, dy.$$

We assumed until now that f is continuous and nonnegative in R. If $f(x, y)$ assumes positive and negative values in the region R, then the double integral will represent the difference between two volumes. Which ones?

Note that a positive value for a double integral will tell us that the volume above the region R is bigger than the volume below, and a negative value for a double integral will tell us that the volume below the region R is bigger than the volume above, and a zero value for a double integral will tell us that both volumes are equal.

In order to determine a double integral's value it is useful to make use of three important properties verified by them. Those properties are

1. $\iint_R cf(x, y)\, dA = c \iint_R f(x, y)\, dA, \ \forall\, c \in \mathbb{R}$,

2. $\iint_R [\, f(x, y) \pm g(x, y)]\, dA = \iint_R f(x, y)\, dA \pm \iint_R g(x, y)\, dA$,

3. if $R = R_1 \cup R_2$ and $R_1 \cap R_2 = \emptyset$, then $\iint_R f(x, y)\, dA = \iint_{R_1} f(x, y)\, dA + \iint_{R_2} f(x, y)\, dA$.

Double integral calculus

If R is a rectangular region of type $[a, b] \times [c, d]$, then we may define the following integrals as **iterated integrals:**

$$\int_c^d \int_a^b f(x, y)\, dx\, dy = \int_c^d \left[\int_a^b f(x, y)\, dx \right] dy,$$

$$\int_a^b \int_c^d f(x, y)\, dy\, dx = \int_a^b \left[\int_c^d f(x, y)\, dy \right] dx,$$

and use Fubini's theorems, given below, to their calculus.

Theorem 7.1. [Fubini's theorem] *(first version)*

Let R be the rectangle defined by $a \le x \le b,\ c \le y \le d$. If $f(x, y)$ is continuous on that rectangle, then

$$\iint_R f(x, y)\, dA = \int_c^d \int_a^b f(x, y)\, dx\, dy$$
$$= \int_a^b \int_c^d f(x, y)\, dx\, dy.$$

If R is not a rectangular region, then we have to distinguish between two types of regions:

- **Regions type I**

$$R = \left\{ (x, y) \in \mathbb{R}^2 : a \le x \le b,\ g_1(x) \le y \le g_2(x) \right\}$$

such that $g_1(x)$ and $g_2(x)$ are continuous in $[a, b]$. See Figs. 7.2 and 7.3.

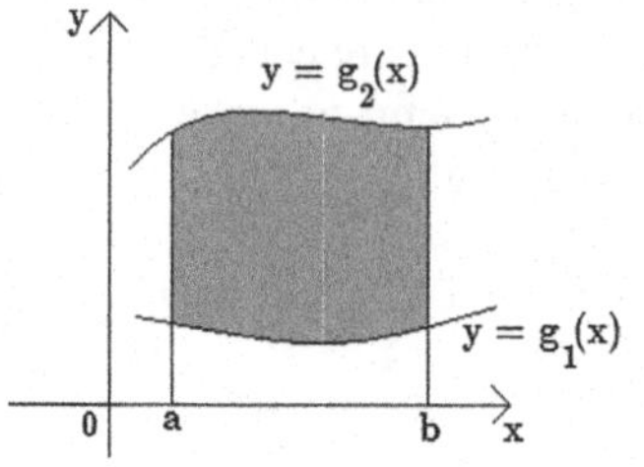

Figure 7.2 Region type I.

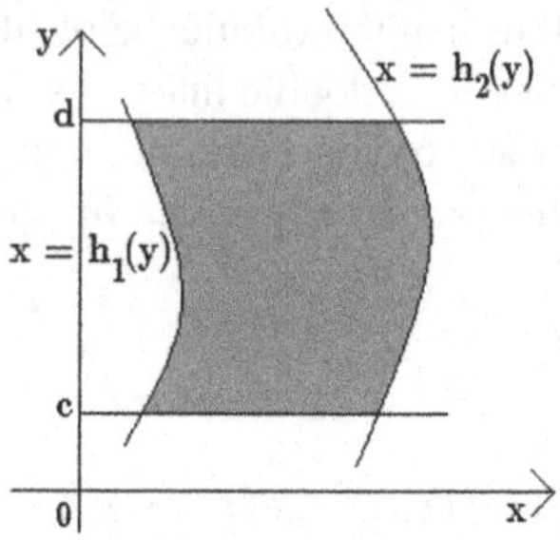

Figure 7.3 Region type II.

- **Regions type II**

$$R=\left\{(x,y)\in\mathbb{R}^2: h_1(y)\le x\le h_2(y),\ c\le y\le d\right\}$$

such that $h_1(y)$ and $h_2(y)$ are continuous in $[c,d]$.

Theorem 7.2. [Fubini's theorem] *(second version – stronger)*
Let $f(x,y)$ be continuous in the region R.

- *If R is a region of type I, then*

$$\iint_R f(x,y)\,dA=\int_a^b\int_{g_1(x)}^{g_2(x)} f(x,y)\,dy\,dx.$$

- *If R is a region of type II, then*

$$\iint_R f(x,y)\,dA=\int_c^d\int_{h_1(y)}^{h_2(y)} f(x,y)\,dx\,dy.$$

Remark. In order to better determine the value of the double integral you should always make a sketch of the region R.

Polar coordinates and their relation with Cartesian coordinates

Observe Fig. 7.4. To define the position of a point P using polar coordinates it is necessary to consider:

- a point O called the origin;
- a line that starts at O and will be called polar axis.

The position, in polar coordinates, of P will be defined by the pair (ρ,θ), where ρ is the distance between O and P and θ is the angle formed by the polar axis and $\dot{O}P$. By convention the polar coordinates of O are $(0,\theta)$. See Fig. 7.5.

Considering both system coordinates and letting the x-axis coincide with the polar axis we can deduce the following relations:

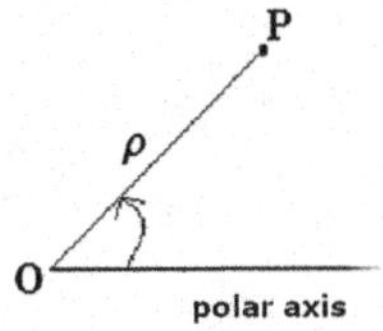

Figure 7.4 Polar representation.

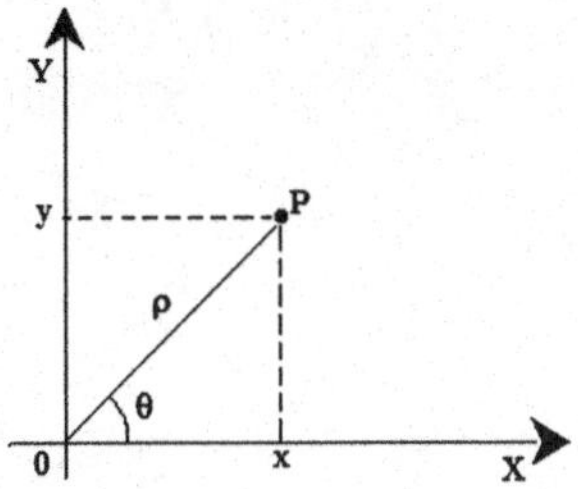

Figure 7.5 Polar and Cartesian coordinates.

If (x, y) are the Cartesian coordinates of the point P at the referential XOY, then its polar coordinates in the polar system can be given using the following equalities:

$$\rho^2 = x^2 + y^2, \qquad \tan(\theta) = \frac{y}{x}.$$

If (ρ, θ) are the polar coordinates P, then the Cartesian coordinates of P are

$$\begin{cases} x = \rho \cos\theta, \\ y = \rho \sin\theta. \end{cases}$$

Remark. To change from Cartesian coordinates to polar coordinates one must have in consideration to which quadrant the point P belongs.

In a system of polar coordinates $\rho = f(\theta)$ represents a curve.

Cylindrical coordinates

In the cylindrical coordinate system, a point P of the space is defined by $P = (\rho, \theta, z)$, where

- $\rho > 0$ and $\theta \in [0, 2\pi[$ (ρ and θ are the polar coordinates of the orthogonal projection of P on the XOY plane);
- z is the vertical Cartesian coordinate sometimes called *height* or *altitude*. See Fig. 7.6.

The relation between cylindrical coordinates and Cartesian coordinates is given by the following equations:

$$\begin{cases} x = \rho \cos\theta, \\ y = \rho \sin\theta, \\ z = z, \end{cases}$$

and thus we also obtain the following relations:

$$\begin{cases} \rho^2 = x^2 + y^2, \\ \tan\theta = \dfrac{y}{x}, \quad x \neq 0, \\ z = z. \end{cases}$$

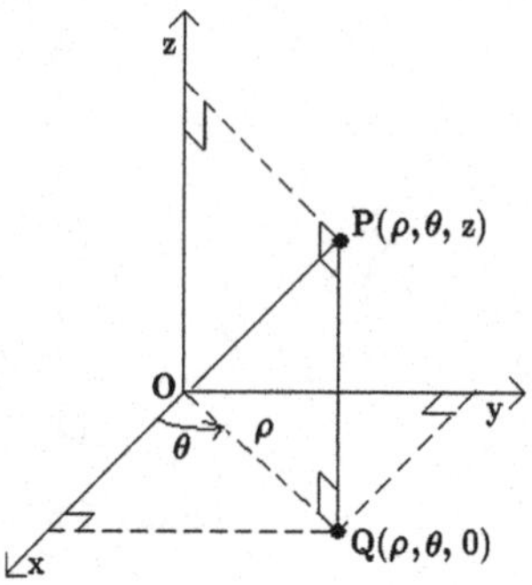

Figure 7.6 Cylindrical coordinates representation.

Spherical coordinates

In the spherical coordinate system, a point P of the space is defined by $P = (\rho, \theta, \phi)$, where

- $\rho \geq 0$ is the distance between P and the origin O;
- $\theta \in [0, 2\pi[$ is equal to θ of the cylindrical coordinates;
- $\phi \in [0, \pi]$ is the angle formed by the vector $\overrightarrow{OP}$ and positive z-axis. See Fig. 7.7.

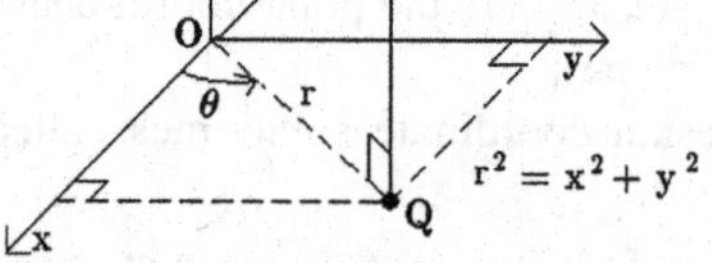

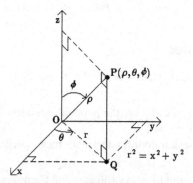

Figure 7.7 Spherical coordinates representation.

The relation between spherical coordinates and Cartesian coordinates is given by the following equations:

$$\begin{cases} x = \rho \sin\phi \cos\theta, \\ y = \rho \sin\phi \sin\theta, \\ z = \rho \cos\phi, \end{cases}$$

and thus we also obtain the following relations:

$$\begin{cases} \rho^2 = x^2 + y^2 + z^2, \\ \phi = \arccos \dfrac{z}{\rho}, \quad \phi \in [0, \pi], \\ \cos\theta = \dfrac{x}{\rho \sin\phi}, \\ \sin\theta = \dfrac{y}{\rho \sin\phi}, \quad \theta \in [0, 2\pi[. \end{cases}$$

Double integrals in polar coordinates

Definition. We call a region of the plane that is limited by $\theta = \alpha$ and $\theta = \beta$ and by the continuous polar curves $\rho = r_1(\theta)$ and $\rho = r_2(\theta)$, where

$$\alpha \leq \beta \quad \text{and} \quad \beta - \alpha \leq 2\pi, \quad 0 \leq r_1(\theta) \leq r_2(\theta),$$

a *simple polar region.*

Examples.

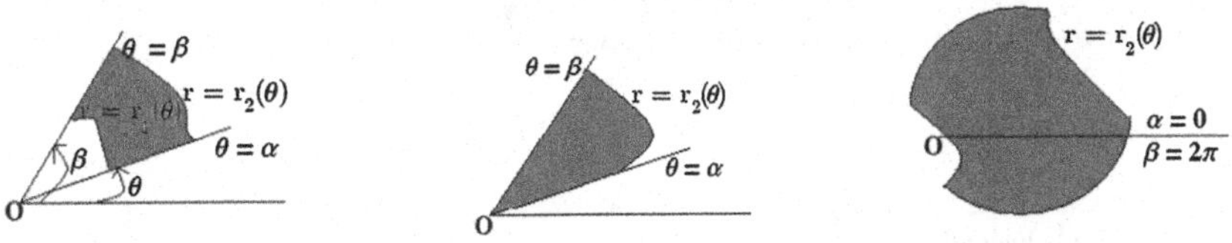

Consider the solid formed by all points that belong to the simple polar region R, of the plane XOY, and under the surface given by the equation, in cylindrical coordinates, $z = f(\rho, \theta)$, where f is continuous in R and $f(\rho, \theta) \geq 0$, $\forall\ (\rho, \theta) \in R$.

The volume of that solid is represented by the following double integral:

$$V = \iint_R f(\rho, \theta)\, dA.$$

Since R is defined through polar coordinates and $f(\rho, \theta)$ is given on cylindrical coordinates, we say that this integral is a **double integral in polar coordinates**. Its calculus can be done using iterated integrals.

Theorem 7.3. *If R is a simple polar region and $f(\rho, \theta)$ is continuous in R, then*

$$\iint_R f(\rho, \theta)\, dA = \int_\alpha^\beta \int_{r_1(\theta)}^{r_2(\theta)} f(\rho, \theta)\, \rho\, d\rho d\theta.$$

Calculus of double integrals in Cartesian coordinates can be hard. Some of those hard to calculate integrals may become easier if we use polar coordinates.

The coordinate change between Cartesian and polar coordinates is done by the following formula:

$$\iint_R f(x, y) dA = \int_\alpha^\beta \int_{g_1(\theta)}^{g_2(\theta)} f(\rho \cos\theta, \rho \sin\theta)\, \rho\, d\rho d\theta,$$

where α, β, $g_1(\theta)$, and $g_2(\theta)$ are obtained using the polar representation of the region R.

Triple integrals

Let G be a limited closed space region and $f(x, y, z)$ a continuous function in G. Dividing G into various small parallelepipeds with faces parallel to the coordinate planes and numbering the ones that are completely contained in G from 1 to n, consider ΔV_k as the volume of the kth parallelepiped and $(x_k^\star, y_k^\star, z_k^\star)$ a point from its interior. If we consider the sum (see Fig. 7.8)

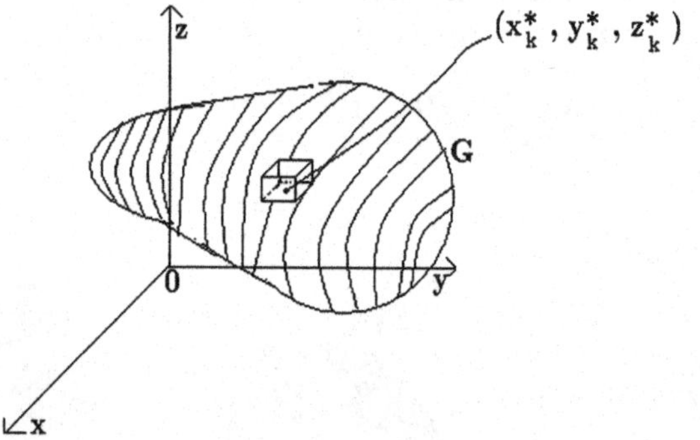

Figure 7.8 Triple integral.

$$S_n = \sum_{k=1}^{n} f(x_k^\star, y_k^\star, z_k^\star)\Delta V_k,$$

then

$$\lim_{n\to+\infty} S_n = \iiint_G f(x, y, z)\, dV$$

is called **triple integral** of $f(x, y, z)$ on the space region G.

If we consider $f(x, y, z) = 1$, then

$$S_n = \sum_{k=1}^{n} \mathbf{1}\, \Delta V_k = \sum_{k=1}^{n} \Delta V_k,$$

thus

$$\lim_{n\to S_n} = \iiint_G dV$$

and we obtain the volume of the space region G.

Triple integrals satisfy properties that are similar to the ones previously referred to for double integrals.

Triple integrals calculus

The simplest case is when G is a parallelepiped with faces parallel to the coordinate axes, i.e.,

$$G = \{(x, y, z) \in \mathbb{R}^3 : a \le x \le b, c \le y \le d, e \le z \le f\}.$$

If $f(x, y, z)$ is continuous in G, then

$$\iiint_G f(x, y, z)\, dV = \int_a^b \int_c^d \int_e^f f(x, y, z)\, dzdydx.$$

The integration order many be switched, which gives us more calculus possibilities.

Suppose that G is a closed space region limited by a line parallel to the z-axis. Consider that the referred line contains an interior point of G and does not intersect more than two points of the G frontier. If we designate by $z = h_1(x, y)$ the upper frontier surface and by R the plane region of XOY in which G is projected, (see Fig. 7.9), then we will have

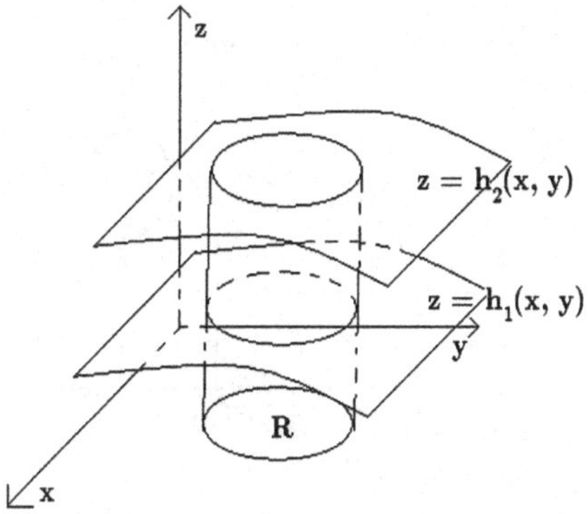

Figure 7.9 R is the projection of G on the XOY-plane.

$$\iiint_G = \iint_R \left[\int_{h_1(x,y)}^{h_2(x,y)} f(x, y, z) dz \right] dA.$$

For certain regions it can be useful to start by calculating the triple integral on the x-variable or on the y-variable. For example, if the solid G is limited at the left side by $y = h_1(x, z)$ and at the right side by $y = h_2(x, z)$ (see Fig. 7.10), then

$$\iiint_G f(x, y, z)\, dV = \iint_R \left[\int_{h_1(x,z)}^{h_2(x,z)} f(x, y, z)\, dy \right] dA,$$

where R is the projection of G on the XOZ-plane.

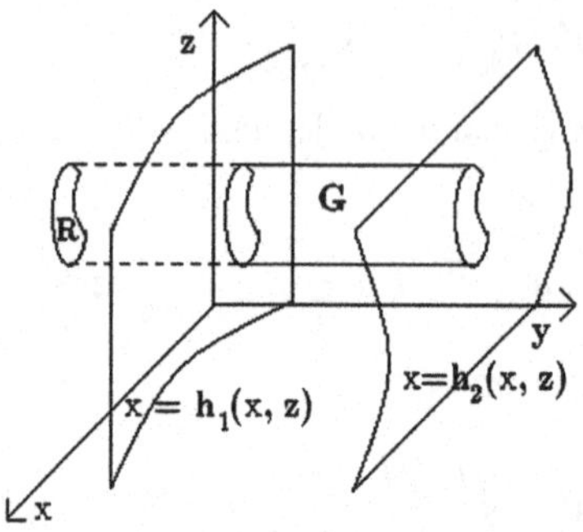

Figure 7.10 R is the projection of G on the XOZ-plane.

If the solid G is limited by $x = h_1(y, z)$ and by $x = h_2(y, z)$ (see Fig. 7.11), then

$$\iiint_G f(x, y, z)\, dV = \iint_R \left[\int_{h_1(y,z)}^{h_2(y,z)} f(x, y, z)\, dx \right] dA,$$

where R is the projection of G on the YOZ-plane.

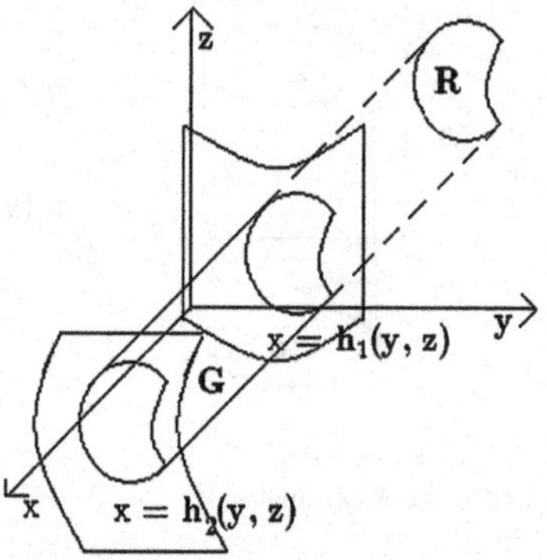

Figure 7.11 R is the projection of G on the YOZ-plane.

Triple integrals in cylindrical and spherical coordinates

Cylindrical coordinates of a point P are given by the triplet (ρ, θ, z), $\rho > 0$, where $\theta \in [0, 2\pi[$. Some surfaces have simple equations in cylindrical coordinates, for example:

- $\rho = r_1$: vertical cylinder with radius r_1, centered at the z-axis;
- $\theta = \theta_1$: vertical semiplane that makes an angle of θ_1 radians with the positive part of the x-axis;
- $z = z_1$: horizontal plane that contains the point $(0, 0, 1)$.

Theorem 7.4. *Let G be a space region whose cylindrical representation is given by the following inequalities:*

$$\begin{aligned} &\theta_1 \le \theta \le \theta_2, \qquad (\theta_2 - \theta_1 \le 2\pi),\\ &0 \le g_1(\theta) \le \rho \le g_2(\theta),\\ &h_1(\rho,\theta) \le z \le h_2(\rho,\theta), \end{aligned}$$

with g_1, g_2, h_1, and h_2 being functions with first-order continuous derivative. If $f(x, y, z)$ is continuous in G, then

$$\iiint_G f(x,y,z)\,dV = \int_{\theta_1}^{\theta_2}\int_{g_1(\theta)}^{g_2(\theta)}\int_{h_1(\rho,\theta)}^{h_2(\rho,\theta)} f(\rho\cos\theta, \rho\sin\theta, z)\,\rho\,dz d\rho d\theta.$$

Spherical coordinates of a point P are given by the triplet (ρ, θ, ϕ), $\rho > 0$, where $\theta \in [0, 2\pi[$ and $\phi \in [0, \pi]$. Some space surfaces that have simple spherical coordinates are:

- $\rho = r_1$: sphere of center (0, 0, 0) and radius r_1;
- $\theta = \theta_1$: vertical semiplane that makes an angle of θ_1 radians with the positive part of the x-axis (as in cylindrical coordinates);
- $\phi = \phi_1$: circular cone with vertices at the origin around the z-axis.

Theorem 7.5. *Let G be a space region whose spherical representation is given by the following inequalities:*

$$\begin{aligned} &\theta_1 \le \theta \le \theta_2, \qquad (\theta_2 - \theta_1 \le 2\pi),\\ &0 \le g_1(\theta) \le \phi \le g_2(\theta) \le \pi,\\ &h_1(\theta,\phi) \le \rho \le h_2(\theta,\phi), \end{aligned}$$

where g_1, g_2, h_1, h_2 are functions with first-order continuous derivative. If $f(x, y, z)$ is continuous in G, then

$$\iiint_G f(x,y,z)\,dV =$$

$$= \int_{\theta_1}^{\theta_2}\int_{g_1(\theta)}^{g_2(\theta)}\int_{h_1(\theta,\phi)}^{h_2(\theta,\phi)} f(\rho\sin\phi\cos\theta, \rho\sin\phi\sin\theta, \rho\cos\phi)\,\rho^2\sin\phi\,d\rho d\phi d\theta.$$

7.2 Applications of multiple integrals

A simple integral finds the area under a curve in one dimension. Applications of multiple integrals arise whenever an area or a volume under a 2D curve (a surface) needs to be measured. Similarly a triple integral can be used to find the sum of some value in a solid (if you are given a density function then the total mass of a solid can be calculated). The basic applications of double integrals is finding volumes. The basic application of the triple integral is finding the mass of a solid. Generally, a solid has some mass but it depends on its density as the density is not constant but varying.

Let $f(x, y)$ be a function defined on a closed region R of the plane XOY. If we intend to determine the area of the portion of the surface $z = f(x, y)$ whose projection on the XOY-plane is the region R, we will be able to do this using double integrals, as guaranteed by the theorem below.

Theorem 7.6. *If $f(x, y)$ has first-order partial derivatives continuous on the closed region R of the plane XOY, then the area S of the portion of the surface $z = f(x, y)$ whose projection on the plane XOY is the region R is given by*

$$S = \iint_R \sqrt{\left(\frac{\partial z}{\partial x}\right)^2 + \left(\frac{\partial z}{\partial y}\right)^2 + 1}\, dA.$$

Multiple integrals are also very useful to determine the mass, moments, and gravity centers of certain plane thin objects. A plane thin object is often called a **blade** (see Fig. 7.12).

Figure 7.12 Blade.

Blades can be homogeneous if the object composition and structure are uniformly distributed or nonhomogeneous in any other case.

The density of a homogeneous blade is defined by $\delta = \frac{M}{A}$, where the object mass is given by M and has area A.

In a homogeneous blade the density at each point is given by a $\delta(x, y)$ function called *density function.*

To calculate the *blade's mass* consider a blade with density function $\delta(x, y)$, continuous, defined in the region R of the XOY-plane. Its total mass is given by

$$M = \iint_R \delta(x, y)\, dA.$$

The moments regarding the different axes are given by:

◇ *First moment blade in relation to the x-axis,*

$$M_x = \iint_R y\delta(x, y)\, dA;$$

◇ *First moment blade in relation to the y-axis,*

$$M_y = \iint_R x\delta(x, y)\, dA;$$

and the *blade's gravity center* is the point $(\overline{x}, \overline{y})$ where

$$\begin{cases} \overline{x} = \dfrac{M_y}{M} = \dfrac{\iint_R x\delta(x, y)\, dA}{\iint_R \delta(x, y)\, dA}, \\ \overline{y} = \dfrac{M_x}{M} = \dfrac{\iint_R y\delta(x, y)\, dA}{\iint_R \delta(x, y)\, dA}. \end{cases}$$

Consider now that we have a *3D solid* (Fig. 7.13) G and $\delta(x, y, z)$ is its continuous density function.

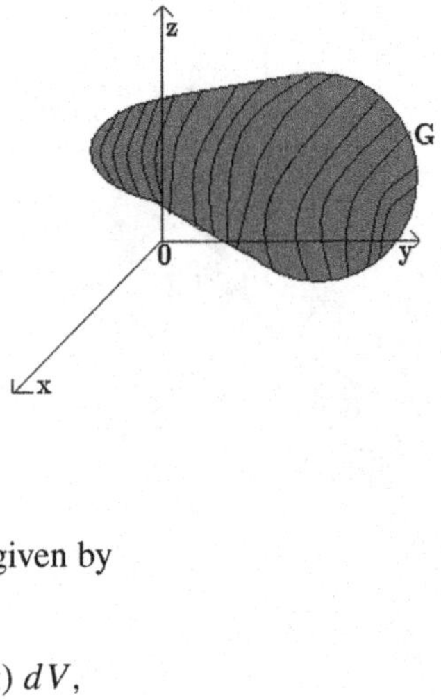

Figure 7.13 Solid.

The *mass* of solid G is given by

$$M = \iiint_R \delta(x, y, z)\, dV,$$

and its *gravity center* is the point $(\overline{x}, \overline{y}, \overline{z})$ where

$$\begin{cases} \overline{x} = \dfrac{1}{M} = \iiint_G x\, \delta(x, y, z)\, dV, \\ \overline{y} = \dfrac{1}{M} = \iiint_G y\, \delta(x, y, z)\, dV, \\ \overline{z} = \dfrac{1}{M} = \iiint_G z\, \delta(x, y, z)\, dV. \end{cases}$$

7.3 Real problems

In this section two real applied examples will be presented: one related to the mass and gravity center of an object and the other related to a viscometer design whose rotation speed has to reach a certain value.

7.3.1 Mass center of an object

Consider an object whose density is given by xz and a tetrahedron with vertices $(0, 0, 0), (1, 1, 0), (0, 1, 0), (0, 1, 1)$. In Fig. 7.14 the surface xz is drawn together with the tetrahedron and the auxiliary plane $x - y + z = 0$ which coincides with one of the tetrahedron's laterals. What we are asked to determine is the mass of the solid that is inside the tetrahedron and also its gravity center.

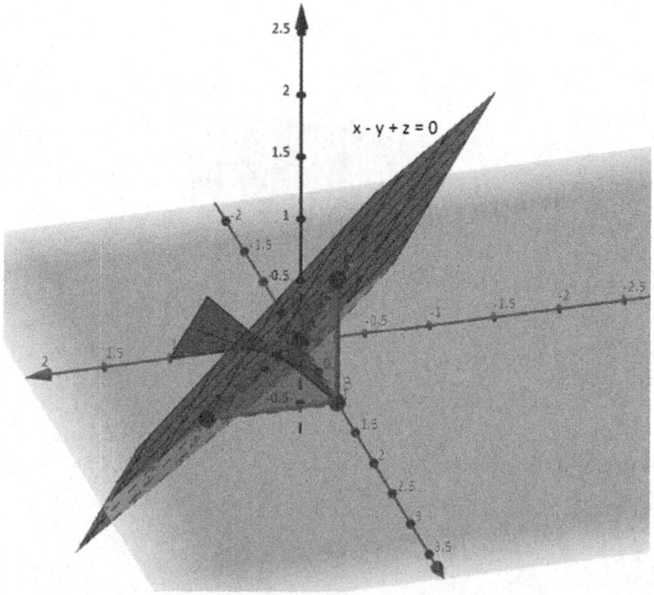

Figure 7.14 Tetrahedron and solid density.

As we have defined above, the object mass will be given by

$$
\begin{aligned}
M &= \int_0^1 \int_x^1 \int_0^{y-x} xz \, dz dy dx \\
&= \int_0^1 \int_x^1 \frac{x(y-x)^2}{2} dy dx
\end{aligned}
$$

$$= \frac{1}{2}\int_0^1 \frac{x(1-x)^3}{3} dy$$
$$= \frac{1}{120},$$

and the gravity center (Fig. 7.15) is the point $GC(\overline{x}, \overline{y}, \overline{z})$ where

$$\begin{cases} \overline{x} = \frac{1}{M} = \iiint_G x\delta(x, y, z)\, dV = 120 \int_0^1 \int_x^1 \int_0^{y-x} xz^2\, dzdydx = \frac{1}{3}, \\ \overline{y} = \frac{1}{M} = \iiint_G y\delta(x, y, z)\, dV = 120 \int_0^1 \int_x^1 \int_0^{y-x} xyz\, dzdydx = \frac{5}{6}, \\ \overline{z} = \frac{1}{M} = \iiint_G z\delta(x, y, z)\, dV = 120 \int_0^1 \int_x^1 \int_0^{y-x} x^2 z\, dzdydx = \frac{1}{3}. \end{cases}$$

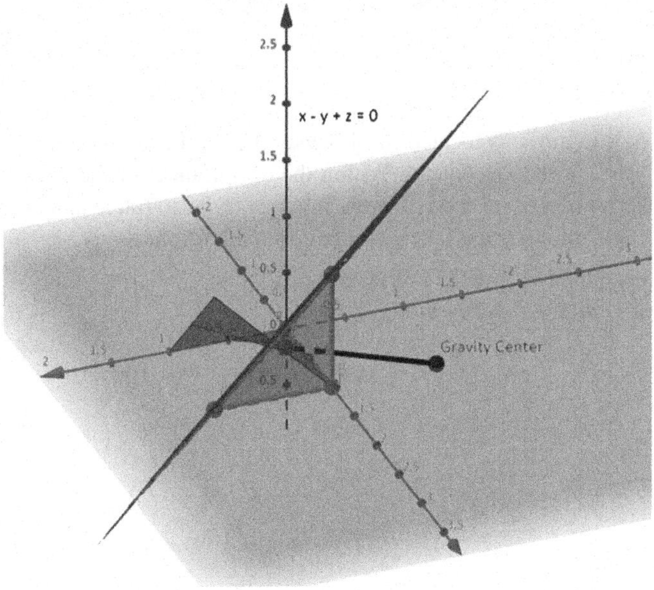

Figure 7.15 Object gravity center.

7.3.2 Viscometer design

A viscometer is a laboratory device used to measure viscosity and other important properties of fluids. A team of engineers specialized in instrumentation equipment intends to design a viscometer consisting of an inner cylinder that rotates with respect to a vessel (stationary cylinder or outer drum), as shown in Fig. 7.16. Both cylinder and vessel are coaxially coupled in such a way that there is a gap between them in which

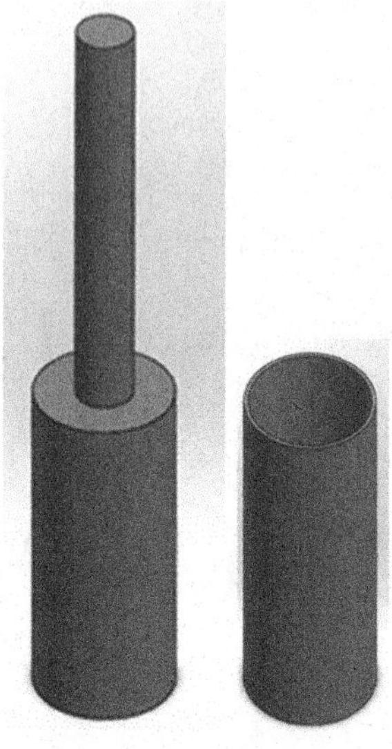

Figure 7.16 Inner and outer cylinders.

the fluid to be examined is located. In this way the fluid is distributed on the lateral sides and bottom of the viscometer. Within the design tasks of the device, engineers want to know the power necessary for the inner cylinder to rotate at a constant speed of 115 rpm. The following are taken as reference data for the study:

- the reference fluid is SAE 30 lubricating oil;
- dimensions are those shown in Fig. 7.17.

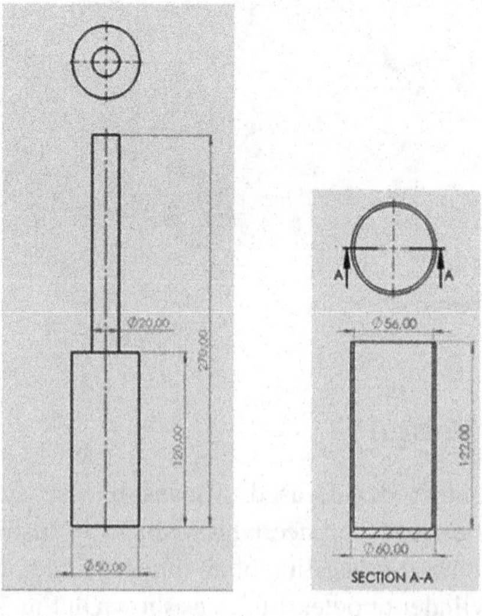

Figure 7.17 Inner and outer cylinders' dimensions.

Viscosity of SAE 30 oil can be consulted by the engineer in the viscosity curves/tables available in the literature (e.g., White (2015)). The dynamic viscosity at 20°C is approximately 0.290 Pa·s. To calculate the required power (P), the previous calculation of torque (T) is needed. Please note that, in this case, as rotation speed is known for this rigid rotatory solid, required power can be computed using the expression $P = wT$. In this problem, to obtain the total torque, the effect of the viscosity of the fluid must be determined in two steps, considering each one of the two surfaces of the moving cylinder in contact with the fluid: lateral surface and bottom surface (Fig. 7.18). For the problem, the radius of the rotating cylinder is known (named r) and the oil film thickness will be named l_i. Based on the geometry of the two parts of the viscometer, the oil film thickness of the lateral surface of the cylinder with respect to the hole (l_1) is 3 mm for the side of the cylinder and 2 mm for the bottom surface of the cylinder (l_2) (Fig. 7.18).

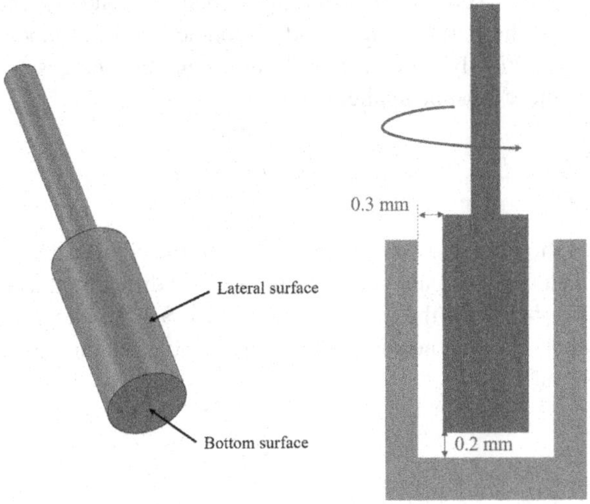

Figure 7.18 (left) Both rotating cylinder surfaces in contact with the oil. (right) The clearance between cylinder and reservoir to be occupied by the oil. As the reader can see, the clearance is not the same on the lateral surface as on the bottom surface.

The dynamic viscosity for a Newtonian fluid can be obtained as $\mu = \frac{\tau}{\frac{dv}{dy}}$, where τ is the shear stress applied to the fluid surface and $\frac{dv}{dy}$ is the velocity gradient normal to the wall.

1. **Lateral surface (side of the rotating cylinder)**
   Torque differential (dT) can be calculated for a surface differential (dA) as $dT = rdF$, while effort differential (dF) can be calculated for a (dA) using $dF = \tau dA$. So, the mathematical integration process will be needed to solve this problem. We have

$$dF = \tau dA = \mu \frac{dv}{dy} dA.$$

In this case, a lineal distribution of velocities is considered. Please note that, in this case, the velocity gradient $\left(\frac{dv}{dy}\right)$ is constant (also considering the nonslip condition): it can be considered as $\frac{dv}{dy} = \frac{v}{l_1}$. Velocity can be included in function of the rotation speed ($v = wr$) and

$$dF_1 = \mu \frac{v}{l_1} dA = \mu \frac{wr}{l_1} dA.$$

It implies that the equation to obtain the torque differential is

$$dT_1 = \mu \frac{wr^2}{l_1} dA.$$

Note that r is really a constant because the velocity does not depend on the radius in this case (all the points of the solid in contact with the fluid have the same velocity). Consequently, to calculate T, an integration process over the lateral surface of cylinder must be applied, i.e.,

$$T_1 = \iint_A \mu \frac{wr^2}{l_1} dA = \mu \frac{wr^2}{l_1} \iint_A dA.$$

As the integration region is the lateral surface of the cylinder in this case, we can consider r as a constant parameter and the surface differential as $dA = dh\ r\ d\theta$, where h is the height of the cylinder (Fig. 7.19). So, the integration process parameters which are then needed to define the whole region are $0 \le \theta \le 2\pi$ and $0 \le h \le 0.12$ m.

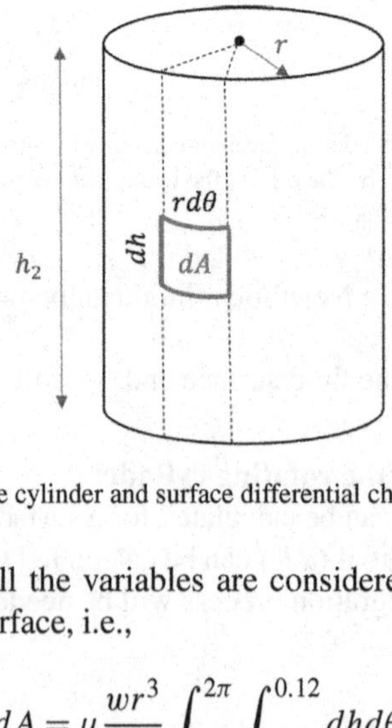

Figure 7.19 Lateral surface of the cylinder and surface differential chosen to parameterize the region.

Numerical values for all the variables are considered to calculate the torque in relation to the lateral surface, i.e.,

$$T_1 = \mu \frac{wr^2}{l_1} \iint_A dA = \mu \frac{wr^3}{l_1} \int_0^{2\pi} \int_0^{0.12} dh d\theta$$

$$= \mu \frac{wr^3}{l_1} \int_0^{2\pi} [h]_0^{0.12} d\theta = \mu \frac{wr^3}{l_1} (0.12)[\theta]_0^{2\pi}$$
$$= \mu \frac{wr^3}{l_1} (0.12)(2\pi)$$

Considering the rest of the parameters, which are known, the torque can be calculated as follows:

$$T_1 = 0.290\ (\text{Pa·s})\ \frac{12.043\ (\text{s}^{-1})\ (0.025\ (\text{m}^3))}{3 \cdot 10^{-3}\ (\text{m})} (0.120\ (\text{m}))[2\pi - 0]$$
$$= 1.371 \cdot 10^{-2}\ \text{Nm}.$$

*Note that rotary speed is included in rad/s.
Please note as conclusion that $\iint dA$ is really the area of the lateral surface of the cylinder in this case.

2. *Inferior surface (bottom)*
Torque also can be computed as $dT = r dF$, while dF can be calculated as $dF = \tau dA$. For this aim, the integration mathematical process is also needed but with some differences that will be indicated below. A lineal distribution of velocities is also considered $\left(\frac{dv}{dy} = \frac{v}{l_1}\right)$. Considering that velocity can be included in the function of the rotation speed ($v = wR$) and the rest of the assumptions also established for the lateral surface, torque differential can be obtained in the following way[1]:

$$dT_2 = \mu \frac{wR^2}{l_2} dA.$$

The difference between this case and the previous one is that the velocity of each fluid particle in contact with the bottom surface of the cylinder now has a different velocity. This velocity varies according to the distance between the point and the center of the circle of the base (Fig. 7.20) (that distance is referred to as $R : v = wR$). Consequently, now the radius is not a constant parameter, but a variable to consider within the integration process $0 \leq R \leq 0.025$ m. To include this parameter in the integral, a new surface differential is established, $dA = R\, d\theta dR$, as shown in Fig. 7.20. We have

$$T_2 = \iint_A \mu \frac{wR^2}{l_2} dA = \mu \frac{w}{l_2} \int_{\theta_1}^{\theta_2} \int_0^r R^3 dR d\theta$$
$$= \mu \frac{wr^2}{l_2} \int_0^{2\pi} \int_0^{0.025} R^3 dR\, d\theta = \mu \frac{w}{l_2} \int_0^{2\pi} \left[\frac{R^4}{4}\right]_0^{0.025} d\theta$$

[1] Please note that in this case, unlike the previous one, R is a variable so it is capitalized, instead of the lowercase r, which is the numerical value of the radius and therefore the integration limit in the following steps.

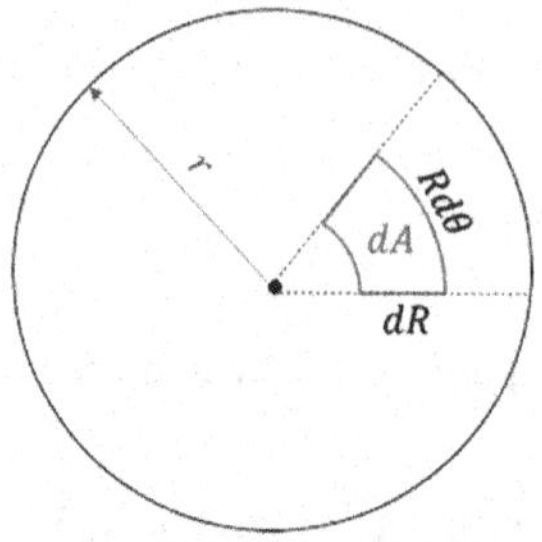

Figure 7.20 Bottom surface of the cylinder to parameterize to solve the integral.

$$= \mu \frac{w}{l_2} \frac{0.025^4}{4} [\theta]_0^{2\pi} = \mu \frac{w}{l_2} \frac{0.025^4}{4} 2\pi.$$

Considering the rest of the parameters which are known, the torque can be calculated (please note that in this case, as the reader can see in Fig. 7.18, the film thickness (l_2) is 2 mm, while for the lateral surface l_1 it was 3 mm). We have

$$T_2 = 0.290\ (\text{Pa}\cdot\text{s}) \frac{12.043\ (\text{s}^{-1})}{2 \cdot 10^{-3}\ (\text{m})} \left(\frac{(0.025\ (\text{m}))^4}{4} \right) 2\pi = 1.071 \cdot 10^{-3}\ \text{Nm}.$$

*Note that rotary speed is included in rad/s.

Total torque can be obtained as the sum of the two partial torques necessary to maintain constant rotary speed ($T = T_1 + T_2$). Finally, the power for this rotating rigid solid can be calculated as the product of total torque and rotary speed ($P = Tw$), i.e.,

$$P = (T_1 + T_2)w = (1.371 \cdot 10^{-2} + 1.071 \cdot 10^{-3})12.043 = 0.178\ \text{W}.$$

As conclusion, the power required for the cylinder to rotate at a constant speed of 115 rpm in the viscometer is 0.178 W.

References

Jerrold, E., Marsden, A.J.T., 2003. Vector Calculus, 5th ed. W. H. Freeman and Company, New York.

Stewart, J., 2016. Calculus. Cengage Learning, Inc.

White, F., 2015. Fluid Mechanics, 8th ed. McGraw–Hill.

Critical forces and collisions. How to solve nonlinear equations and their systems

8

Monika Kováčová, Daniela Richtáriková
Slovak University of Technology in Bratislava, Bratislava, Slovakia

8.1 Preliminaries

Definition 8.1. A function $f : \mathbb{R}^n \to \mathbb{R}$ is called *nonlinear* when it does not satisfy the superposition principle or property of homogeneity, which means

$$f(ax_1 + bx_2 + \ldots) \neq af(x_1) + bf(x_2) + \ldots.$$

To solve an equation with one variable $f(x) = 0$ in $\mathbb{R}$ means to find all numbers $p \in \mathbb{R}$ for which $f(p) = 0$.

A *system of nonlinear equations* is a system of two or more equations with two or more variables containing at least one equation that is *not linear*.

Definition 8.2. A *system of nonlinear equations* is a set of equations

$$\begin{aligned} f_1(x_1, x_2, \ldots, x_n) &= 0 \\ f_2(x_1, x_2, \ldots, x_n) &= 0 \\ &\vdots \\ f_n(x_1, x_2, \ldots, x_n) &= 0, \end{aligned} \tag{8.1}$$

where $(x_1, x_2, \ldots, x_n) \in \mathbb{R}^n$ and at least one f_i is a nonlinear real function, $i = 1, 2, \ldots, n$.

Here is an example of a nonlinear system of two equations:

$$\begin{aligned} f_1(x_1, x_2) &= 0 \\ f_2(x_1, x_2) &= 0, \end{aligned} \qquad \begin{aligned} x_1^2 - 4x_1 - x_2 + 2 &= 0 \\ 2x_1^2 + x_2^2 - 3 &= 0, \end{aligned} \qquad \begin{aligned} x^2 - 4x - y + 2 &= 0 \\ 2x^2 + y^2 - 3 &= 0. \end{aligned}$$

To denote the final result of solving an equation or a system of equations we will use the term root or solution.

Definition 8.3. A *solution of the system of equations* (8.1) with n variables is a vector $(a_1, \ldots, a_n) \in \mathbb{R}^n$ such that $f_1(a_1, \ldots, a_n) = \cdots = f_n(a_1, \ldots, a_n) = 0$.

Nonlinear equations and their systems cannot be solved by a finite number of arithmetic operations; usually iterative methods are used.

Calculus for Engineering Students. https://doi.org/10.1016/B978-0-12-817210-0.00015-1

Definition 8.4. An *iterative method* is a mathematical procedure that uses an initial guess to generate a sequence of improving approximate solutions, in which the nth approximation is derived from the previous ones.

We can simply say that an iterative method is a procedure that is repeated over and over again where the goal is to find the root of an equation or a system of equations.

Definition 8.5. Let $\boldsymbol{\varphi}$ be a vector of real functions from $D \subset \mathbb{R}^n$ to $\mathbb{R}^n$. If $\boldsymbol{\varphi}(\mathbf{p}) = \mathbf{p}$, for some $\mathbf{p} \in D$, then $\mathbf{p}$ is said to be a fixed point of $\boldsymbol{\varphi}$.

A fixed point is a special point in the space of possible solutions. Later, we will show the close connection between fixed points and solutions of system equations.

8.1.1 General principles for iterative methods – or how to jump from one equation to the system of equations

Any nonlinear equation $f(x) = 0$ can be expressed as $x = g(x)$. If x_0 is a proper initial guess for an iterative method, then the solution p can be reached by the numerical sequence

$$x_{i+1} = g(x_i), \quad i = 0, 1, 2, ..., \quad \text{and}$$
$$\lim_{i\to\infty} x_{i+1} = \lim_{i\to\infty} x_i = p.$$

This iteration is termed *Piccard process* and p, the limit of the sequence $\{x_i\}$, is called a *fixed iterative point.* To find a root of the equation, it is necessary that the sequence $\{x_i\}$ converges (see Chapter 4) to its solution. The order of convergence tells us "how fast we will get the solution."

Theorem 8.1. *If $x = g(x)$ has a solution p within an interval I and if $g(x)$ satisfies the Lipschitz condition ($|f(x_i) - f(x_j)| \leq s|x_i - x_j|$, $s \in (0, 1)$, $x_i, x_j \in I$), then for each initial point x_0 from interval I, all iterated values x_i lie in interval I and converge towards p. The solution p is unique within interval I. The theorem represents a* sufficient condition for convergence *and is known as the Banach fixed point theorem.*

Definition 8.6. Let $\{x_i\}$ be a sequence that converges to p, where $x_i \neq p$. If for constants λ and α

$$\lim_{i\to\infty} \frac{|x_{i+1} - p|}{|x_i - p|^\alpha} = \lambda,$$

then it is said that $\{x_i\}$ converges to p of order α with constant λ.

We distinguish three main orders of sequence convergence: linear convergence, quadratic convergence, and superlinear convergence.

To construct the *algorithm that leads to an appropriate fixed point method* in the 1D case, a function φ with the property

$$g(x) = x - \varphi(x) \cdot f(x) \tag{8.2}$$

can be used. It gives a quadratic convergence to the fixed point p of the function g. If we choose $\varphi(x) = 1/f'(x)$, assuming that $f'(x) \neq 0$, it tends to the well-known *Newton's method,*

$$x_{i+1} = x_i - \frac{1}{f'(x_i)} \cdot f(x_i). \tag{8.3}$$

To explore convergence in the n-dimensional case, a similar approach can be used as in Theorem 8.2, or we can use Theorem 8.3.

Theorem 8.2. *Let $\boldsymbol{\varphi}(\mathbf{x})$ satisfy $\|\boldsymbol{\varphi}(\mathbf{x}) - \boldsymbol{\varphi}(\mathbf{y})\| \leq K\|\mathbf{x}-\mathbf{y}\|$ for all vectors $\mathbf{x}$, $\mathbf{y}$ such that $\|\mathbf{x}-\mathbf{x}^{(0)}\| \leq \varepsilon$, $\|\mathbf{y}-\mathbf{y}^{(0)}\| \leq \varepsilon$ with the Lipschitz constant, K, satisfying $0 < K < 1$.*

Let the initial iterate $\mathbf{x}^{(0)}$ satisfy $\|\boldsymbol{\varphi}(\mathbf{x}^{(0)}) - \mathbf{x}^{(0)}\| \leq (1-K)\varepsilon$. Then all iterates $\mathbf{x}^{(i+1)} = \boldsymbol{\varphi}(\mathbf{x}^{(i)})$ hold $\|\mathbf{x}^{(i)} - \mathbf{x}^{(0)}\| \leq \varepsilon$ and $\lim_{i\to\infty} \mathbf{x}^{(i)} = \boldsymbol{\alpha}$, which is a unique root of $\mathbf{x} = \boldsymbol{\varphi}(\mathbf{x})$.

Theorem 8.3. *Let $D = \{(x_1, x_2, ..., x_n)^T, a_i \leq x_i \leq b_i\}$, $i = 1, 2, ..., n$. Suppose $\boldsymbol{\varphi}$ is a continuous function from $D \subset \mathbb{R}^n$ into $\mathbb{R}^n$, $\boldsymbol{\varphi}(\mathbf{x}) \in D$, and $\mathbf{x} \in D$. Then $\boldsymbol{\varphi}$ has a fixed point in D. Moreover, if all component functions of $\boldsymbol{\varphi}$ have continuous partial derivatives, for which*

$$\left|\frac{\partial \varphi_i(\mathbf{x})}{\partial x_j}\right| \leq \frac{K}{n},$$

where constant $K < 1$, then the sequence $\{\mathbf{x}^{(i)}\}_{i=0}^{\infty}$ determined by an arbitrarily selected $\mathbf{x}^{(0)}$ in D and generated by $\mathbf{x}^{(i+1)} = \boldsymbol{\varphi}(\mathbf{x}^{(i)})$, $i \geq 0$, converges to the unique fixed point $\mathbf{p} \in D$ and

$$\left\|\mathbf{x}^{(i)} - \mathbf{p}\right\| \leq \frac{K^i}{1-K}\left\|\mathbf{x}^{(i)} - \mathbf{x}^{(0)}\right\|.$$

Remark. One way to accelerate convergence of the fixed point iteration is to use the latest estimates $x_1^{(i)}, ..., x_{j-1}^{(i)}$ instead of $x_1^{(i-1)}, ..., x_{j-1}^{(i-1)}$ to compute $x_j^{(i)}$, as in the Gauss–Seidel method for linear systems.

The system of equations (8.1) can be expressed in matrix form $\mathbf{F}(x_1, x_2, ..., x_n) = \mathbf{0}$. Hence the desired function $\varphi(\mathbf{x})$ can be defined through the inverse Jacobian matrix $J^{-1}(\mathbf{x})$ and the corresponding Newton's method can be derived as

$$\mathbf{x}^{(i+1)} = \mathbf{x}^{(i)} - J\left(\mathbf{x}^{(i)}\right)^{-1} \cdot \mathbf{F}\left(\mathbf{x}^{(i)}\right), \tag{8.4}$$

where $i = 1, 2, ..., n$ represents an iteration number, $\mathbf{x}$ is a vector in $\mathbb{R}^n$, $\mathbf{F}(\mathbf{x})$ is a vector function, and $J(\mathbf{x})^{-1}$ is the inverse matrix of the Jacobian matrix.

Too complicated? Let us go step by step.

8.1.2 Newton's method for systems of nonlinear equations again and properly

First, we will discuss Eq. (8.4) and explain all its components.

- Let $\mathbf{F}$ be a function such that $\mathbf{F}: \mathbb{R}^n \to \mathbb{R}^n$ and

$$\mathbf{F}(\mathbf{x}) = \mathbf{F}(x_1, x_2, ..., x_n) = \begin{bmatrix} f_1(x_1, x_2, ..., x_n) \\ f_2(x_1, x_2, ..., x_n) \\ \vdots \\ f_n(x_1, x_2, ..., x_n) \end{bmatrix},$$

  where $f_i : \mathbb{R}^n \to \mathbb{R}$.
- Let $\mathbf{x} \in \mathbb{R}^n$. Then $\mathbf{x}$ represents a vector

$$\mathbf{x} = [x_1, x_2, ..., x_n]^{\mathrm{T}}.$$

- Let $J(\mathbf{x})$ be the Jacobian matrix. Thus its inverse $J(\mathbf{x})^{-1}$ is

$$J(\mathbf{x})^{-1} = \begin{bmatrix} \frac{\partial f_1}{\partial x_1}(\mathbf{x}) & \frac{\partial f_1}{\partial x_2}(\mathbf{x}) & \dots & \frac{\partial f_1}{\partial x_n}(\mathbf{x}) \\ \frac{\partial f_2}{\partial x_1}(\mathbf{x}) & \frac{\partial f_2}{\partial x_2}(\mathbf{x}) & \dots & \frac{\partial f_2}{\partial x_n}(\mathbf{x}) \\ \vdots & \vdots & \dots & \vdots \\ \frac{\partial f_n}{\partial x_1}(\mathbf{x}) & \frac{\partial f_n}{\partial x_2}(\mathbf{x}) & \dots & \frac{\partial f_n}{\partial x_n}(\mathbf{x}) \end{bmatrix}^{-1}.$$

  Let us describe each step in Newton's method.

Step 1: Take an initial guess.
Let $\mathbf{x}^{(0)} = [x_1^{(0)}, x_2^{(0)}, ..., x_n^{(0)}]^{\mathrm{T}}$ be a given initial vector, $i = 0$.

Step 2: Calculate $\mathbf{F}(\mathbf{x}^{(0)})$ and $J(\mathbf{x}^{(0)})^{-1}$ or (due to computational problems) go to step 4.

Step 3: Denote

$$\mathbf{x}^{(1)} = \mathbf{x}^{(0)} - J\left(\mathbf{x}^{(0)}\right)^{-1} \cdot \mathbf{F}\left(\mathbf{x}^{(0)}\right)$$

and in general

$$\mathbf{x}^{(i+1)} = \mathbf{x}^{(i)} - J\left(\mathbf{x}^{(i)}\right)^{-1} \cdot \mathbf{F}\left(\mathbf{x}^{(i)}\right).$$

Skip to step 6.
In practice, usually explicit computation of $J(\mathbf{x})^{-1}$ is avoided by performing the operation in the following two-step manner.

Step 4: Calculate vector $\mathbf{y}^{(0)}$, where $\mathbf{y} = [y_1, y_2, ..., y_n]^{\mathrm{T}}$, $\mathbf{y}^{(0)} = -J(\mathbf{x}^{(0)})^{-1}\mathbf{F}(\mathbf{x}^{(0)})$.
In order to find $\mathbf{y}^{(0)}$, we solve the linear system $J(\mathbf{x}^{(0)}).\mathbf{y}^{(0)} = -\mathbf{F}(\mathbf{x}^{(0)})$, using, e.g., Gaussian elimination or another proper method.

Step 5: From (8.4)

$$\mathbf{x}^{(1)} = \mathbf{x}^{(0)} + \mathbf{y}^{(0)} = \begin{bmatrix} x_1^{(0)} \\ x_2^{(0)} \\ \vdots \\ x_n^{(0)} \end{bmatrix} + \begin{bmatrix} y_1^{(0)} \\ y_2^{(0)} \\ \vdots \\ y_n^{(0)} \end{bmatrix}$$

and in general

$$\mathbf{x}^{(i+1)} = \mathbf{x}^{(i)} + \mathbf{y}^{(i)}.$$

Step 6: The process is repeated until $\mathbf{x}^{(i)}$ converges to $\mathbf{p}$. Usually it stops when a stopping condition holds, for instance, $\|\mathbf{y}^{(i)}\| = \|\mathbf{x}^{(i+1)} - \mathbf{x}^{(i)}\| < \varepsilon$, supposing that $\lim_{i\to\infty} \|\mathbf{y}^{(i)}\| = 0$. Then the root $\mathbf{p}$ is approximated by the output $\mathbf{x}^{(i+1)}$ with respect to a tolerance ε, where $\varepsilon > 0$ represents the required volume of accuracy. The norm can stand for Euclidean norm

$$\|\mathbf{x}^{(i+1)} - \mathbf{x}^{(i)}\| = \sqrt{\left(x_1^{(i+1)} - x_1^{(i)}\right)^2 + \ldots + \left(x_n^{(i+1)} - x_n^{(i)}\right)^2}.$$

This indicates we have reached the solution/numerical solution of $\mathbf{F}(x) = 0$, where $\mathbf{p}$ is the root of the system (8.1).

Remark. The iteration $x^{(i+1)} = x^{(i)} + y^{(i)}$, where $y^{(i)} = -J(\mathbf{x}^{(i)})^{-1} \cdot \mathbf{F}(\mathbf{x}^{(0)})$, is known as the local Newton's method.

8.1.3 Convergence of Newton's method. Advantages and disadvantages of Newton's method

The advantage of Newton's method is its speed of convergence once a sufficiently accurate initial guess is known. The method converges quadratically (Burden and Faires, 2011). The second advantage is that it is not too complicated in form and it can be used to solve a variety of problems.

On the other hand, a weakness of this method is that initial approximation has to be relatively close to the solution in order to converge. The other major disadvantage associated with Newton's method in the case of systems is the fact that $J(\mathbf{x})$, as well as its inversion, has to be calculated at each iteration. Calculating the Jacobian matrix and its inverse can be quite time consuming, depending on the size of the system. Another problem, which we may be confronted with, is that the method may still fail to converge, for instance, the linear systems in step 3 may be ill-conditioned or singular.

In the case Newton's method fails to converge and it results in an oscillation between points, one can think about three possible reasons: an error in computations has occurred, a wrong initial point was chosen, or a solution may not exist.

Generally, three conditions can be formulated as a convergence assumption. The method requires

1. nonsingular $J(\mathbf{x})$,
2. sufficiently smooth first derivatives,
3. sufficiently close initial guess $\mathbf{x}^{(0)}$ to the root $\mathbf{p}$.

If the initial iterate $\mathbf{x}^{(0)}$ is chosen far from the root $\mathbf{p}$, then Newton's algorithm may either require too many iterations to converge, or it may even not converge at all.

Another major disadvantage of Newton's method is that the algorithm is *locally convergent.* One can test on the examples in Section 8.3.2 that if a different initial guess is chosen, a different result or no result is reached. Since the location of solution $\mathbf{p}$ may not be known, it is generally difficult to choose an initial iterate $\mathbf{x}^{(0)}$ close to $\mathbf{p}$. There are several modifications of the local Newton's method to make it globally convergent. Such modifications of Newton's algorithm are known as *globalized Newton's method*, also called *Broyden's method.* In subsection 8.1.4 we explain one possible way how the local Newton's algorithm should be modified so that it is possible to set any arbitrary initial iterate $\mathbf{x}^{(0)}$. We will speak about the simpler one, where step lengths are used.

In the case of single nonlinear equations $f(x)=0$, the situation is less complicated. For convergence of the iteration process (8.3) we require determination of an interval (a,b), where

1. for any x: $a<x<b$, $f(x)$ and its derivative $f'(x)$ are continuous, $f'(x)\neq 0$, $f''(x)\neq 0$ ($f(x)$ is strictly monotone and strictly convex/concave). The initial guess $a<x_0<b$;
2. a root p exists, $a<p<b$, $f(p)=0$.

The convergent iteration process can be terminated when a stopping condition holds (for instance, $|x_{i+1}-x_i|<\varepsilon$, $|f(x_{i+1})|<\varepsilon$, or estimation of error $\frac{|f(x_{i+1})|}{m}<\varepsilon$, $m=\min|f'(x)|$, $a<x<b$, $\varepsilon>0$).

Remark. In the case the root lies in close vicinity of an inflex point ($f''(x)=0$), the condition of strict convexity/concavity cannot hold, and we have to be aware of possible cycling between values.

8.1.4 Modifications of Newton's algorithm. Globalization using step lengths

The iteration $x^{(i+1)}=x^{(i)}+y^{(i)}$, where $y^{(i)}=-J(\mathbf{x}^{(i)})^{-1}\cdot\mathbf{F}(\mathbf{x}^{(0)})$, known as the local Newton's method, is modified to the term

$$x^{(i+1)}=x^{(i)}+\alpha^{(i)}\cdot y^{(i)},$$

where $\alpha^{(i)}$ is known as a step length to be determined at each iteration. It works (sufficient decrease condition) if each new iterate is determined in such a way that

$$\|\mathbf{F}(\mathbf{x}^{(i+1)})\|^2<(1-\sigma\alpha^{(i)})\|\mathbf{F}(\mathbf{x}^{(i)})\|^2, \tag{8.5}$$

where σ is a given constant. This requirement represents a sufficient decrease condition which guarantees that the norm $\|\mathbf{F}(\mathbf{x})\|$ reduces sufficiently from iteration to iteration, until the process converges to the root with given tolerance.

Note that if we define $\mathbf{f}(\mathbf{x}) = \frac{1}{2}\|\mathbf{F}(\mathbf{x})\|^2$, then

$$\begin{aligned}\mathbf{f}(x^{(i)} + \alpha^{(i)} y^{(i)}) &= \frac{1}{2}\|\mathbf{F}(\mathbf{x}^{(i+1)})\|^2 \le \frac{1}{2}(1 - \sigma\alpha^{(i)})\|\mathbf{F}(\mathbf{x}^{(i)})\|^2 \\ &\le \frac{1}{2}\|\mathbf{F}(\mathbf{x}^{(i)})\|^2 = \mathbf{f}(\mathbf{x}^{(i)}).\end{aligned}$$

And so $\mathbf{f}(x^{(i)} + \alpha^{(i)} y^{(i)}) \le \mathbf{f}(\mathbf{x}^{(i)})$.
Hence $\mathbf{y}^{(i)}$ is a decent direction for $\mathbf{f}(\mathbf{x}) = \frac{1}{2}\|\mathbf{F}(\mathbf{x})\|^2$.

The question is how to choose the step length $\alpha^{(i)}$. If $\alpha^{(1)} = 1$ and $\mathbf{x}^{(i+1)} = \mathbf{x}^{(i)} + \mathbf{y}^{(i)}$ satisfies the requirement (8.5), then $\alpha^{(i)} = 1$ is a good choice. Otherwise, choose an $\alpha^{(i)}$ so that $0 < \alpha^{(i)} < 1$.

In the case the requirement

$$\|\mathbf{F}(\mathbf{x}^{(i+1)})\|^2 \le (1 - 2\sigma\alpha^{(i)})\|\mathbf{F}(\mathbf{x}^{(i)})\|^2, \quad \sigma \in \left(0, \frac{1}{2}\right),$$

the modified Newton's method is known as *Armijo rule method*.
Commonly used constants are $\alpha = 0.5$ and either $\sigma = 10^{-4}$ or $\sigma = 10^{-2}$.

8.1.4.1 *Properties of the modified Newton's method*

If the vector function $\mathbf{F}(\mathbf{x})$ is two times differentiable and the Jacobi matrix $J(\mathbf{x})$ is invertible at each point $\mathbf{x}$, then an arbitrary initial iterate $\mathbf{x}^{(0)}$ can be chosen. The sequence of iterates $\{\mathbf{x}^{(i)}\}$ converges to the solution $\mathbf{p}$ superlinearly. After a certain number of iterations, it is possible to choose a step length $\alpha^{(i)} = 1$, i.e., the modified Newton's method behaves just like Newton's method with initial guess near the solution.

There are also disadvantages. First, the Jacobian matrix $J(\mathbf{x})$ is required to have an analytic expression or to be exactly computable. Second, the good convergence properties are guaranteed only if the Jacobian matrix $J(\mathbf{x})$ is invertible. And after all, at each iteration step (i), the system of linear equations $J(\mathbf{x}^{(i)}) \cdot \mathbf{y}^{(i)} = -\mathbf{F}(\mathbf{x}^{(i)})$ should be exactly solved in order to obtain $\mathbf{y}^{(i)}$. These are serious issues when solving large-scale systems of nonlinear equations in practical engineering applications. Mainly for large-scale systems of nonlinear equations, the direct computations of Jacobian could take enormous CPU time. Three frequently used methods to approximately compute the Jacobian follow finite difference approximation, quasi-Newton's methods (with variants including the Sherman–Morrison formula), and automatic differentiation.

Finally, let us note there exist also other strategies of solving nonlinear systems of equations. The advantage of the Newton's and quasi-Newton's methods for solving systems of nonlinear equations is their speed of convergence once a sufficiently accurate approximation is known. A weakness of these methods is that a very close initial guess to the solution is needed to ensure convergence. To compare, for instance, the steepest descent method converges only linearly to the solution, but it will usually converge even for poor initial approximations. As a consequence, this method is used to find sufficiently accurate starting approximations for the Newton-based techniques in the same way the bisection method is used for a single equation.

8.2 Why the nonlinear systems of equations are important. How important it is to study nonlinear systems

Naturally, almost all processes are in their essence nonlinear. They are present in different areas of everyday life including nature, industrial processes, economic data, biology, life sciences, medicine and health care, etc. Exploring them, we notice a big group of processes that are usually linearized and then solved as linear systems without losing their main character. But there is also a not smaller group that represent complex events with often unpredictable behaviors in natural surroundings whose nonlinear character cannot be omitted. From a mathematical point of view, many complex problems in science and engineering require the solution of nonlinear equations and their systems. Here are a few properties that are not found in linear systems but they occur in nonlinear systems:

- *Finite escape time*: The state of a nonlinear system can go to infinity in finite time.
- *Multiple equilibria*: A nonlinear system can have multiple isolated equilibrium points.
- *Limit cycles*: Nonlinear systems can go into an oscillation of fixed amplitude and frequency independent of the initial conditions or linearized eigenvalues of the system.
- *Periodic oscillations*: nonlinear systems may undergo periodic excitation and they can oscillate with frequencies that are completely unrelated to the input excitation frequency.
- *Chaos/chaotic dynamics*: A nonlinear system with sensitivity to initial conditions that diverge very rapidly is called chaotic.

The study of nonlinear processes belongs to the biggest challenges of science. The nonlinear characteristic of the systems can cause abrupt changes due to some slight modifications of valid parameters, including time, and it is not easy to control them. Moreover, it is difficult to develop general modeling methods due to the enormous variety of problems, which are subject to uncertainties and disturbances. Anyway, numerical methods can help us to cross these uncertainties. Once adequate system identification is achieved, the resulting model can be used for various purposes, such as for control design, state variable estimation, modeling and prediction, finding the fixed point of the system, modeling stable states, or looking for existence and uniqueness conditions.

For the modeling and solving of linear systems, well-structured theories, methodologies, and algorithms were developed. In the case of nonlinear systems, the situation is more complex. In this sense, the chosen method can be only as good as the mathematical model it utilizes, depending on system input–output relations which are measured. The result depends on the selection of numerical method and starting points. In addition, all these facts must be considered already in the process of choosing the proper method, and before the analysis of the engineering problem that is to be solved.

Many problems in science and engineering need a formulation through nonlinear equations or through a system of nonlinear equations. Numerical methods are used when exact solutions cannot be determined via algebraic methods. They construct successive approximations that aim to converge to the exact solution of equations or their systems. The principle consists in starting from arbitrary points – the points as close as possible to the solution sought – and involves arriving at the solution gradually through successive tests.

The following two criteria need to be taken into account when choosing a method for solving a nonlinear system:

- method of convergence (conditions of convergence, speed of convergence, etc.),
- the cost of calculating of the method.

We will show in the next section how these simple principles can be used for solving real problems in mechanical engineering.

8.3 Nonlinear equations and their systems in applications

8.3.1 Problems of nonlinear equations

Problem 8.1. Determination of *Euler's critical load* is one of the basic topics in the study of elasticity and strength. The deflection curve of a beam is described by the Euler–Bernoulli second-order differential equation of bending moment $M = -E \cdot J \cdot y''$ from which the equation for bending curve $y''(x) = -M/(E \cdot J)$ follows, resulting in the general solution $y = y_0 + c_1 \cos(kx) + c_2 \sin(kx)$, where E is Young's modulus of elasticity, J is the inertia moment, x is the distance from the bottom of the beam, and $k^2 = \frac{F}{E \cdot J}$, from which the critical load is the lowest possible value of F.

Reflecting various types of beam constructions and possibilities of fixing, its solution often leads to nonlinear equations where trigonometric functions appear. Consider the beam fixed only in a bottom which is composed of two parts of lengths l_1 and l_2 and diameters d_1 and d_2. Taking into account two segments, we solve two equations which produce two general solutions. After installing five boundary conditions (in a fixed end $y(x_1) = 0$, $y'(x_1) = 0$, in a connection point $y_1(l_1) = y_2(l_1)$, $y_1'(l_1) = y_2'(l_1)$, and in a free end $y_2(l_1 + l_2) = y_0$) we come to the equation

$$\tan(k_1 l_1) = \frac{k_2}{k_1} \cdot \cot(k_2 l_2),$$

with $k_2/k_1 = \sqrt{J_1/J_2}$, from which $k_2 = \sqrt{J_1/J_2} \cdot k_1$ and the explored equation will have the form $\tan(kl_1) - \sqrt{\frac{J_1}{J_2}} \cdot \cot\left(\sqrt{\frac{J_1}{J_2}} k l_2\right) = 0$, where the unknown is $k = k_1$.

Let parameters for a given beam be $l_1 = 3$ m; $l_2 = 2$ m; $d_1 = 0.08$ m; $d_2 = 0.05$ m; $E = 2.1 \cdot 10^{11}$ Pa; $J_1 = \frac{\pi d_1^4}{64} = 2.01062 \cdot 10^{-6}$ m^4; $J_1 = \frac{\pi d_1^4}{64} = 2.01062 \cdot 10^{-6}$ m^4. Then we solve the equation

$$-2.56 \cot(5.12k) + \tan(3k) = 0 \tag{8.6}$$

with $f(k) = -2.56 \cot(5.12k) + \tan(3k)$. To determine endpoints of the interval for an iteration process we can plot the graph and estimate the endpoints from the graph, or we can use the principle of changing sign of $f(k)$ when the graph crosses the k axis.

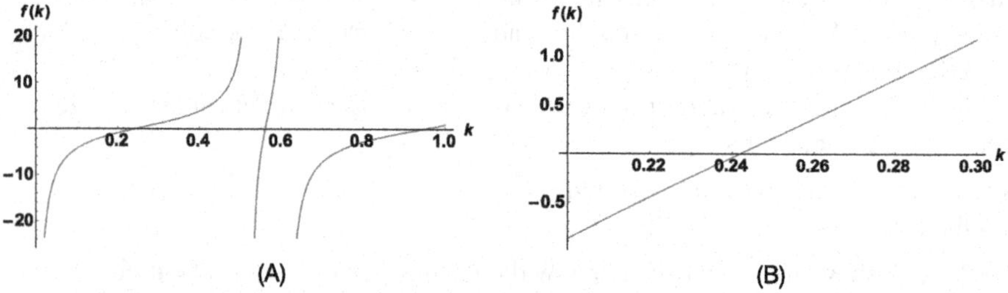

Figure 8.1 Graph of $f(k)$. Estimation of interval (a, b).

Trigonometric functions are periodic functions, so there are infinitely many roots of (8.6) (see Fig. 8.1(A)). We are interested in the smallest positive root, which will determine the critical force. From Fig. 8.1(B) we suggest $(a, b) = (0.19, 0.3)$. For initial guess $k_0 = a = 0.2$, the iteration process (8.3) converges to $k = 0.241707978$ in four iterations. The critical force $F = k^2 \cdot E \cdot J = 24\,667.84$ N.

Problem 8.2. The compressor increases the density of the air or fuel mixture supplied to the turbocharged or spark-ignition engine, increasing its performance. The formula for specific polytrophic work of compressor is expressed as

$$a_{pol} = \frac{n}{n-1} rT \left(\left(\frac{p_2}{p_1} \right)^{\frac{n-1}{n}} - 1 \right).$$

We would like to calculate the value of the *polytrophic index* n, which represents the compressor power if specific polytrophic work $a_{pol} = 67\,800$ J kg^{-1}, the gas constant $r = 288$ J kg^{-1} K^{-1}, static temperature of gas flowing in the suction pipe $T = 298.25$ K, static pressure of gas flowing in suction pipe $p_1 = 1.018 \cdot 10^5$ Pa, and pressure of gas after polytrophic compression $p_2 = 2.032 \cdot 10^5$ Pa. Then we solve the equation

$$\frac{n}{n-1} \cdot 85\,896 \cdot \left(-1 + 1.99607^{\frac{n-1}{n}}\right) = 0$$

with $f(n) = \frac{n}{n-1} \cdot 85\,896 \cdot \left(-1 + 1.99607^{\frac{n-1}{n}}\right)$.

From Fig. 8.2 we suggest $(a, b) = (1.1, 1.8)$ and the stopping condition $|f(x_{i+1})| < 10^{-10}$. For initial guess $n_0 = a = 1.2$, iteration process (8.3) stops

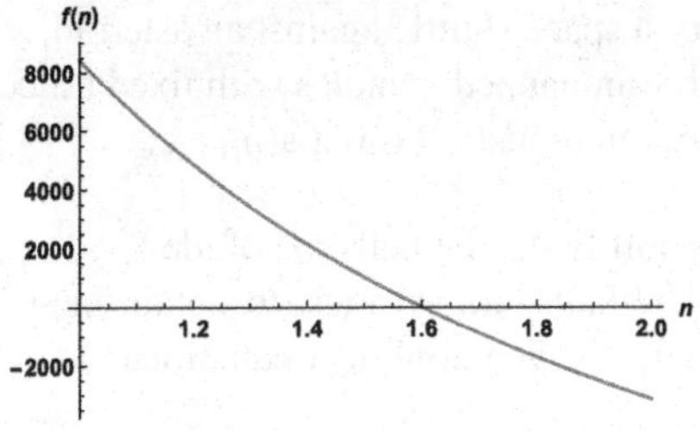

Figure 8.2 Graph of $f(n)$. Estimation of interval (a, b).

for $n \doteq n_5 = 1.60274$ after five iterations. Estimation of error $\frac{|f(x_5)|}{m} = \frac{|f(x_5)|}{|f'(1.8)|} = 1.86756 \times 10^{-15}$.

8.3.2 Problems on systems of nonlinear equations

8.3.2.1 A simple numerical example of Newton's method for systems of nonlinear equations. Collision avoidance for unmanned vehicles with fixed trajectories

Example (hypothetical): Suppose, two balls are moving in the (x, y)-plane along fixed trajectories (Fig. 8.3):

- trajectory of ball 1: $\{[x, y] \mid x^2 - 4x - y + 2 = 0\}$,
- trajectory of ball 2: $\{[x, y] \mid 2x^2 + y^2 - 3 = 0\}$.

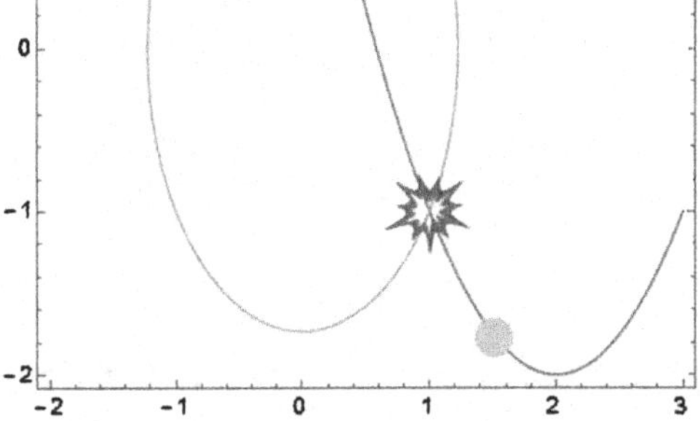

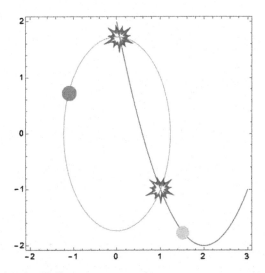

Figure 8.3 Trajectories of two balls.

The said mathematical description may aim at several practical applications:

- collision avoidance of a space shuttle against an asteroid,
- collision avoidance for unmanned vehicles with fixed trajectories,
- collision avoidance for an industrial robot arm, etc.

Question: Is there a possibility for the balls to collide?

Taking into account the balls' trajectories, to obtain their intersection points, we solve the following system of given nonlinear equations:

$$x^2 - 4x - y + 2 = 0,$$
$$2x^2 + y^2 - 3 = 0.$$

We choose an initial guess (the graphics can help to choose the point not far from the supposed intersection points),

$$\mathbf{x}^{(0)} = \begin{bmatrix} 0.2 \\ 0.4 \end{bmatrix}.$$

Solution

Step 1: Take initial vector $x^{(0)} = [0.2, 0.4]^{\mathrm{T}}$.

Step 2: Define $\mathbf{F}(\mathbf{x})$ and $J(\mathbf{x})$,

$$\mathbf{F}(\mathbf{x}) = \begin{bmatrix} x^2 - 4x - y + 2 \\ 2x^2 + y^2 - 3 \end{bmatrix},$$
$$J(\mathbf{x}) = \begin{bmatrix} -4 + 2x & -1 \\ 4x & 2y \end{bmatrix}.$$

In our case we can easily compute $J(\mathbf{x})^{-1}$, i.e.,

$$J(\mathbf{x})^{-1} = \begin{bmatrix} \frac{2y}{4x-8y+4xy} & \frac{1}{4x-8y+4xy} \\ -\frac{4x}{4x-8y+4xy} & \frac{-4+2x}{4x-8y+4xy} \end{bmatrix}.$$

We calculate $\mathbf{F}(\mathbf{x}^{(0)})$ and $J(\mathbf{x}^{(0)})^{-1}$, where $x^{(0)} = [0.2, 0.4]^{\mathrm{T}}$. We have

$$\mathbf{F}\left(\mathbf{x}^{(0)}\right) = \begin{bmatrix} 0.2^2 - 4 * 0.2 - 0.4 + 2 \\ 2 * 0.2^2 + 0.4^2 - 3 \end{bmatrix} = \begin{bmatrix} 0.84 \\ -2.76 \end{bmatrix}$$

and

$$J\left(\mathbf{x}^{(0)}\right)^{-1} = \begin{bmatrix} \frac{2*0.4}{4*0.2-8*0.4+4*0.2*0.4} & \frac{1}{4*0.2-8*0.4+4*0.2*0.4} \\ -\frac{4*0.2}{4*0.2-8*0.4+4*0.2*0.4} & \frac{-4+2*0.2}{4*0.2-8*0.4+4*0.2*0.4} \end{bmatrix}$$
$$= \begin{bmatrix} -0.384615 & -0.480769 \\ 0.384615 & 1.73077 \end{bmatrix}.$$

Step 3: In case *we are able* to compute $J(\mathbf{x})^{-1}$ in its exact form, the easiest way to compute $\mathbf{x}^{(i+1)}$ is to proceed with iteration scheme (8.4),

$$\mathbf{x}^{(i+1)} = \mathbf{x}^{(i)} - J\left(\mathbf{x}^{(i)}\right)^{-1} \cdot \mathbf{F}\left(\mathbf{x}^{(i)}\right).$$

In our case,

$$\mathbf{x}^{(1)} = \mathbf{x}^{(0)} - J\left(\mathbf{x}^{(0)}\right)^{-1} \cdot \mathbf{F}\left(\mathbf{x}^{(0)}\right),$$

$$\mathbf{x}^{(1)} = \begin{bmatrix} 0.2 \\ 0.4 \end{bmatrix} - \begin{bmatrix} -0.384615 & -0.480769 \\ 0.384615 & 1.73077 \end{bmatrix} \cdot \begin{bmatrix} 0.84 \\ -2.76 \end{bmatrix}$$
$$= \begin{bmatrix} -0.803846 \\ 4.85384615 \end{bmatrix}.$$

Skip to step 6.

If *we are not able* to compute $J(\mathbf{x})^{-1}$ in its exact form, we use the two-step technique.

Step 4: Solve the linear system $J(\mathbf{x}^{(0)}) \cdot \mathbf{y}^{(0)} = -\mathbf{F}(\mathbf{x}^{(0)})$ and compute $\mathbf{y}^{(0)}$. We have

$$\begin{bmatrix} -3.6 & -1 \\ 0.8 & 0.8 \end{bmatrix} \cdot \begin{bmatrix} y_1^{(0)} \\ y_2^{(0)} \end{bmatrix} = -\begin{bmatrix} 0.84 \\ -2.76 \end{bmatrix}.$$

It yields the result

$$\mathbf{y}^{(0)} = \begin{bmatrix} -1.00385 \\ 4.45385 \end{bmatrix}.$$

Step 5: Compute

$$\mathbf{x}^{(1)} = \mathbf{x}^{(0)} + \mathbf{y}^{(0)} = \begin{bmatrix} 0.2 \\ 0.4 \end{bmatrix} + \begin{bmatrix} -1.00385 \\ 4.45385 \end{bmatrix} = \begin{bmatrix} -0.803846 \\ 4.85385 \end{bmatrix}.$$

Step 6: Compute the next iteration $\mathbf{x}^{(i+1)}$, $i = 1, 2, \ldots$, using the same procedure.

In general, we repeat the process until $\mathbf{x}^{(i+1)}$ converges to some fixed point (see Table 8.1).

If we set the tolerance $\varepsilon = 10^{-15}$, then $\|x^{(8)} - x^{(7)}\| = 2.23773 \times 10^{-16} < \varepsilon$. Taking into consideration the conventional computer machine precision, we can state practically $\|x^{(i+1)} - x^{(i)}\| = 0$. This indicates that our system $\mathbf{F}(x) = 0$ has converged to the solution after eight iterations. Therefore its solution

$$\mathbf{x} = \mathbf{p} = \begin{bmatrix} 0.0688575 \\ 1.72931 \end{bmatrix}$$

which is one of collision points of balls trajectories.

Remark. As shown in Fig. 8.3, there are two possible solutions (collisions). To find the second root we will repeat the same process with a different initial guess, e.g., $[1.2, 1.4]^{\mathrm{T}}$, with result $\mathbf{x} = [1, -1]^{\mathrm{T}}$.

Table 8.1 Collision point of two balls. Table of iterations.

i	$x_1^{(i)}$	$x_2^{(i)}$	$\|\mathbf{x}^{(i+1)} - \mathbf{x}^{(i)}\|$
0	0.2	0.4	–
1	−0.803846	4.85385	4.56557
2	−0.255139	2.78457	2.14079
3	−0.00414241	1.95359	0.868066
4	0.0636412	1.74489	0.219429
5	0.068828	1.7294	0.0163373
6	0.0688575	1.72931	0.0000921701
7	0.0688575	1.72931	2.94122×10^{-9}
8	0.0688575	1.72931	2.23773×10^{-16}

8.3.2.2 Robot arm example

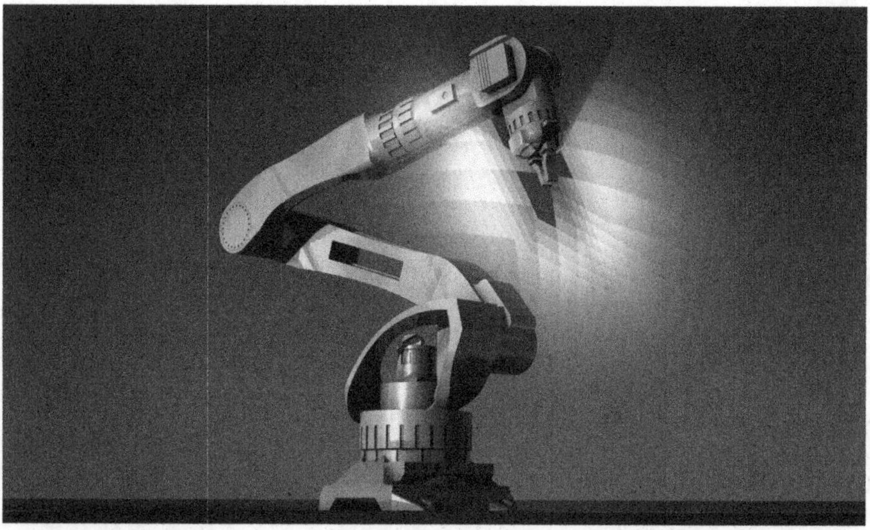

Figure 8.4 Robot arm.
Credit: Free image from https://pixabay.com/illustrations/cybernetics-robot-robot-arm-4216899/.

In mechanical engineering and in industrial robotic applications, engineers often solve problems dealing with different configurations of automatic robot arms (Fig. 8.4). Let us take a simple case aiming to configure a robot arm so that it rests at a prescribed steady-state position. Let an idealized two-link robot arm consist of two rods, $l_1 = 4[m]$ and $l_2 = 2.5[m]$, that are joined end-to-end in the plane. The robot arm is free to rotate with respect to the joints. The base of the robot arm is assumed to be fixed at the origin $[0, 0]$ of a coordinate plane.

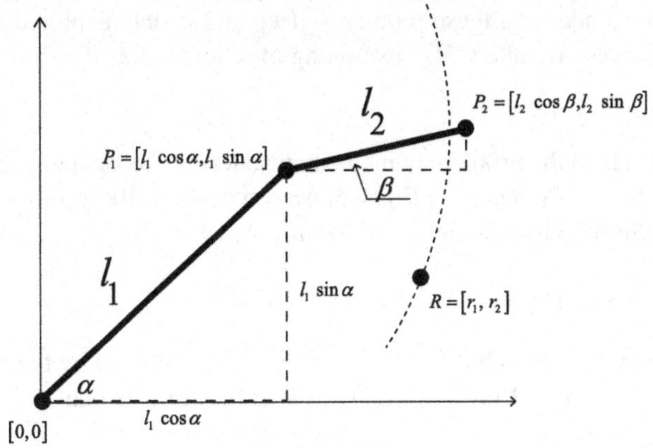

Figure 8.5 An idealized two-link robot arm.

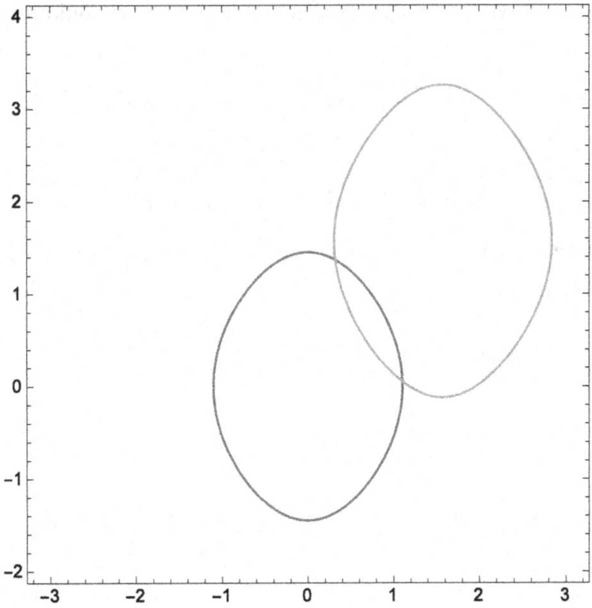

Figure 8.6 Graph of equations (two curves) in the system (8.7).

Question: Is it possible that the robot's arm comes to a desired position $R = [r_1, r_2]$ (Fig. 8.5) with given coordinates (e.g., $R = [4.3, 3.7]$)? Find a spatial configuration of rods for this case.

In real applications, a realistic industrial robot arm has more than two links. As a consequence, there should be many possible configurations. Here, the robot arm has only two possible configurations and we would like to determine at least one of them.

Let us note that concerning these problems, deep and widely explored theory exists, especially in the case we allow 3D positioning of robot arms.

Solution

First, let us formulate the problem in mathematical terms. Recognizing the right angle triangle with vertices P_1 and P_2 in Fig. 8.5, we can describe the terminal points of the rod P_1 and P_2 through trigonometric functions

$$P_1 = [l_1 \cos\alpha, l_1 \sin\alpha] \quad \text{and} \quad P_2 = [l_2 \cos\beta, l_2 \sin\beta].$$

The coordinates of the robot arm free end P_2 are determined by the sum $P_1 + P_2$, defining also the distance from the origin of the 2D coordinate plane,

$$P_1 + P_2 = [l_1 \cos\alpha + l_2 \cos\beta, l_1 \sin\alpha + l_2 \sin\beta].$$

To answer if is it possible for the robot's arm to come to the given position $R = [r_1, r_2]$, we need to solve the following system of equations for unknown α and β (see Fig. 8.6):

$$\begin{aligned} l_1 \cos\alpha + l_2 \cos\beta &= r_1, & \quad 4\cos\alpha + 2.5\cos\beta &= 4.3, \\ l_1 \sin\alpha + l_2 \sin\beta &= r_2, & \quad 4\sin\alpha + 2.5\sin\beta &= 3.7. \end{aligned} \tag{8.7}$$

Let us follow step by step the algorithm from Section 8.1.2.

Step 1: Set the initial guess $\mathbf{x}^{(0)} = [\alpha, \beta]^{\mathrm{T}} = [1, 3]^{\mathrm{T}}$.

Step 2: Define $\mathbf{F}(\mathbf{x})$ and $J(\mathbf{x})$,

$$\mathbf{F}(\mathbf{x}) = \begin{bmatrix} 4\cos\alpha + 2.5\cos\beta - 4.3 \\ 4\sin\alpha + 2.5\sin\beta - 3.7 \end{bmatrix},$$

$$J(\mathbf{x}) = \begin{bmatrix} -4\sin\alpha & -2.5\sin\beta \\ 4\cos\alpha & 2.5\cos\beta \end{bmatrix}.$$

The Jacobian inverse $J(\mathbf{x})^{-1}$ is computable, but too complicated,

$$J(\mathbf{x})^{-1} = \begin{bmatrix} \frac{2.5\cos\beta}{-10\cdot\cos\beta\cdot\sin\alpha + 10\cdot\cos\alpha\cdot\sin\beta} & \frac{2.5\sin\beta}{-10\cdot\cos\beta\cdot\sin\alpha + 10\cdot\cos\alpha\cdot\sin\beta} \\ \frac{4\cos\alpha}{-10\cdot\cos\beta\cdot\sin\alpha + 10\cdot\cos\alpha\cdot\sin\beta} & \frac{4\sin\alpha}{-10\cdot\cos\beta\cdot\sin\alpha + 10\cdot\cos\alpha\cdot\sin\beta} \end{bmatrix}.$$

Calculate $\mathbf{F}(\mathbf{x}^{(0)})$,

$$\mathbf{F}(\mathbf{x}^{(0)}) = \begin{bmatrix} 4\cos\alpha + 2.5\cos\beta - 4.3 \\ 4\sin\alpha + 2.5\sin\beta - 3.7 \end{bmatrix} = \begin{bmatrix} -4.61377 \\ 0.018684 \end{bmatrix}.$$

The inverse Jacobi matrix has complicated form $J(\mathbf{x})^{-1}$, so we will use the alternative.

Skip to step 4.

Step 4: In order to find $\mathbf{y}^{(0)}$, solve the linear system $J(\mathbf{x}^{(0)}) \cdot \mathbf{y}^{(0)} = -\mathbf{F}(\mathbf{x}^{(0)})$,

$$\begin{bmatrix} -3.36588 & -0.352800 \\ 2.16121 & -2.47498 \end{bmatrix} \cdot \begin{bmatrix} y_1^{(0)} \\ y_2^{(0)} \end{bmatrix} = - \begin{bmatrix} -4.61377 \\ 0.018684 \end{bmatrix}.$$

It yields

$$\mathbf{y}^{(0)} = \begin{bmatrix} -1.25653 \\ -1.08968 \end{bmatrix}.$$

Step 5: Compute $\mathbf{x}^{(1)}$

$$\mathbf{x}^{(1)} = \mathbf{x}^{(0)} + \mathbf{y}^{(0)} = \begin{bmatrix} 1 \\ 3 \end{bmatrix} + \begin{bmatrix} -1.25653 \\ -1.08968 \end{bmatrix} = \begin{bmatrix} -0.25653 \\ 1.91032 \end{bmatrix}.$$

Step 6: Compute the next iteration $\mathbf{x}^{(i+1)}$, $i = 1, 2, \ldots$, using the same procedure. Repeat the process until the stopping condition is true (see Table 8.2).

Table 8.2 The robot rod configuration angles. Table of iterations.

i	$x_1^{(i)}$	$x_2^{(i)}$	$\lVert\mathbf{x}^{(i+1)} - \mathbf{x}^{(i)}\rVert$
0	1	3	–
1	−0.256530	1.91032	1.25653
2	0.287916	1.608731	0.544445
3	0.299453	1.37954	0.229192
4	0.317856	1.36956	0.0184028
5	0.317987	1.36922	0.000340118
6	0.317987	1.36922	4.87645×10^{-8}

If we set the tolerance $\varepsilon = 10^{-7}$, then $\lVert\mathbf{x}^{(6)} - \mathbf{x}^{(5)}\rVert = 4.87645 \times 10^{-8} < \varepsilon$. Based on Table 8.2, the norm reaches a value less than the set tolerance at the sixth iteration. We can state the root

$$\mathbf{p} \doteq \mathbf{x}^{(6)} = \begin{bmatrix} 0.317987 \\ 1.36922 \end{bmatrix}.$$

To reach the desired position of the arm $R = [4.3, 3.7]$, the rod configuration angles have to be $\alpha = 0.317987$ and $\beta = 1.36922$.

8.3.3 The two-bar truss problem/buckling problem for two bars

Buckling is an instability that leads to structural failure. The failure modes can be, in simple cases, found by nonlinear systems of equations. For complex structures, the failure modes are found by more complicated optimization numerical tools. We will solve the buckling problem for two bars.

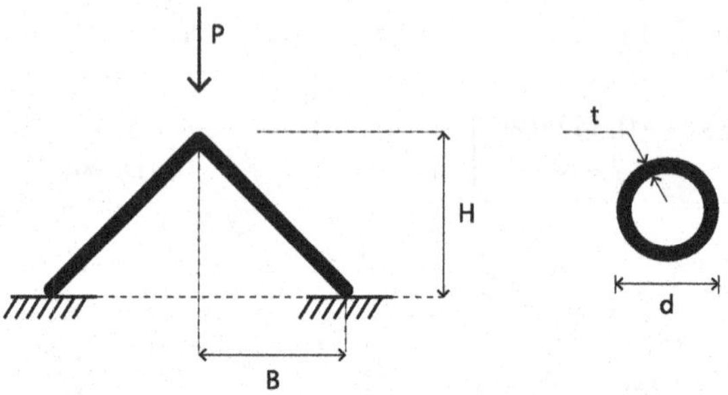

Figure 8.7 The tubular members of a plane two-bar truss.

When a structure is subjected to compressive axial load, buckling may occur. Buckling is characterized by a sudden sideways deflection of a structural member. This may occur even though the stresses developed in the structure are well below the critical values that are needed to cause failure of the material of which the structure is composed.

In the example, bars are subjected under a concentric axial compression, so the load exhibits the characteristic deformation of buckling.

Let the height H and the diameter d of the tubular members of a plane two-bar truss (Fig. 8.7) vary in order to minimize the volume of the material. The volume is proportional to the total weight – so in applied mechanics, it is very important to find an optimized proportion. With respect to given technical application, the following requirements (technical constraints) have to be satisfied:

1. the member stress should be less than the yield stress F, and
2. the members should not buckle.

Let us take real parameters of a given structure,

E	Young's modulus	1.99948×10^{11} [Pa], [Pa] $= \mathrm{kg\,m^{-1}\,s^{-2}}$,
B	half-distance between supports	2.54 [m],
σ_y	yield stress of material	2.482×10^8 [Pa],
t	wall thickness of tube	0.00635 [m],
P	applied load	100 000 [N], [N] $= \mathrm{kg\,m\,s^{-2}}$.

Then the objective function and the mechanical constraints can be expressed in mathematical forms.

The member force ([N] $= \mathrm{kg\,m\,s^{-2}}$) is

$$F = \frac{P}{2}\frac{\sqrt{B^2 + H^2}}{H}.$$

Member stress ([Pa] $= \mathrm{kg\,m^{-1}\,s^{-2}}$) is

$$\sigma = \frac{F}{A}.$$

Buckling stress ($[\mathrm{Pa}] = \mathrm{kg\,m^{-1}\,s^{-2}}$) is

$$\sigma_{cr} = \frac{\pi^2 EI}{L^2}\frac{1}{A}.$$

The second moment of inertia ($[\mathrm{m}^4]$) is

$$I = \frac{\pi}{64}\left[(d+t)^4 - (d-t)^4\right] = \frac{\pi t d}{8}(d^2+t^2).$$

The volume of material is simply the bar area A multiplied by the total length of two bars. Using the approximation $A = \pi t d$ – valid for thin-wall tubes – the volume is

$$V = 2AL = 2(\pi t d)\sqrt{H^2+B^2}.$$

Hence we can write the objective function f as

$$f = 2(\pi d t)\sqrt{H^2+B^2},$$

and the constraints g_1, g_2 as

$$g_1: \qquad \frac{P}{2}\frac{\sqrt{H^2+B^2}}{H}\frac{1}{dt\pi} - \sigma_y \le 0,$$

$$g_2: \qquad \frac{P}{2}\frac{\sqrt{H^2+B^2}}{H}\frac{1}{dt\pi} - \frac{\pi^2 E(d^2+t^2)}{8(H^2+B^2)} \le 0.$$

In the case we look at the problem from an optimization point of view, the scalar function f is to be optimized with respect to zero or more constraints g_i. On the other hand, there are two nonlinear equations $g_1 = 0$ and $g_2 = 0$. To find critical points, the following nonlinear equation system can be solved:

$$g_1(d,H) = 0: \quad \frac{P}{2}\frac{\sqrt{H^2+B^2}}{H}\frac{1}{dt\pi} - F_y = 0,$$

$$g_2(d,H) = 0: \quad \frac{P}{2}\frac{\sqrt{H^2+B^2}}{H}\frac{1}{dt\pi} - \frac{\pi^2 E(d^2+t^2)}{8(H^2+B^2)} = 0.$$

The constraint equations are modeled in 3D as surfaces. We are looking for a point on their intersection curve (see Fig. 8.8, created in software Mathematica), where its value is equal to zero.

Solution

Let us follow the described algorithm.

Step 1: Take initial guess $\mathbf{x}^{(0)} = [d, H]^{\mathrm{T}} = [0.1, 0.1]^{\mathrm{T}}$.

Step 2: Define $\mathbf{F}(\mathbf{x})$ and $J(\mathbf{x})$,

$$\mathbf{F}(\mathbf{x}) = \begin{bmatrix} g_1(d,H) \\ g_2(d,H) \end{bmatrix}$$

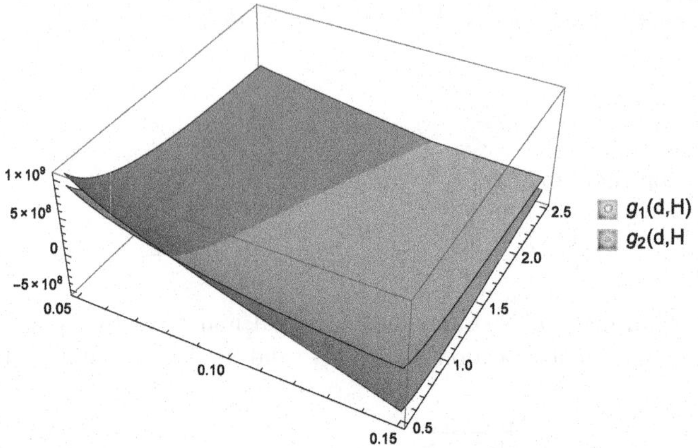

Figure 8.8 Graph of the constraint equations in 3D.

$$= \begin{bmatrix} -2.482 \times 10^8 + \frac{2\,506\,377.057\sqrt{6.4516+H^2}}{d.H} \\ -\frac{2.46676\times 10^{11}(0.0000403225+d^2)}{6.4516+H^2} + \frac{2\,506\,377.057\sqrt{6.4516+H^2}}{d.H} \end{bmatrix},$$

$$J(\mathbf{x}) = J(d,H) = \begin{bmatrix} \frac{\partial g_1}{\partial d}(\mathbf{x}) & \frac{\partial g_1}{\partial H}(\mathbf{x}) \\ \frac{\partial g_2}{\partial d}(\mathbf{x}) & \frac{\partial g_2}{\partial H}(\mathbf{x}) \end{bmatrix}.$$

Here, $J(\mathbf{x})^{-1}$ is computable only if aided by some CAS system.
Compute $\mathbf{F}(\mathbf{x}^{(0)})$,

$$\mathbf{F}\left(\mathbf{x}^{(0)}\right) = \begin{bmatrix} g_1(d,H) \\ g_2(d,H) \end{bmatrix} = \begin{bmatrix} 3.88913 \times 10^8 \\ 2.53817 \times 10^8 \end{bmatrix}.$$

Since the inverse Jacobi matrix has a complicated form, we will use the alternative. Skip to step 4.

Step 4: In order to find $\mathbf{y}^{(0)}$, solve the linear system $J(\mathbf{x}^{(0)}) \cdot \mathbf{y}^{(0)} = -\mathbf{F}(\mathbf{x}^{(0)})$,

$$\begin{bmatrix} -6.37113 \times 10^9 & -6.36127 \times 10^9 \\ -1.40063 \times 10^{10} & -6.34941 \times 10^9 \end{bmatrix} \cdot \begin{bmatrix} y_1^{(0)} \\ y_2^{(0)} \end{bmatrix} = -\begin{bmatrix} 3.88913 \times 10^8 \\ 2.53817 \times 10^8 \end{bmatrix}.$$

It yields

$$\mathbf{y}^{(0)} = \begin{bmatrix} -0.0175717 \\ 0.0787365 \end{bmatrix}.$$

Be careful in the real model, and do not forget to check if the linear system is well conditioned.

Step 5: Compute $\mathbf{x}^{(1)}$

$$\mathbf{x}^{(1)} = \mathbf{x}^{(0)} + \mathbf{y}^{(0)} = \begin{bmatrix} 0.1 \\ 0.1 \end{bmatrix} + \begin{bmatrix} -0.0175717 \\ 0.0787365 \end{bmatrix} = \begin{bmatrix} 0.0824283 \\ 0.178737 \end{bmatrix}.$$

Step 6: Compute the next iteration, and repeat the process until the stopping condition holds (see Table 8.3).

Table 8.3 Critical point. Table of iterations.

i	$d^{(i)}$	$H^{(i)}$	$\|\mathbf{x}^{(i+1)} - \mathbf{x}^{(i)}\|$
0	0.1	0.1	–
1	0.0824283	0.178737	0.0787365
2	0.0807248	0.259151	0.0804145
3	0.0808962	0.307981	0.0488297
4	0.0809528	0.318924	0.0109435
5	0.0809551	0.319329	0.000404801
6	0.0809551	0.319330	5.181785×10^{-7}

If we take the tolerance $\varepsilon = 10^{-6}$, then $\|x^{(6)} - x^{(5)}\| = 5.181785 \times 10^{-7} < \varepsilon$. Based on Table 8.3, the norm is less than the tolerance at the sixth iteration. We can state the root

$$\mathbf{p} \doteq \mathbf{x}^{(6)} = \begin{bmatrix} 0.0809551 \\ 0.319330 \end{bmatrix}$$

giving the sought values $d = 0.0809551$ and $H = 0.319330$.

We are not at the end yet! The task in the example was to compute the minimum of the objective function under given constraints

$$f = 2(\pi dt)\sqrt{H^2 + B^2} = 0.00826870.$$

This problem has a unique solution (not all problems do) and the executed method converges for a chosen starting point. Testing different starting points we can see that not all points are good choices. You can test also other methods like the modified Newton's algorithm with different step lengths.

Remark. CAS systems provide for solution of nonlinear equations and their system inbuilt commands. In Wolfram Mathematica one can use FindRoot[f, $\{x, x_0\}$] or FindRoot[$\{eq_1, eq_2, \ldots\}, \{\{x, x_0\}, \{y, y_0\}, \ldots\}$] to apply Newton–Ralphston's method with small modifications included.

The nonlinear systems methods in the IMSL and NAG libraries use the *Levenberg–Marquardt method*, which is a weighted average of Newton's method and the steepest descent method. The weight is biased toward the steepest descent method until convergence is detected, at which time the weight is shifted toward the more rapidly convergent Newton's method. In either routine a finite difference approximation to the Jacobian can be used or a user-supplied subroutine enters to compute the Jacobian.

Very easy and friendly for reading is the book Numerical Analysis from Burden and Faires (2011). A comprehensive treatment of methods for solving nonlinear systems of equations can be found in Ortega and Rheinboldt (1970) and in Dennis and Schnabel (1983). Recent developments on iterative methods can be found in Argyros and Szidarovszky (1993), and information on the use of continuation methods is available in Allgower and Georg (1990).

References

Allgower, E., Georg, K., 1990. Numerical Continuation Methods: An Introduction. Springer-Verlag, New York. 388 pp.

Argyros, I.K., Szidarovszky, F., 1993. The Theory and Applications of Iteration Methods. CRC Press, Boca Raton, FL. 355 pp.

Burden, R.L., Faires, D.J., 2011. Numerical Analysis, 9th edition. Brooks/Cole, Cengage Learning.

Dennis Jr., J.E., Schnabel, R.B., 1983. Numerical Methods for Unconstrained Optimization and Nonlinear Equations. Prentice-Hall, Englewood Cliffs, NJ. 378 pp.

Ortega, J.M., Rheinboldt, W.C., 1970. Iterative Solution of Nonlinear Equations in Several Variables. Academic Press, New York. 572 pp.

Shortest path problem and computer algorithms

9

Deolinda M.L. Dias Rasteiro
Coimbra Polytechnic – ISEC, Coimbra, Portugal

9.1 Background

In the optimal path problem, a real function is considered which assigns a value to each path that can be defined between a given pair of nodes in a given network; a path with the best value in a subset of paths between that pair of nodes is what has to be determined (Martins et al., 1999).

For an extensive and complete characterization of network problems we suggest the book *Network Flows* by Ravindra K. Ahuja et al. (Ahuja et al., 1993). We will start with some definitions, notation, and properties that we consider necessary to fully understand the problems and their solutions.

9.1.1 Notation, definitions, and properties

Let $(\mathcal{N}, \mathcal{A})$ be a given network, where $\mathcal{N} = \{v_1, \ldots, v_n\}$ is the set of nodes and $\mathcal{A} = \{a_1, \ldots, a_m\} \subseteq \mathcal{N} \times \mathcal{N}$ is the set of arcs. In what follows, node v_i is sometimes denoted by i. Each arc $a_k \in \mathcal{A}$, $k \in \{1, \ldots, m\}$, is represented by the pair (i, j), where $i, j \in \{1, \ldots, n\}$. We assume that $\mathcal{N}$ and $\mathcal{A}$ are finite sets.

Consider two nodes i and j of $\mathcal{N}$. If all pairs $(i, j) \in \mathcal{A}$ are ordered, i.e., if all the arcs of the network are directed, we will say that $(\mathcal{N}, \mathcal{A})$ is a directed network; otherwise we will say that $(\mathcal{N}, \mathcal{A})$ is an undirected network.

In what follows, without loss of generality, we assume that $(\mathcal{N}, \mathcal{A})$ is a directed network and that there are no arcs of the form (i, i), $i \in \mathcal{N}$.

Definition. Let s and t be two nodes of $\mathcal{N}$ such that $s \neq t$. We denote by s the initial node and by t the terminal node.

A *path*[1] p from s to t in $(\mathcal{N}, \mathcal{A})$ is an alternating sequence of nodes and arcs of the following form:

$$p = \langle s = v'_1, a'_1, v'_2, \ldots, a'_{l-1}, v'_l = t \rangle,$$

where $l \geq 2$ and the following conditions hold:

- $v'_k \in \mathcal{N}, \forall k \in \{1, \ldots, l\}$,
- $a'_k = (v'_k, v'_{k+1}) \in \mathcal{A}, \forall\, k \in \{1, \ldots, l-1\}$.

[1] For convenience sometimes we will represent a path by only one node.

Calculus for Engineering Students. https://doi.org/10.1016/B978-0-12-817210-0.00016-3

Given two nodes $i, j \in \mathcal{N}$, we denote by $\mathcal{P}_{ij}$ the set of paths from node i to node j in $(\mathcal{N}, \mathcal{A})$; $\mathcal{P}$ is the set of paths from node s to t ($\mathcal{P}_{st}$). In what follows, we also assume that the sets $\mathcal{P}_{si}$ and $\mathcal{P}_{it}$ are nonempty for each $i \in \mathcal{N}$.

A *loopless path* from s to t is a path from s to t such that all nodes are distinct. A *cycle* (or a loop) is a path from a node to itself whose nodes are all distinct except for the first, which is also the last.

The symbol "$\diamond$" denotes the *concatenation* of paths where the terminal node of the first path is also the initial node of the second path, i.e., $p_{ij} \diamond p_{jk} = p_{ik}$.

Definition. Let $a_k = (i, j) \in \mathcal{A}$ be an arc from $(\mathcal{N}, \mathcal{A})$. Associated to a_k there will be at least one real value d_{ij} that will be called arc parameter(s).

In what follows the parameters associated to arcs may be called distance, cost, or time, depending on what the problem is about. For the sake of exposition we will refer to the parameter d_{ij} as the length of the arc $(i, j) \in \mathcal{A}$.

Definition. The length of a directed path $p \in \mathcal{P}_{st}$ will be defined as the sum of the lengths of the arcs in the path p. That is, $d(p) = \sum_{(i,j)\in p} d_{ij}$.

Important network assumptions

1. **The parameters d_{ij} associated to each arc are integer values.** Note that if that is not the case, one can always transform rational numbers into integers and that it is necessary to convert irrational numbers into rational ones if we want to represent them in a computer. Thus this is not really a restriction of the problem.
2. **The network contains a directed path from node s to every node j in the network.** Fictitious arcs may be added to the original network if associated to them we considered a very high parameter value.
3. **The network that represents our problem does not contain negative cycles.** In the case that happens when trying to obtain the solution we will necessarily obtain an unbounded solution.
4. **The network is directed.** In the case of nonnegative parameters this is easily done by considering two directed arcs with opposite ways but the same direction and the same value parameters.

The network optimization problems are, in general, solved by using labeling algorithms (Martins et al., 1999). These algorithms are improvements of the exhaustive search by the optimality principle definition that is given below.

Definition (Strong optimality principle). Let i and j be two nodes of $\mathcal{N}$. We say that problem P1 satisfies the strong optimality principle if every optimal path from i to j in $(\mathcal{N}, \mathcal{A})$ is formed by optimal subpaths.

Definition (Weak optimality principle). We say that problem P1 satisfies the weak optimality principle if for any $i, j \in \mathcal{N}$ there exists at least an optimal path from i to j in $(\mathcal{N}, \mathcal{A})$ that is formed by optimal subpaths.

According to these principles, an optimal solution can be determined by finding successive optimal subsolutions. As a consequence, when an optimization problem satisfies the strong optimality principle, a labeling algorithm can be used to solve it. However, labeling algorithms can be applied only if there is an optimal path under those conditions, and they do really determine this optimal path. Then it can be concluded that if a problem satisfies the weak optimality principle, even then it still can be solved using a labeling algorithm. It can be easily established that an optimal path which satisfies the strong optimality principle also satisfies the weak optimality one.

Under the 1.–4. important network assumptions made above, we can guarantee that the shortest path problem verifies the strong optimality principle and therefore the weak optimality principle. As a curiosity, the maximum capacity path problem only verifies the weak optimality principle; see Martins et al. (1999), pages 6–8, for a detailed explanation and example.

Before detailing labeling algorithms we will first explain the data structure that is efficient to deal with network problems.

9.1.2 Data structure and labeling correcting algorithms

In order to store a network we will consider at least three vectors (more will be necessary if the number of parameters associated to each arc is bigger than 1). The network can be stored in considering the arcs that go out from each node (forward star form) or the arcs that get in on a node (reverse forward star form) (Dial et al., 1979). For the representation in the forward star form, the first step is numbering the network's arcs in lexicographic order of the outgoing arcs, that is, to numerate the arcs in increasing order such that the number of the arc (i, j) is smaller than the number of arc (i', j') if and only if $i < i'$. The vectors used will be called $pointer$, suc, and $parameter$ and their dimensions are $n+1$, m, and m, respectively, where n is the number of nodes and m is the number of arcs in the network $(\mathcal{N}, \mathcal{A})$. The vector *pointer* is constructed as

$$pointer[1] = 1;$$

$$pointer[i+1] = pointer[i] + \text{ number of outgoing arcs of i;}$$

$$\forall i = 2, \ldots, n-1$$

$$pointer[n+1] = m+1.$$

The vector *suc* will store the successor of each arc, that is, $suc[k] = j$ if the arc $k = (i, j) \in \mathcal{A}$. The vector *parameter* will store the value associated to each arc, i.e., $parameter[k] = x$ if x is, for example, the distance associated to arc $k = (i, j) \in \mathcal{A}$.

For the representation in the reverse forward star form, the first step is numbering the network's arcs in lexicographic order of the incoming arcs that is to numerate the arcs in increasing order such that the number of the arc (i, j) is smaller than the number of arc (i', j') if and only if $j < j'$. The vectors used will be called *rpointer*, *ant*, and *rparameter* and their dimensions are $n+1$, m, and m, respectively, where n is the number of nodes and m is the number of arcs in the network $(\mathcal{N}, \mathcal{A})$. The vector *rpointer* is constructed as

$$rpointer[1] = 1;$$
$$rpointer[i+1] = rpointer[i] + \text{ number of incoming arcs of i};$$
$$\forall i = 2, \ldots, n-1$$
$$rpointer[n+1] = m+1.$$

The vector *ant* will store the antecessor of each arc, that is, $ant[k] = i$ if the arc $k = (i, j) \in \mathcal{A}$. The vector *parameter* will store the value associated to each arc, i.e., $parameter[k] = x$ if x is, for example, the distance associated to arc $k = (i, j) \in \mathcal{A}$.

If the network has more parameters associated to each arc the network storage will be two plus the number of vectors needed to store the parameters. It is worth to mention that generally the arcs' forward star form numeration is different from the arcs' numeration in the reverse star form numeration. Although we may use only one of the representations, for huge problems it is worthy to know both of them because the solution might be obtained applying algorithms that start at the same time at the initial and terminal node and the optimal path determined by concatenating both subsolutions in the "middle" of the network.

In order to efficiently solve the type of network problems defined in the previous section one may use labeling correcting algorithms. These algorithms make use of common sense resolution and generalize a greedy approach to the solution starting from the initial node $s \in \mathcal{A}$. An implementation of the algorithms presented in this chapter may be consulted in Chapter 15.

Labeling correcting algorithm

```
Initialize vectors that are used in the labeling procedure
while the list of used labeled L is not empty do
        Remove node i from the list L
        For each arc k starting at node i
                Consider node j as the successor of arc k
                If (existent path from s to j distance is not better than the one
                    obtained concatenating the path from s to i with arc k) then
                        the path from s to j is actualized
                        the information that node j was labeled by node i is saved
                        If node j does not belong to the list L then
                                the node j is added to the list L
                        end if
                end if
        end for
end while
```

If we want to determine the *shortest path,* then an existent path to a node $j \in \mathcal{N}$ is not better than the one obtained concatenating the path from s (initial node) to i with arc $k = (i, j)$, if the parameter associated to arc k plus the label of i is smaller than the actual label of j, i.e.,

$$label[j] > label[i] + parameter[k].$$

If we want to determine the *maximum capacity path,* then an existent path to a node $j \in \mathcal{N}$ is not better than the one obtained concatenating the path from s (initial node) to i with arc $k = (i, j)$, if the minimum between the parameter associated to arc k and the label of i is bigger than the actual label of j, i.e.,

$$label[j] \geq \min\{label[i], parameter[k]\}.$$

In the example below it is shown how a network can be stored using this type of data structure.

Example. Consider the following network:

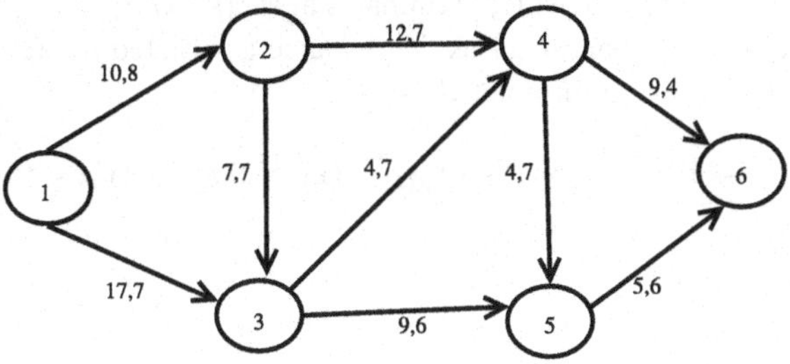

One possible arc numeration in the forward star form (blue, light gray in print version) is

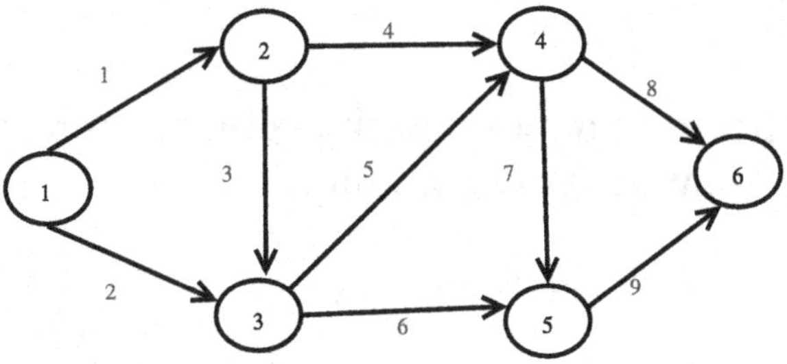

The data that will represent the above network will be

pointer =	1	3	5	7	9	10	10

suc =	2	3	3	4	4	5	5	6	6

*parameter*1 =	10	17	7	12	4	9	4	9	5

*parameter*2 =	8	7	7	7	7	6	7	4	6

In the reverse forward star form numeration the representation will be

$$rpointer = \boxed{1 \mid 1 \mid 2 \mid 4 \mid 6 \mid 8 \mid 10}$$

$$ant = \boxed{1 \mid 1 \mid 2 \mid 2 \mid 3 \mid 3 \mid 4 \mid 4 \mid 5}$$

Since, for this network, the numerations are the same, in the reverse star form the vectors corresponding to parameters will both be the same. With this representation it is very easy to know how many arcs begin/end at each node and also to which nodes they are connected. As an example, the number of arcs that begin at node 3 is given by $pointer[4] - pointer[3] = 7 - 5 = 2$ i.e., arc 5, which terminates in the node $suc[pointer[3]] = 4$, and arc 6, which terminates at $suc[pointer[3] + 1] = 5$.

If we consider the first parameter as the distance associated to arc (i, j), then the shortest path problem can be formulated as

$$\begin{aligned}
&\min \quad 10x_{12} + 17x_{13} + 7x_{23} + 12x_{24} + 4x_{34} + 9x_{35} + 4x_{45} + 9x_{46} + 5x_{56} \\
&\text{subject to} \\
&x_{12} + x_{13} = 1, \\
&x_{12} - x_{23} - x_{24} = 0, \\
&x_{13} + x_{23} - x_{34} - x_{35} = 0, \\
&x_{24} + x_{34} - x_{45} - x_{46} = 0, \\
&x_{35} + x_{45} - x_{56} = 0, \\
&x_{46} + x_{56} = 1, \\
&x_{ij} \in \{0, 1\}.
\end{aligned}$$

9.2 Description of general path problems and areas where they are very common

Shortest path problems arise frequently in practice when one wants to send some material/good between two specific points in a network, the first one called the initial point and the last one called the terminal point. That material might be a message, water, a vehicle, or a package of any kind. One system applying these algorithms that almost all of us make use of is navigation by the Global Positioning System (commonly known as GPS). Often each one of us uses this tool to get to know the best route from one place to another. The best route may be determined as being the fastest or the shortest (sometimes they coincide but that is not always the case). Optimizing energy as part of growing preoccupation with the creation of a sustainable world and also environmental protection is another area where the knowledge of these algorithms is very useful: dimensioning an efficient electric network to deliver the amount of energy needed to a certain city and dimensioning a water deliver plan are some examples that may be solved using optimal path problems and their derivations. Since our objective is to send material as quickly as possible or through the shortest distance possible or even as cheaply as possible, the criteria for choosing the so-called shortest path do not always involve distance between a pair of nodes, as will be seen in the sequel.

The following linear problem is the one that has to be solved in order to obtain the *shortest path between two given nodes*, the initial and the terminal node, usually denoted by s and t, respectively:

$$
\begin{aligned}
&\min \textstyle\sum_{(i,j)\in\mathcal{A}} d_{ij}x_{ij}\\
&\text{subject to}\\
&\sum_{\{j:(s,j)\in\mathcal{A}\}} x_{sj} - \sum_{\{j:(j,s)\in\mathcal{A}\}} x_{js} = 1,\\
&\sum_{\{j:(i,j)\in\mathcal{A}\}} x_{ij} - \sum_{\{j:(j,i)\in\mathcal{A}\}} x_{ji} = 0, \quad \forall i \in \mathcal{N}\setminus\{s,t\},\\
&\sum_{\{j:(t,j)\in\mathcal{A}\}} x_{tj} - \sum_{\{j:(j,t)\in\mathcal{A}\}} x_{jt} = 1,\\
&x_{ij} \in \{0,1\},
\end{aligned}
\tag{P1}
$$

where $(\mathcal{N}, \mathcal{A})$ is a directed network where associated to each arc $(i, j) \in \mathcal{A}$ is a parameter d_{ij}. Assume that d_{ij} is the length of arc (i, j) and x_{ij} is a binary variable where $x_{ij} = 1$ if arc $(i, j) \in \mathcal{A}$ belongs to the solution and $x_{ij} = 0$ otherwise.

Note, as we mentioned above, that the parameter d_{ij} may be distance, in which case the shortest path is indeed a shortest path, cost, in which case the shortest path will be the cheapest path, or time, in which case we may call it the fastest path. If any other function is required to be optimized, then the optimal path is named accordingly.

Problem P1 may also be viewed as the problem of sending 1 unit of flow as fast/cheaply as possible (with arc flow parameters d_{ij} being time or cost) from node s to each of the nodes in $\mathcal{N}\setminus\{s\}$ in an uncapacitated directed network. Then the problem may be stated as

$$
\begin{aligned}
&\min \textstyle\sum_{(i,j)\in\mathcal{A}} d_{ij}x_{ij}\\
&\text{subject to}\\
&\sum_{\{j:(s,j)\in\mathcal{A}\}} x_{sj} - \sum_{\{j:(j,s)\in\mathcal{A}\}} x_{js} = n-1,\\
&\sum_{\{j:(i,j)\in\mathcal{A}\}} x_{ij} - \sum_{\{j:(j,i)\in\mathcal{A}\}} x_{ji} = -1, \quad \forall i \in \mathcal{N}\setminus\{s\},\\
&x_{ij} \in \{0,1\}.
\end{aligned}
\tag{P2}
$$

The parameter d_{ij} may also be considered as arc $(i, j) \in \mathcal{A}$ capacity. Therefore we may be interested in determining the path between two nodes which has the maximum capacity. The formulation for this problem will be

max v
subject to

$$\begin{aligned}
&\sum_{\{j:(s,j)\in\mathcal{A}\}} x_{sj} - \sum_{\{j:(j,s)\in\mathcal{A}\}} x_{js} &&= v, \\
&\sum_{\{j:(i,j)\in\mathcal{A}\}} x_{ij} - \sum_{\{j:(j,i)\in\mathcal{A}\}} x_{ji} &&= 0, \qquad \forall i \in \mathcal{N} \setminus \{s,t\}, \\
&\sum_{\{j:(t,j)\in\mathcal{A}\}} x_{tj} - \sum_{\{j:(j,t)\in\mathcal{A}\}} x_{jt} &&= -v, \\
& && 0 \le x_{ij} \le d_{ij}.
\end{aligned} \tag{P3}$$

Note that the optimal solution for problem P3 is the maximum value of the function $d(p) = \min\limits_{(i,j)\in p} d_{ij}$. In order to obtain the optimal solution of either problem P1, P2, or P3, the 1.–4. important network assumptions imposed in Section 9.1 must be verified before labeling algorithms can be applied.

9.3 Real problems

In this section, two examples of very common network problems will be presented and solved. The context in which the referred problems are defined may easily be adapted to other realities. In Problem 1 the considered parameter is the distance between cities or convenient places, but the problem can also be defined in a similar way if one considers the parameter as being time between places or even another quantitative variable that one needs to optimize (prices, amount of goods to be delivered, ...). If the reader is an informatics engineering student, time or distance between two players in any kind of game that we need to program on a computer may also be a challenging application.

9.3.1 Problem 1: traveling in Portugal

Suppose you arrive to Porto (in Portugal) and rent a car. Then you want to get to Faro (Algarve) driving by the shortest possible path. At your disposal you have a network where some cities (represented by its nodes) and roads (represented by its directed arcs) are marked. An international industrial company has to do hundreds of travels a year between Porto and Faro. Since one of the company's major expenses is fuel and car maintenance, they always want to do it following the shortest path which, in some cases, does not coincide with the path given by *Google Maps* or other *GPS* software. Usually those softwares combine distance with time or other relevant travel information. The distance between each pair of cities is marked. What you are asked to determine is the shortest path between Porto and Faro. Fig. 9.1 shows a Portuguese map with one possible undirected network on top of it.

Considering the suggested network we may complete it using the distance, in kilometers, from one city to another. For the sake of simplicity only integer values were

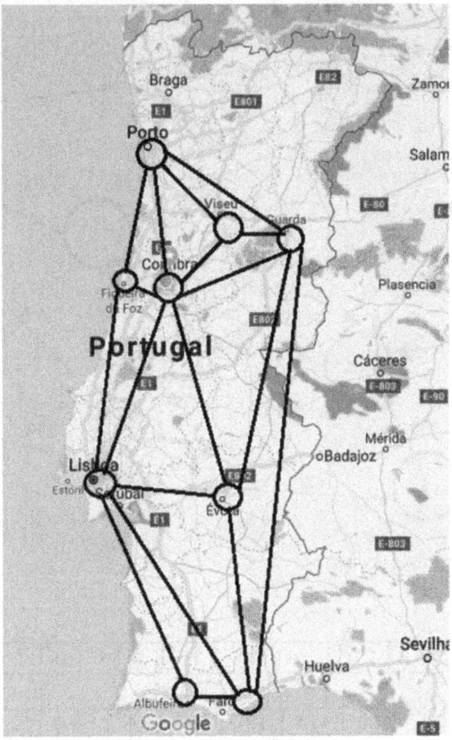

Figure 9.1 Draft of the undirected network.

considered but, as we have seen before with the assumptions made to these types of problems, that is not something that forbids their usage. In Fig. 9.2 the complete directed network is presented. The task we have now is to determine the shortest path from Porto to Faro. How will we do it?

The data structure defined in the previous section will allow us to store the network. Thus, for the above problem, if we use the forward star form, we will have

pointer =

1	5	8	10	12	14	17	18	19	19

suc =

2	3	4	5	5	7	9	2	5	5	6	6	7	7	9	8	9	9

parameter =

134	80	115	107	108	222	397	57	68	35	162	169

187	107	216	199	170	30

If we use the reverse star form, we will have

rpointer =

1	1	3	4	5	9	11	14	15	19

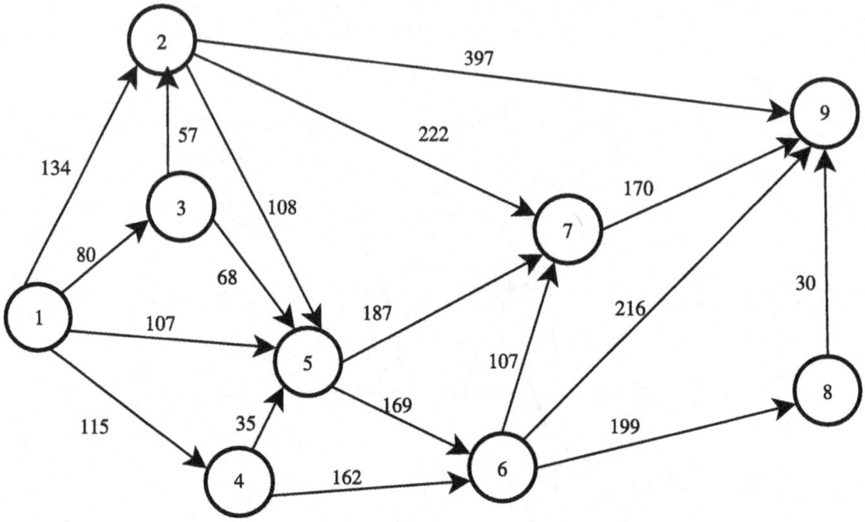

Figure 9.2 Problem 1, directed network.

ant =

1	3	1	1	1	2	3	4	4	5	2	5	6	6	2	6	7	8

rparameter =

134	57	80	115	107	108	68	35	162	169	222	187

107	199	397	216	170	30

Note that the lexicographic order is not the same in both star forms although sometimes it may coincide.

The shortest path problem within the above network can be formulated as

$$\begin{aligned} \min \quad & 134x_{12} + 80x_{13} + 115x_{14} + 107x_{15} + 108x_{25} + 222x_{27} + 397x_{29} \\ & + 57x_{32} + 68x_{35} + 35x_{45} + 162x_{46} + 169x_{56} + 187x_{57} + 107x_{67} \\ & + 199x_{68} + 216x_{69} + 170x_{79} + 30x_{89} \end{aligned}$$

subject to

$$\begin{aligned} & x_{12} + x_{13} + x_{14} + x_{15} = 1, \\ & x_{12} + x_{13} - x_{25} - x_{27} - x_{29} = 0, \\ & x_{13} - x_{32} - x_{35} = 0, \\ & x_{14} - x_{45} - x_{46} = 0, \\ & x_{15} + x_{25} + x_{35} + x_{45} - x_{56} - x_{57} = 0, \\ & x_{46} + x_{56} - x_{67} - x_{68} - x_{69} = 0, \\ & x_{27} + x_{57} + x_{67} - x_{79} = 0, \\ & x_{68} - x_{89} = 0, \\ & x_{29} + x_{69} + x_{79} + x_{89} = 1, \\ & x_{ij} \in \{0, 1\}. \end{aligned}$$

The solution we obtain using labeling correcting algorithms, considering as arc parameters the distance values, is the following (Fig. 9.3).

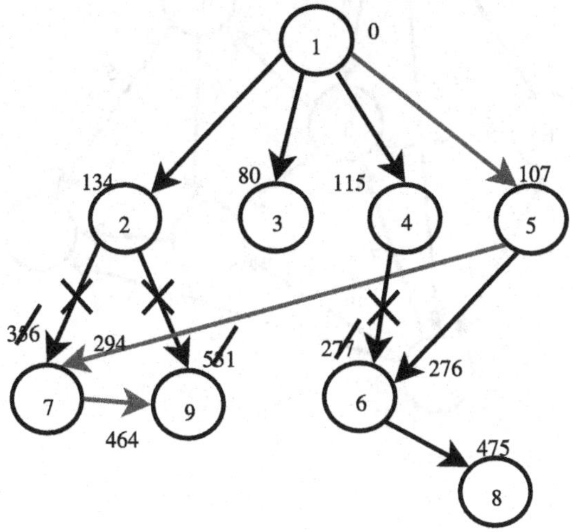

Figure 9.3 Problem 1, solution.

The list L referred to when the algorithm was present, in the previous section, is $L = \{1, 2, 3, 4, 5, 7, 9, 6, 5\}$. Note that nodes may not enter the list in their natural order.

The reader is challenged to observe that the previous solution is not the one given by *Google Maps*. To obtain it we should consider the time of travel between all network pairs of cities.

The shortest path problem verifies the strong optimality principle, which means that each subpath obtained from Porto to Faro is also optimal. Thus if the company driver changes his mind and wants to go to any of the cities that belong to the optimal path, he is certain that there is no other shortest path from Porto to that city.

9.3.2 *Problem 2: maximum capacity path*

An international company has one store to which it must deliver goods to be sold. Along the years this company has built a network of distributors to whom it sends the merchandise and these, subsequently, forward it to other distributors or to the store. Not all distributors have the same capacity to ship the goods. The figure below reflects the network of distributors and the company's final store as well as the capacity to dispose them. What needs to be determined is the set of maximum capacity paths on this international company network (see Fig. 9.4).

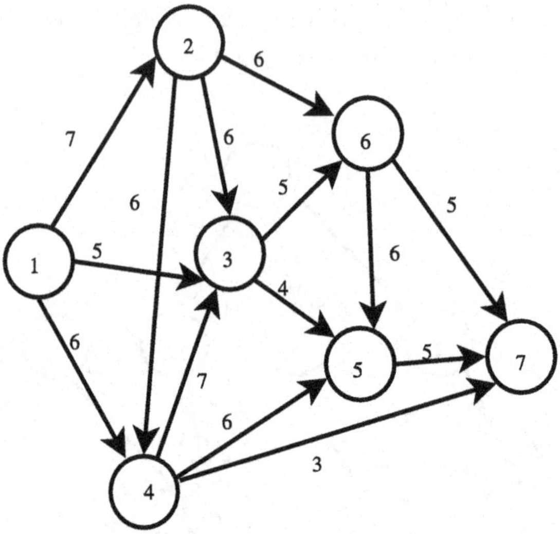

Figure 9.4 Problem 2, capacity company network.

The representation of this network in the forward star form is

pointer = | 1 | 4 | 7 | 9 | 12 | 13 | 15 | 15 |

suc = | 2 | 3 | 4 | 3 | 4 | 6 | 5 | 6 | 3 | 5 | 7 | 7 | 5 | 7 |

parameter = | 7 | 5 | 6 | 6 | 6 | 6 | 4 | 5 | 7 | 6 | 3 | 5 | 6 | 5 |

The formulation for this maximum capacity problem is the following:

$$
\begin{aligned}
&\max \quad v \\
&\text{subject to} \\
&x_{12} + x_{13} + x_{14} = v, \\
&x_{12} - x_{23} - x_{24} - x_{26} = 0, \\
&x_{13} + x_{23} + x_{43} - x_{35} - x_{36} = 0, \\
&x_{14} + x_{24} - x_{43} - x_{45} - x_{47} = 0, \\
&x_{35} + x_{45} + x_{65} - x_{57} = 0, \\
&x_{46} + x_{56} - x_{67} - x_{68} - x_{69} = 0, \\
&x_{26} + x_{36} - x_{65} - x_{67} = 0, \\
&x_{47} + x_{57} + x_{67} = -v, \\
&0 \le x_{ij} \le d_{ij}.
\end{aligned}
$$

Applying the labeling correcting algorithm to this problem we obtain the following solution, see Fig. 9.5.

Observe the existence of dashed arrows in the optimal maximum capacity path. Those arrows represent alternative paths. The maximum capacity path only satisfies

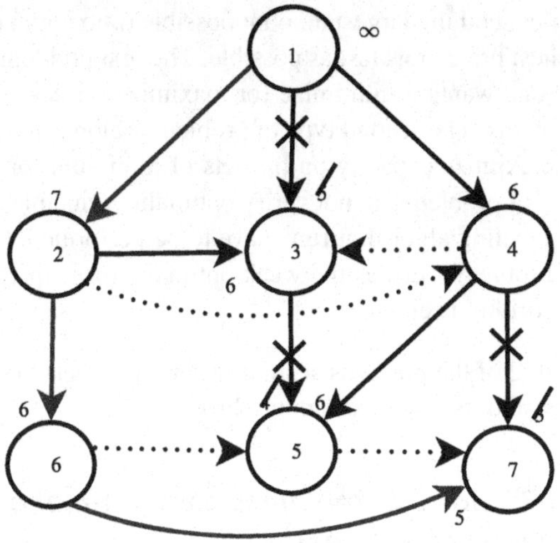

Figure 9.5 Problem 2, solution.

the weak optimal principle and we are able to see it in the above problem since there exists at least one maximum capacity path which is not constituted by optimal sub-paths, e.g., $(\langle 1, (1,2), 2, (2,3), 3, (3,6), 6, (6,5), 5, (5,7), 7\rangle)$. On the referred path, node 5 is labeled with value 5 and there exists another path from node 1 to node 7, e.g., $(\langle 1, (1,2), 2, (2,6), 6, (6,5), 5, (5,7), 7\rangle)$, where node 5 is labeled with the value 6. This is very important when we want to combine both problems, that is, to determine the shortest path with maximum capacity, or the maximum capacity shortest path. Although the last two problems may seem the same, they are indeed different. The shortest path with maximum capacity implies that one is looking for the maximum capacity path on the set of shortest paths and so, since the initial problem (shortest path) verifies the strong optimality principle, the combined problem may be solved using only one passage in the network. The maximum capacity shortest path implies that one is looking for the shortest path on the set of maximum capacity paths. Since the primary problem only verifies the weak optimality principle, in order to solve the combined problem we have to run twice the labeling correcting algorithm and to perform in between a slight network modification. These combined problems and possible applications will be discussed in the next section.

9.4 Combined network problems

A class of combined network problems often arise when, for example, one enterprise wants to know the path that has the lowest or highest price to buy or sell its items on the set of paths that have the highest capacity of shipping their products. The problem may also be considered as sending products the fastest as possible and the largest amount as possible as well. Although similar but not the same is the problem where one wants to obtain or send the largest amount possible (maximum capacity problem) at the lowest or highest price or as fast as possible. The first problem belongs to the set of problems where one wants to minimize (or maximize) prices (time) on the set of maximum capacity paths. The second type of problems belongs to the set of problems where we want to maximize capacity on the sets of minimum (or maximum) prices (time). The first type of problems do not verify optimality principles and thus, in order to solve them, two labeling algorithm runs have to be performed. On the other hand, the second type of problems verifies the weak optimality principle and thus only one run of a labeling algorithm is enough.

Consider Problem 2 of the previous section, where we determined the set of maximum capacity paths, and its modification into Problem 3.

9.4.1 Problem 3: shortest (minimize cost or time) path on the set of maximum capacity paths

An international company has one store to which it must deliver goods to be sold. Along the years this company has built a network of distributors to whom it sends the merchandise and this, subsequently, is forwarded to other distributors or to the store. Not all distributors have the same capacity to ship the goods. The time associated to each path is another parameter associated to each network arc. The figure below reflects the network of distributors and the company final store as well as the time and capacity to ship and dispose of the goods. What needs to be determined is the fastest path (which will imply minimized time) on the set of maximum capacity paths (Fig. 9.6).

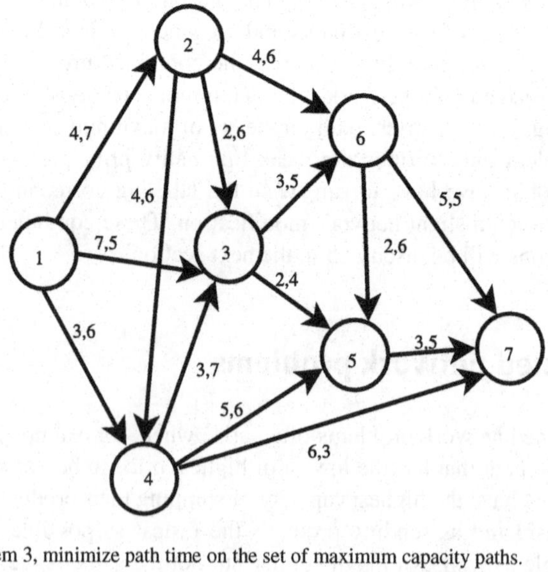

Figure 9.6 Problem 3, minimize path time on the set of maximum capacity paths.

The representation of this network in the forward star form is similar to the one made for Problem 2 in the previous section, but, since each arc has two parameters, we must have two vectors to store this information. Thus, together with vectors *pointer* and *suc* we will have the vectors *parameter_time* and *parameter_capacity*:

parameter_time =

4	7	3	2	4	4	2	3	3	5	6	3	2	5

parameter_capacity =

7	5	6	6	6	6	4	5	7	6	3	5	6	5

Since capacity problems do not verify any of the optimal path principles, Problem 3 has to be solved in two moments and the algorithm to obtain its solution is the following.

Algorithm for combined problems which do not verify any of the optimality principles

Step 1: Considering parameter capacity, determine the set of maximum capacity paths;
Step 2: Remove from the network all arcs where the capacity parameter is smaller than the maximum obtained in step 1;
Step 3: Considering parameter time, determine the set of shortest paths from 1 to 7.

Applying this algorithm to Problem 3 we obtain the following solution, see Fig. 9.7.

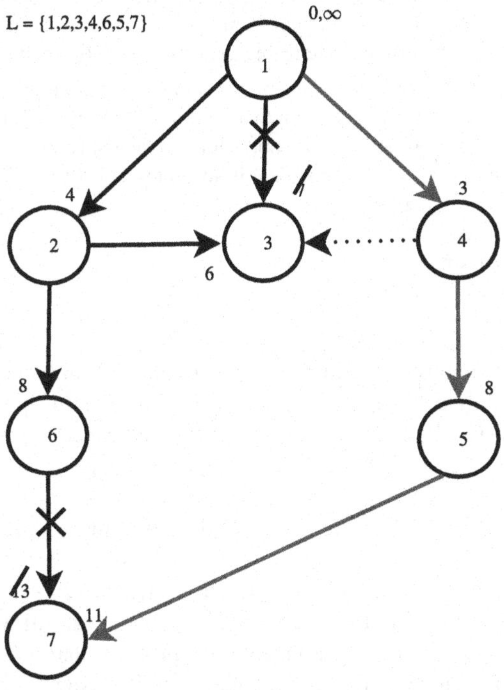

Figure 9.7 Problem 3, solution.

If one wants to solve the second type of problem referred to at the beginning of this section (the problem where one wants to obtain or send the largest amount possible [maximum capacity problem] at the lowest or highest price or as fast as possible) a labeling algorithm may be applied since the minimum path problem verifies the strong optimality principle. Thus only one passage through the network is necessary and only some small adjustments are needed to perform in the general labeling algorithm. The adequate algorithm to solve this type of problems is the following.

Algorithm for combined problems that verify optimality principles

```
Initialize vectors that are used in the labeling procedure
while the list of used labeled L is not empty do
    Remove node i from the list L
    For each arc k starting at node i
        Consider node j as the successor of arc k
        If (existent path from s to j distance is not better than the one
            obtained concatenating the path from s to i with arc k) then
                the path from s to j is actualized regarding distance and capacity
                the information that node j was labeled by node i is saved
                If node j does not belong to the list L then
                    the node j is added to the list L
                end if
        end if
        If (existent path from s to j distance is equal to the one
            obtained concatenating the path from s to i with arc k) then
                If (existent path from s to j capacity is not better than the one
            obtained concatenating the path from s to i with arc k) then
                the path from s to j is actualized regarding capacity
                the information that node j was labeled by node i is saved
                If node j does not belong to the list L then
                    the node j is added to the list L
                end if
                end if
        end if
    end for
end while
```

When applying the above algorithm to the network of Problem 3 in order to solve the problem of maximizing capacity into the set of minimum time paths (fastest paths), that is, the path for which the company is sure that the capacity distribution is maximized into the set of fastest delivery paths, we obtain the following solution, see Fig. 9.8.

Observe that there exists a dashed arc, which represents an alternative path from node 1 to node 7.

More applied problems exist around us! What we tried to show the reader with the examples exposed above is that to solve them one must be careful with the optimality principles and always verify their appliance to the problem that must be solved. Once that is verified, labeling algorithms are most helpful to obtain an online straightforward solution.

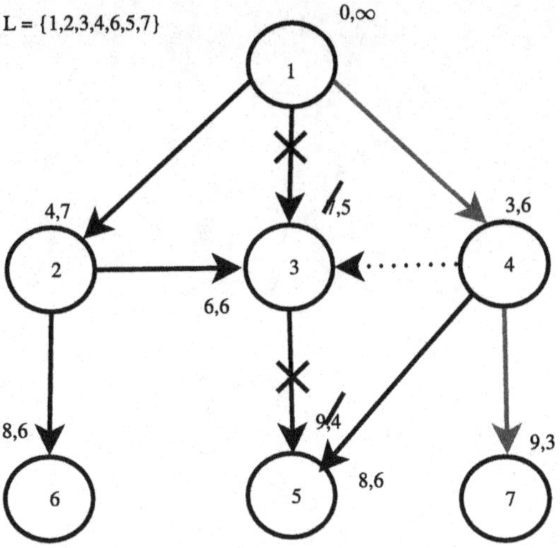

Figure 9.8 Problem 4, solution.

References

Ahuja, Ravindra K., Magnanti, T.L., Orlin, J.B., 1993. Network Flows: Theory, Algorithms, and Applications. Prentice Hall, Upper Saddle River, New Jersey 07458.

Bollobás, B., 2001. Random Graphs. Cambridge University Press, United Kingdom.

Dial, R., Glover, F., Karney, D., Klingman, D., 1979. A computational analysis of alternative algorithms and labeling techniques for finding shortest path trees. Networks 9 (3), 215–2482. https://onlinelibrary.wiley.com/doi/abs/10.1002/net.32300903042.

Martins, E., Pascoal, M., Rasteiro, D.D., Santos, J., 1999. The optimal path problem. Investigação Operacional 19-1, 43–60.

Random variables as arc parameters when solving shortest path problems

10

Deolinda M.L. Dias Rasteiro[a], *Nelson Chibeles-Martins*[b]
[a]Coimbra Polytechnic – ISEC, Coimbra, Portugal, [b]University NOVA of Lisbon, Caparica, Portugal

10.1 Background

A random graph, Bollobás (2001), is a collection of vertices and edges randomly connecting them pairwise. Starting with the theory of Paul Erdős and Albert Rényi (1959, 1960, 1961) random graph theory developed into one of the most important fields of modern discrete mathematics. In all of the studies we made the assumption that the presence or absence of any edge between two vertices is independent of the presence or absence of any other edge, so that each edge may be considered to be present with independent probability p. If there are N vertices in a graph and each is connected to an average of k vertices, then $p = \frac{k}{N-1}$. The number of edges connected to any particular vertex is called the degree d of that vertex, and has a probability distribution of p_d given by

$$p_d = \binom{N}{d} p^d (1-p)^{N-d} \approx \frac{k^d e^{-k}}{k!},$$

where the second equality becomes exact in the limit of large N. This is the Poisson distribution, i.e., the normal random graphs has a Poisson distribution of vertex degrees. Since random graphs were employed as models of real-world networks of various types such as epidemiology, networks of telephone calls, and airline timetables, as well as networks of physical systems (including neural networks), and it is found that the distribution, Aiello and Lu (2000), Amaral et al. (2002) of vertex degrees in many of these networks is substantially different from the Poisson distribution, which suggests that there are features of such networks that will the lost if we approximate them by a normal random graph. Newman et al. (2001) developed the theory of random graphs with arbitrary degree distributions and in addition to simple undirected graphs they examined the properties of directed and bipartite graphs.

Probabilistic networks

Different types of optimal path problems over random probabilistic networks are considered in the literature. The first publication belongs to Frank (1969), who determined

Calculus for Engineering Students. https://doi.org/10.1016/B978-0-12-817210-0.00017-5

the shortest path on a random graph and presented a process for obtaining the probability distribution of the shortest path.

One of the most utilized criteria to determine the optimal path is the criterion that maximizes the expected value of a utility function. This criterion stems from the von Neumann–Morgenstern formulation of how evaluations of preferences should be made under uncertainty Loui (1983).

Another criterion of optimization is to maximize the probability the optimal path length does not exceed some specified value Frank (1969), Andreatta et al. (1985). This criterion was utilized by Henig (1990) for the stochastic knapsack problem. Sigal et al. (1980) considered a problem where the optimal path was the one with the highest probability of being the shortest. The same problem was later studied by Kamburowski (1985) and by Adlakha (1986), where approximated solutions were obtained using Monte Carlo simulations.

Related with the stochastic optimal problem is the definition given by Jaillet (1992), who determined the path that minimizes the expected distance. He allows the existence of failure nodes, i.e., nodes that have some probability of failure. In 1991 Bard and Bennett (1991) developed heuristic methods based on Monte Carlo simulations for the stochastic optimal path with nonincreasing utility function. Their computational results regarded networks with 20 up to 60 nodes. Mirchandani and Soroush (1985) presented an algorithm for the same problem but with a quadratic utility function; more recently Murthy and Sarkar (1996, 1997, 1998) presented not only an algorithm for the quadratic case but also for the linear nonincreasing and quasilinear concave.

In this chapter we present an algorithm which determines the path maximizing the expected value of a utility function over a probabilistic network. The considered utility functions are linear, quadratic, and exponential. Our algorithm is based on multiobjective theory and is very efficient in terms of memory and time.

Notation, definitions, and properties

Let $(\mathcal{N}, \mathcal{A})$ be a given probabilistic network, where $\mathcal{N} = \{v_1, \ldots, v_n\}$ is the set of nodes and $\mathcal{A} = \{a_1, \ldots, a_m\} \subseteq \mathcal{N} \times \mathcal{N}$ is the set of arcs. In what follows, node v_i is sometimes denoted by i. Each arc $a_k \in \mathcal{A}$, $k \in \{1, \ldots, m\}$, is represented by the pair (i, j), where $i, j \in \{1, \ldots, n\}$. We assume that $\mathcal{N}$ and $\mathcal{A}$ are finite sets.

Consider two nodes i and j of $(\mathcal{N}, \mathcal{A})$. If all pairs (i, j) are ordered, i.e., if all the arcs of the network are directed, we will say that $(\mathcal{N}, \mathcal{A})$ is a directed network; otherwise we will say that $(\mathcal{N}, \mathcal{A})$ is an undirected network.

In what follows, without loss of generality, we assume that $(\mathcal{N}, \mathcal{A})$ is a directed network and that there are no arcs of the form (i, i), $i \in \mathcal{N}$.

Definition. Let s and t be two nodes of $\mathcal{N}$ such that $s \neq t$. We denote by s the initial node and by t the terminal node.

A *path*[1] p from s to t in $(\mathcal{N}, \mathcal{A})$ is an alternating sequence of nodes and arcs of the following form:

$$p = \langle s = v'_1, a'_1, v'_2, \ldots, a'_{l-1}, v'_l = t \rangle,$$

[1] For convenience sometimes we will represent a path by only one node.

where $l \geq 2$ and the following conditions hold:

- $v'_k \in \mathcal{N}, \forall k \in 1, \ldots, l,$
- $a'_k = (v'_k, v'_{k+1}) \in \mathcal{A}, \forall\, k \in 1, \ldots, l-1.$

Given two nodes $i,\ j \in \mathcal{N}$, we denote by $\mathcal{P}_{ij}$ the set of paths from node i to node j in $(\mathcal{N}, \mathcal{A})$; $\mathcal{P}$ is the set of paths from node s to t $(\mathcal{P}_{st})$. In what follows, we also assume that the sets $\mathcal{P}_{si}$ and $\mathcal{P}_{it}$ are nonempty for each $i \in \mathcal{N}$.

A *loopless path* from s to t is a path from s to t such that all nodes are distinct. A *cycle* (or a loop) is a path from a node to itself whose nodes are all distinct except for the first, which is also the last.

The symbol "$\diamond$" denotes the *concatenation* of paths where the terminal node of the first path is also the initial node of the second path, i.e., $p_{ij} \diamond p_{jk} = p_{ik}$.

Definition. A random variable X has *normal* distribution with the mean μ and the variance σ^2 $(\mu \in \mathbb{R},\ \sigma \in \mathbb{R}^+)$ and we write $X \sim N(\mu, \sigma^2)$ if X is continuous and its density function has the form

$$g(x) = \frac{1}{\sqrt{2\pi}\sigma} \exp\left(-\frac{1}{2}\left(\frac{x-\mu^2}{\sigma}\right)\right), \qquad \text{for each } x \in \mathbb{R}. \tag{10.1}$$

Remark 10.1. If $X_1, \ldots, X_n,\ n \in \mathbb{N}$, are n independent real random variables with normal distributions with parameters μ_i and σ_i^2, then the real random variable $\sum_{i=1}^{n} X_i$ also has normal distribution with the mean value $\sum_{i=1}^{n} \mu_i$ and the variance $\sum_{i=1}^{n} \sigma_i^2$. In what follows, this property is called stability of the normal distribution with respect to the sum.

10.2 Description of general problems and areas where they are very common

In the stochastic shortest path problem a directed probabilistic network $(\mathcal{N}, \mathcal{A})$ is given where each arc $(i, j) \in \mathcal{A}$ is associated with the mean value and the variance of real random variable X_{ij}, which is called the random cost of the arc $(i, j) \in \mathcal{A}$. We assume that the real random variables X_{ij} are normally distributed and independent, and have the mean values μ_{ij} and the variances σ_{ij}^2.

Associated to the path p, we define the real random variable $X_p = \sum_{(i,j)\in p} X_{ij}$ representing the random cost of the loopless path $p \in \mathcal{P}$. Using the stability of the normal distribution with respect to the sum, we obtain

$$X_p \sim N\left(\sum_{(i,j)\in p} \mu_{ij}, \sum_{(i,j)\in p} \sigma_{ij}^2\right).$$

We denote by μ_p and σ_p^2 the mean value and the variance of X_p, respectively.

If an appropriate utility is assigned to each possible consequence and the expected utility of each alternative is calculated, then the best action is to consider the alternative with the highest expected utility. Different axioms that imply the existence of utilities with the property that expected utility is an appropriate guide to consistent decision making are presented in von Neumann and Mogenstern (1947), Savage (1954), Luce and Raiffa (1957), Pratt et al. (1965), Fishburn (1970). The choice of the adequate utility function for a specific type of problem can be taken using direct methods presented in Keeney and Raiffa's book.

With the objective of determining the optimal path, we consider a real function $\mathcal{U}: \mathcal{P} \longrightarrow \mathbb{R}$, called *utility function*, such that for each loopless path p, $\mathcal{U}(p)$ depends on the random variables associated to the arcs of p.

In the stochastic shortest path, we pretend to determine the loopless path $p^\star \in \mathcal{P}$ that maximizes the expected value of the utility function. The loopless path $p^\star$ is called *optimal solution* of the referred problem.

10.2.1 Linear utility function

There are problems were the utility of the path p is a nonincreasing function of d, which represents the amount of available resource. The value of each loopless path $p \in \mathcal{P}$ is defined as follows:

$$\mathcal{U}\left(X_p\right) = \begin{cases} a - bX_p & X_p \le d, \\ 0 & X_p > d, \end{cases}$$

where $a,\ b,\ d \in \mathbb{R}^+$ are such that $a - bd \ge 0$ (then $\mathcal{U}\left(X_p\right)$ is nonnegative).

The probability density function associated to the loopless path p is

$$g_p(x) = \frac{1}{\sqrt{2\pi}\sigma_p}\, e^{\left(-\frac{1}{2}\left(\frac{x-\mu_p}{\sigma_p}\right)^2\right)}, \quad \text{for each } x \in \mathbb{R}. \tag{10.2}$$

The expected value of the utility function at $p \in \mathcal{P}$, $\mathrm{E}\left(\mathcal{U}(X_p)\right)$, is then

$$\begin{aligned} \mathrm{E}\left(\mathcal{U}(X_p)\right) &= \int_{\mathbb{R}} \mathcal{U}(x) g_p(x)\, dx \\ &= \int_{-\infty}^{d} (a - bx)\frac{1}{\sqrt{2\pi}\sigma_p}\, e^{\left(-\frac{1}{2}\left(\frac{x-\mu_p}{\sigma_p}\right)^2\right)} dx \\ &= \int_{-\infty}^{\frac{d-\mu_p}{\sigma_p}} (a - b\sigma_p y - b\mu_p)\frac{1}{\sqrt{2\pi}}\, e^{\left(\frac{y^2}{2}\right)} dy \end{aligned}$$

$$= (a - b\mu_p)P\left(Z \leq \frac{d-\mu_p}{\sigma_p}\right) + b\sigma_p \frac{1}{\sqrt{2\pi}} e^{\left(-\frac{1}{2}\left(\frac{d-\mu_p}{\sigma_p}\right)^2\right)}$$

$$= (a - b\mu_p)G\left(\frac{d-\mu_p}{\sigma_p}\right) + b\sigma_p g\left(\frac{d-\mu_p}{\sigma_p}\right),$$

where $G(\cdot)$ and $g(\cdot)$ are the probability density and distribution functions of the random variable $Z \sim N(0, 1)$, respectively.

Therefore, in the case of linear utility function, the stochastic shortest path is

$$\underset{p\in\mathcal{P}}{\mathrm{m\ x}} \left[(a - b\mu_p)G\left(\frac{d-\mu_p}{\sigma_p}\right) + b\sigma_p g\left(\frac{d-\mu_p}{\sigma_p}\right)\right]. \tag{10.3}$$

The network optimization problems are, in general, solved by using labeling algorithms Martins et al. (1999). These algorithms are improvements of the exhaustive search by the optimality principle definition. We define the weak and the strong optimality principles only in $\mathbb{R}^2$ since that is sufficient for this problem. Therefore, consider the network optimization problem (P).

Definition (*strong optimality principle*). Let i and j be two nodes of $\mathcal{N}$. We say that problem (P) satisfies the strong optimality principle if any nondominated path from i to j in $(\mathcal{N}, \mathcal{A})$ is formed by nondominated subpaths.

Definition (*weak optimality principle*). We say that problem (P) satisfies the weak optimality principle if, for any $i, j \in \mathcal{N}$ and any pair $\left(\mu_q, \sigma_q^2\right) \in \left\{\left(\mu_p, \sigma_p^2\right), p \in \mathcal{P}_{ij}\right\}$, there exists at least a nondominated path, with random cost X_q from i to j in $(\mathcal{N}, \mathcal{A})$, that is formed by nondominated subpaths.

If a problem satisfies at least the weak optimality principle, then it can be solved by labeling algorithms, but unfortunately that does not happen for the stochastic optimal path problem, as one can observe in the example shown in Fig. 10.1.

In this example, we can observe that the path from node 1 to node 4 that maximizes the expected value of the utility function is the path $\langle 1, (1,3), 3, (3,4), 4\rangle$, with optimal value 89.249. Nevertheless, this path is not formed by optimal subpaths once the path $\langle 1, (1,3), 3\rangle$ has value 88.686 but there exists another path $\langle 1, (1,2), 2, (2,3), 3\rangle$ with value 98.

Thus, in order to solve problem (10.3), we need a different approach. The partial derivatives of the objective function of problem (10.3) are

$$\frac{\partial \mathrm{E}\left(\mathcal{U}(X_p)\right)}{\partial \mu_p} = -bG\left(\frac{d-\mu_p}{\sigma_p}\right) - \frac{a-bd}{\sigma_p} g\left(\frac{d-\mu_p}{\sigma_p}\right),$$

$$\frac{\partial \mathrm{E}\left(\mathcal{U}(X_p)\right)}{\partial \sigma_p^2} = \frac{g\left(\frac{d-\mu_p}{\sigma_p}\right)}{2\left(\sigma_p^2\right)^{\frac{3}{2}}}\left[(a-bd)\mu_p + b\sigma_p^2 - d(a-bd)\right].$$

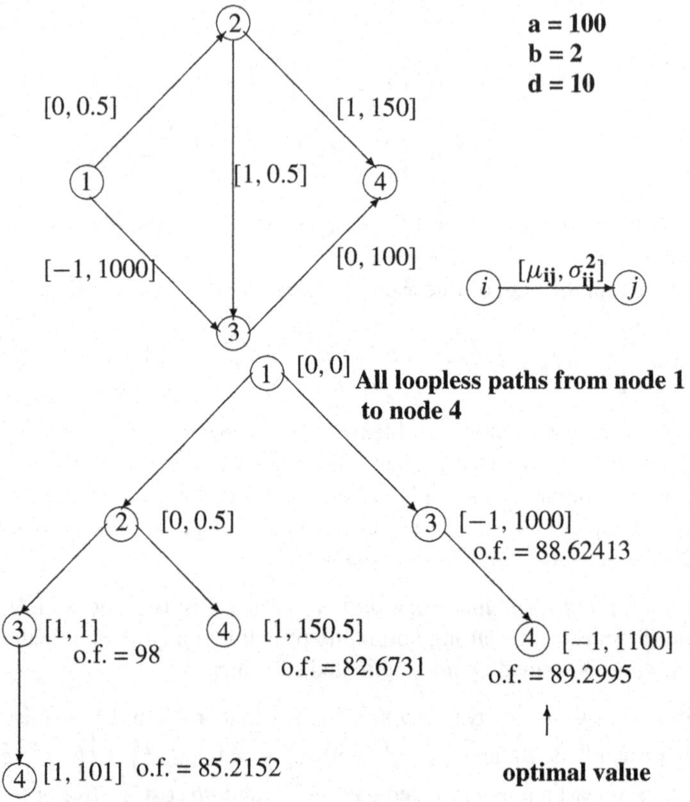

Figure 10.1 Shortest path loopless paths tree.

Since we have assumed that $a - bd \geq 0$ and $b > 0$ and, by definition, the density and distribution functions are nonnegative, we see that

$$\frac{\partial \mathrm{E}\left(\mathcal{U}(X_p)\right)}{\partial \mu_p} \leq 0.$$

Therefore, for the *same variance value, it is preferable a loopless path with smaller mean value*. The sign of $\dfrac{\partial \mathrm{E}\left(\mathcal{U}(X_p)\right)}{\partial \sigma_p^2}$ depends on the sign of the function

$$D\left(\mu_p, \sigma_p^2\right) = (a - bd)\mu_p + b\sigma_p^2 - d(a - bd).$$

Thus, if the random cost of the loopless path $p \in \mathcal{P}$ is such that $D\left(\mu_p, \sigma_p^2\right)$ is nonpositive, then the expected value of $\mathcal{U}(X_p)$ is a nonincreasing function of μ_p and σ_p^2.

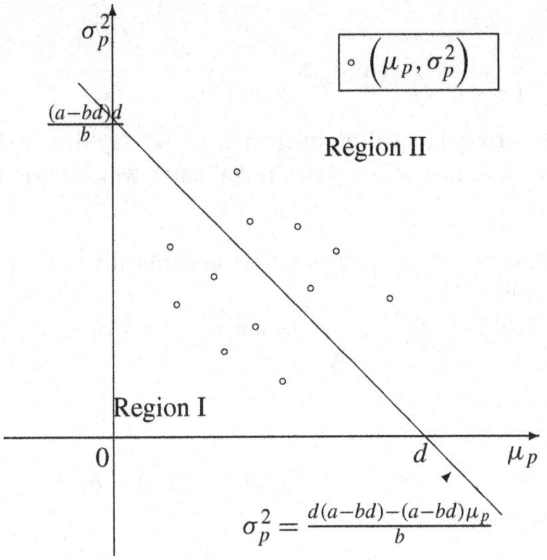

Figure 10.2 Admissible searching regions.

The optimal loopless path, $p^\star \in \mathcal{P}$, for problem (10.3) is such that

$$X_{p^\star} \sim N\left(\min \sum_{(i,j)\in p} \mu_{ij}, \min \sum_{(i,j)\in p} \sigma_{ij}^2\right). \tag{10.4}$$

If $D\left(\mu_p, \sigma_p^2\right)$ is positive, then the expected value of $\mathcal{U}(X_p)$ is a nonincreasing function of μ_p and a nondecreasing function of σ_p^2. Therefore, in this case, we will search the path $p^\star$ such that

$$X_{p^\star} \sim N\left(\min \sum_{(i,j)\in p} \mu_{ij}, \mathrm{m\,x} \sum_{(i,j)\in p} \sigma_{ij}^2\right). \tag{10.5}$$

In conclusion, we can state that $D\left(\mu_p, \sigma_p^2\right) = 0$ separates the referential $\mu_p O \sigma_p^2$ in two search regions of $p^\star$. (See Fig. 10.2.)

With the goal of determining the optimal solution of problem (10.3), some results and definitions are necessary Martins et al. (1999), Santos (1997). Let $\mathcal{T}$ be the tree of loopless paths from node s to node t, the dominance relation on $\mathcal{T}$ as the following definition.

Definition. Let $(\mathcal{N}, \mathcal{A})$ be a network. Let $i, j \in \mathcal{N}$ and let p and q be two paths of $\mathcal{T}_{ij}$ (tree of the paths from i to j in $(\mathcal{N}, \mathcal{A})$). The *dominance relation* in $\bigcup_{i,j\in\mathcal{N}} \mathcal{T}_{ij}$, denoted by "$\preceq$," is defined as follows:

$$p \preceq q \Leftrightarrow \left(\mu_p, \sigma_p^2\right) \preceq \left(\mu_q, \sigma_q^2\right),$$

i.e.,

$$p \preceq q \Leftrightarrow \mu_p \leq \mu_q \ \wedge \ \sigma_p^2 \leq \sigma_q^2,$$

with at least one strong inequality. In such case, we say that p *dominates* q or q *is dominated* by p. Note that we only can relate paths which have the same initial and terminal nodes.

The "$\preceq$" relation in $\bigcup_{i,j \in \mathcal{N}} \mathcal{T}_{ij}$ is reflexive and transitive, i.e., it is a partial preorder relation Santos (1997).

If all the loopless paths extensions p_{si} and q_{si} of node $i \in \mathcal{N}$ in $\mathcal{P}$ belong to region I, then p_{si} dominates q_{si} if and only if

$$\mu_{p_{si}} \leq \mu_{q_{si}} \ \wedge \ \sigma_{p_{si}}^2 \geq \sigma_{q_{si}}^2,$$

with at least one strong inequality. Similarly, if p_{si} and q_{si} belong to region II, then p_{si} dominates q_{si} if and only if

$$\mu_{p_{si}} \leq \mu_{q_{si}} \ \wedge \ \sigma_{p_{si}}^2 \leq \sigma_{q_{si}}^2,$$

with at least one strong inequality.

Theorem 10.1. *Let (P) be the problem of determining the loopless path $p^\star \in \mathcal{P}$ that is a nondominated solution of $(f_1(p), f_2(p))$, where $f_i(p)$, $i = 1, 2$, is the maximum or the minimum of $\sum_{(i,j)\in p} \mu_{ij}$ or $\sum_{(i,j)\in p} \sigma_{ij}^2$.*

The problem (P) satisfies the weak optimality principle if the network $(\mathcal{N}, \mathcal{A})$ has no positive or negative cycles.

Proof. Without loss of generality, consider

$$(f_1(p), f_2(p)) = \left(\min \sum_{(i,j)\in p} \mu_{ij}, \mathrm{m\ x} \sum_{(i,j)\in p} \sigma_{ij}^2 \right).$$

We prove that the problem (P) satisfies the weak optimality principle.

Let $p^\star$ be a nondominated loopless path from i to $j \in \mathcal{N}$. Suppose that there exists a node $k \in p^\star$ such that the loopless subpath $p_{ik}^\star \subset p^\star$ is a dominated path from i to k. Then there also exists a loopless path $q \in \mathcal{P}_{ik}$ such that

$$\mu_q \leq \mu_{p_{ik}^\star} \ \wedge \ \sigma_q^2 \geq \sigma_{p_{ik}^\star}^2,$$

with at least one strong inequality. Therefore,

$$f_1(q) \leq f_1(p_{ik}^\star) \wedge f_2(q) \geq f_2(p_{ik}^\star)$$

also with at least one strong inequality. Since $p^\star = p_{si}^\star \diamond p_{ik}^\star \diamond p_{kt}^\star$, it follows that $\mu_{p^\star} = \mu_{p_{si}^\star} + \mu_{p_{ik}^\star} + \mu_{p_{kt}^\star} \geq \mu_{p_{si}^\star} + \mu_q + \mu_{p_{kt}^\star}$ and $\sigma_{p^\star}^2 = \sigma_{p_{si}^\star}^2 + \sigma_{p_{ik}^\star}^2 + \sigma_{p_{kt}^\star}^2 \leq \sigma_{p_{si}^\star}^2 + \sigma_q^2 + \sigma_{p_{kt}^\star}^2$,

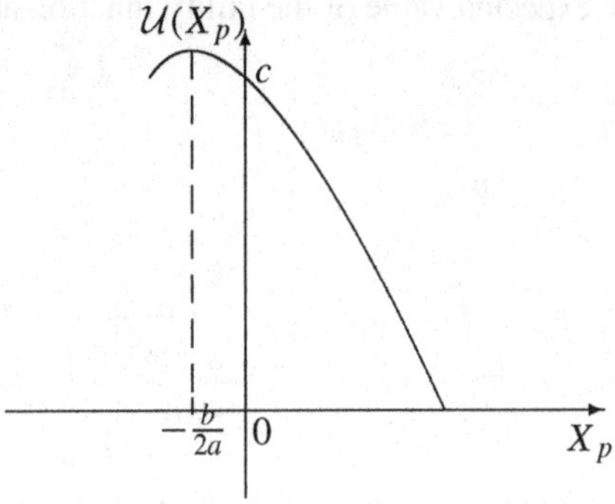

Figure 10.3 Quadratic utility function.

with at least one strong inequality, and thus the loopless path $p^\star$ is dominated, which is impossible. Since k was chosen arbitrarily in the set of nodes of $p^\star$, it follows that all the loopless subpaths of $p^\star$ are nondominated.

Similarly, the remaining cases of $(f_1(p), f_2(p))$ can be easily proved. Note that the network $(\mathcal{N}, \mathcal{A})$ must not contain negative (positive) cycles in order for the minimization (maximization) problem to satisfy the weak optimality principle. □

Theorem 10.1 implies that the biobjective optimal path problem defined above satisfies the weak optimality principle in each of the regions I or II defined above. Therefore, its optimal solution in the linear case is a nondominated solution of one of the biobjective problems from region I or II. Thus, the following algorithm is proposed:

- obtain all nondominated solutions of region I;
- obtain all nondominated solutions of region II;
- among the solutions obtained above, determine the solution corresponding to the loopless path $p^* \in P$ such that X_{p^*} maximizes

$$\left\{(a - b\mu_p)\mathrm{G}\left(\frac{d-\mu_p}{\sigma_p}\right) + b\sigma_p g\left(\frac{d-\mu_p}{\sigma_p}\right)\right\}.$$

10.2.2 Quadratic utility function

In this subsection, we assume that the utility function of the path p is a quadratic nonincreasing function. Moreover, $\mathcal{U}(X_p)$ is defined by the formula

$$\mathcal{U}\left(X_p\right) = -aX_p^2 - bX_p + c,$$

where $a,\ c \in \mathbb{R}^+$ and $b \in \mathbb{R}_0^+$. Its graphic representation is shown in Fig. 10.3.

In this case, the expected value of the utility function at $p \in \mathcal{P}$, $\mathrm{E}\left(\mathcal{U}(X_p)\right)$, is

$$\begin{aligned}
\mathrm{E}\left(\mathcal{U}(X_p)\right) &= \int_{\mathbb{R}} \mathcal{U}(x) g_p(x)\, dx \\
&= \int_{-\infty}^{+\infty} (-ax^2 - bx + c) \frac{1}{\sqrt{2\pi}\sigma_p} e^{\left(-\frac{1}{2}\left(\frac{x-\mu_p}{\sigma_p}\right)^2\right)} dx \\
&= c - b\mu_p - a\mu_p^2 - a\sigma_p^2.
\end{aligned}$$

Hence, one can define the stochastic shortest path as follows:

$$\underset{p \in \mathcal{P}}{\mathrm{m\ x}} \left[c - b\mu_p - a\mu_p^2 - a\sigma_p^2\right]. \tag{10.6}$$

Determining the partial derivatives of the objective function of problem (10.6), we obtain

$$\frac{\partial \mathrm{E}\left(\mathcal{U}(X_p)\right)}{\partial \mu_p} = -b - 2a\mu_p, \qquad \frac{\partial \mathrm{E}\left(\mathcal{U}(X_p)\right)}{\partial \sigma_p^2} = -a.$$

By the assumption $a \in \mathbb{R}^+$, we have

$$\frac{\partial \mathrm{E}\left(\mathcal{U}(X_p)\right)}{\partial \sigma_p^2} < 0, \ \forall p \in \mathcal{P}.$$

Moreover, if for a given loopless path p we have $\mu_p > -\dfrac{b}{2a}$, then

$$\frac{\partial \mathrm{E}\left(\mathcal{U}(X_p)\right)}{\partial \mu_p} < 0.$$

Thus, an optimal loopless path $p^\star \in \mathcal{P}$ for problem (10.7) is such that

$$X_{p^\star} \sim N\left(\min \sum_{(i,j)\in p} \mu_{ij}, \min \sum_{(i,j)\in p} \sigma_{ij}^2\right). \tag{10.7}$$

For a given node $i \in \mathcal{N}$ and two loopless paths p_{si} and q_{si} in $\mathcal{P}$, we say that p_{si} dominates q_{si} if and only if

$$\mu_{p_{si}} \leq \mu_{q_{si}} \ \wedge \ \sigma_{p_{si}}^2 \leq \sigma_{q_{si}}^2,$$

with at least one strong inequality.

Therefore, the optimal solution for the problem in the quadratic case is a nondominated solution of the biobjective problem

$$\left(\min_{p \in \mathcal{P}} \sum_{(i,j) \in \mathcal{P}} \mu_{ij}, \min_{p \in \mathcal{P}} \sum_{(i,j) \in \mathcal{P}} \sigma_{ij}^2 \right), \tag{10.8}$$

which satisfies the weak optimality principle if there are no negative cycles in the network.

In this case the proposed algorithm is as follows:

- obtain all nondominated solutions of problem (10.8);
- among the solutions obtained above, select the solution corresponding to the loopless path $p^* \in P$ such that X_{p^*} maximizes

$$\left\{ c - b\mu_p - a\mu_p^2 - a\sigma_p^2 \right\}.$$

10.2.3 Exponential utility function

In this subsection, we assume that the utility function of the path p is nonincreasing. Moreover, $\mathcal{U}(X_p)$ is defined by the formula

$$\mathcal{U}\left(X_p\right) = a\, e^{-bX_p},$$

where $a \in \mathbb{R}^+$ and $b \in \mathbb{R}_0^+$.

In this case, the expected value of the utility function at $p \in \mathcal{P}$, $\mathrm{E}\left(\mathcal{U}(X_p)\right)$, is

$$\begin{aligned}
\mathrm{E}\left(\mathcal{U}(X_p)\right) &= \int_{\mathbb{R}} \mathcal{U}(x) g_p(x)\, dx \\
&= \int_{-\infty}^{+\infty} (a\, e^{-bx}) \frac{1}{\sqrt{2\pi}\sigma_p}\, e^{\left(-\frac{1}{2}\left(\frac{x-\mu_p}{\sigma_p}\right)^2\right)} dx \\
&= a\, e^{\dfrac{-\mu_p^2 + (\mu_p - \sigma_p^2 b)^2}{2\sigma_p^2}} \\
&\quad \times \int_{-\infty}^{+\infty} \frac{1}{\sqrt{2\pi}\sigma_p}\, e^{\left(-\frac{1}{2}\left(\frac{x-(\mu_p - \sigma_p^2 b)}{\sigma_p}\right)^2\right)} dx \\
&= a\, e^{\left(-\mu_p b + \frac{\sigma_p^2 b^2}{2}\right)}.
\end{aligned}$$

Hence, one can define the stochastic shortest path as follows:

$$\underset{p\in\mathcal{P}}{\mathrm{m\ x}}\left[a\,e^{\left(-\mu_p b+\frac{\sigma_p^2 b^2}{2}\right)}\right]. \tag{10.9}$$

Determining the partial derivatives of the objective function of problem (10.9), we obtain

$$\frac{\partial \mathrm{E}\left(\mathcal{U}(X_p)\right)}{\partial \mu_p}=-ab\,e^{\left(-\mu_p b+\frac{\sigma_p^2 b^2}{2}\right)}$$

$$\frac{\partial \mathrm{E}\left(\mathcal{U}(X_p)\right)}{\partial \sigma_p^2}=a\,\frac{b^2}{2}\,e^{\left(-\mu_p b+\frac{\sigma_p^2 b^2}{2}\right)}.$$

By the assumptions $a\in\mathbb{R}^+$ and $b\in\mathbb{R}_0^+$, we have

$$\frac{\partial \mathrm{E}\left(\mathcal{U}(X_p)\right)}{\partial \mu_p}<0,\quad \frac{\partial \mathrm{E}\left(\mathcal{U}(X_p)\right)}{\partial \sigma_p^2}>0,\ \forall p\in\mathcal{P}.$$

Thus, an optimal loopless path $p^\star\in\mathcal{P}$ for problem (10.9) is such that

$$X_{p^\star}\sim N\left(\min\sum_{(i,j)\in p}\mu_{ij},\mathrm{m\ x}\sum_{(i,j)\in p}\sigma_{ij}^2\right). \tag{10.10}$$

For a given node $i\in\mathcal{N}$ and two loopless paths p_{si} and q_{si} in $\mathcal{P}$, we say that p_{si} dominates q_{si} if and only if

$$\mu_{p_{si}}\le\mu_{q_{si}}\ \wedge\ \sigma_{p_{si}}^2\ge\sigma_{q_{si}}^2,$$

with at least one strong inequality.

Therefore, the optimal solution for the problem in the exponential case is a non-dominated solution of the biobjective problem

$$\left(\min_{p\in\mathcal{P}}\sum_{(i,j)\in\mathcal{P}}\mu_{ij},\underset{p\in\mathcal{P}}{\mathrm{m\ x}}\sum_{(i,j)\in\mathcal{P}}\sigma_{ij}^2\right), \tag{10.11}$$

which satisfies the weak optimality principle if there are no negative cycles in the network.

In this case, the proposed algorithm is as follows:

- obtain all nondominated solutions of problem (10.11);
- among the solutions obtained above, select the solution corresponding to the loopless path $p^* \in P$ such that X_{p^*} maximizes

$$\left\{ a\, e^{\left(-\mu_p b + \frac{\sigma_p^2 b^2}{2}\right)} \right\}.$$

Observe that the nondominated solutions of the various biobjective problems can all be obtained by using a Dijkstra-like algorithm that respects the dominance relations defined above.

10.3 Real problems

The police patrols problem

What is the police patrols problem? Using police slang, the "beat cop" is a police agent to whom a set of streets (the beat) is assigned that he will patrol during his shift, and it is one weapon used in crime fighting. The patrolling objective is the prevention of felonies, due to the presence of authority agents in the streets. On an ideal situation, it would be possible to have a police agent on every city corner, but that is actually unpractical.

It is necessary to choose which streets should be assigned to each agent in the most useful way possible, i.e., the patrolling scheme should prevent the highest number of crimes.

Nowadays, in Portugal, cities are divided in squads, and the task of assigning streets to agents is responsibility of the squad chief. As the number of available agents is insufficient, knowing which streets should be patrolled in a efficient way is fundamental. How to evaluate the efficiency of a patrol system?

• Crime potential

To evaluate the quality of a patrol we need to create an index that, in a way, indicates what is the potential for crime occurrence in that street. A statistical analysis was applied to historical data extracted from police databases. The evolution of criminality through the last years was studied by Chibeles-Martins (1999). According to this report, the expected number of crimes declared to police in a given time unit is proposed as a measure for a street crime potential. In the present study we considered the month as the time unit. Because of the used algorithm it was also necessary to estimate the variance (and the standard deviation) of the number of crimes in each street. Thus, we can define *crime potential of a patrol* as the sum of the crime potential of every street patrolled. An urban zone can be considered as a graph G, where its streets are the edges and streets' extremities are the vertices. In the particular case of walking patrols, street direction is irrelevant, so G is an undirected graph. Between two adjacent vertices there is one and only one edge. A beat, i.e., the set of streets assigned to one agent, is a subgraph of G. In this study a beat is defined as a path between two vertices, not necessarily different (it can be a closed path). Suppose that there are k agents available. The goal is to find k beats, maximizing the crime potential of the set of streets patrolled by the k beats. The algorithm presented here will be used to fulfill this goal.

• **The algorithm**

The algorithm used to solve this problem, known as ORCA, was proposed by Rasteiro and Anjo (2004). Let G be a directed network where the cost of each arc a_{ij} is random with the $Normal(\mu_{ij}, \sigma_{ij})$ distribution. The objective is to find the shortest path between any two vertices. The algorithm proposes the maximization of the expected value of an adequate utility function. However, the optimization of that expected value does not verify the strong or weak optimality principles, so it is necessary to define a dominating relation between two paths. This relation is constructed using the expected values and the variances of paths' costs Rasteiro and Anjo (2004). The best path is the one which maximizes the expected value of the utility function, chosen in the set of nondominated paths. To apply the ORCA algorithm to determine optimal beats it is necessary to make some adaptations to our problem:

1. Build an adequate utility function.
2. Transform the undirected graph $G(V, A)$ into a directed $G'(V, A')$. The vertices of G' will be the same of G. Let n be the number of vertices. To each edge of G two arcs will be created in G', one with the inverse direction of the other. If $a_{ij} \in A \Rightarrow a_{ij} \in A' \bigwedge a_{ji} \in A'$. Obviously, if G has m edges, G' will have $2m$ arcs.
3. Create in G' two fictitious super-vertices s and t, that are the initial super-vertex and the final super-vertex, respectively. $2n$ fictitious arcs are also created, with null expected cost and variance. Each node became the end of a fictitious arc successor of s and the beginning of another fictitious arc that ends in t. So G' has now $n+2$ vertices and $m+2n$ arcs.
4. Find the optimal path between s and t, using the ORCA algorithm. This path is the first beat t_1.
5. Remove every arc belonging to t_1 from G'.
6. Repeat 4 and 5 until k beats are obtained.

• **Adaptations to the problem**

Definition of criminality utility function

It seems natural to consider a crescent utility function for criminality. Streets with higher criminality should be considered for being patrolled with a higher priority. And, as the monthly expected number of felonies is obviously nonnegative, the utility of negative values should be null. In this study a linear utility function is considered:

$$U(X) = \begin{cases} 0 & \text{if } X < 0, \\ X & \text{if } X \geqslant 0. \end{cases} \tag{10.12}$$

A graphical representation of $U(X)$ is presented in Fig. 10.4.

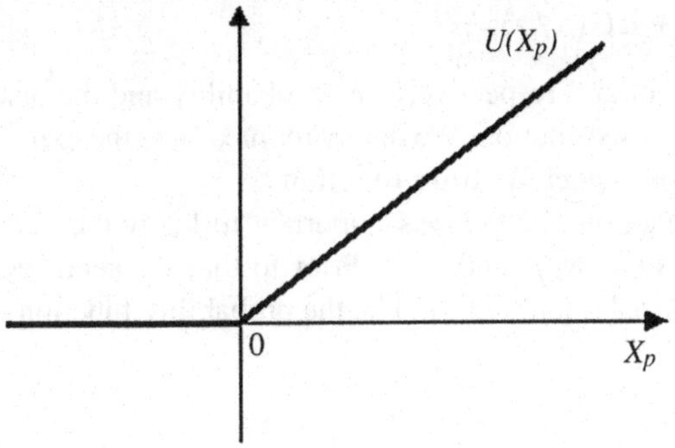

Figure 10.4 Linear utility function.

The expected value of utility function is the following integral:

$$E(U(x)) = \int_{-\infty}^{+\infty} U(x) f(x) dx, \tag{10.13}$$

where $f(x)$ if the probability function of the normal distribution with expected value μ and variance σ^2.

The integral (10.13) can be written as

$$E(U(X)) = \int_{0}^{+\infty} x \frac{e^{\frac{-(x-\mu)^2}{2\sigma^2}}}{\sqrt{2\pi\sigma^2}} dx. \tag{10.14}$$

Substituting x for the variable $y = \frac{x-\mu}{\sigma}$, we have

$$= \int_{-\mu/\sigma}^{+\infty} (\sigma y - \mu) \frac{e^{\frac{-y^2}{2}}}{\sqrt{2\pi}} dy = \tag{10.15}$$

$$= \frac{\sigma}{\sqrt{2\pi}} \int_{-\mu/\sigma}^{+\infty} y e^{\frac{-y^2}{2}} dy - \mu \int_{-\mu/\sigma}^{+\infty} \frac{e^{\frac{-y^2}{2}}}{\sqrt{2\pi}} dy = \tag{10.16}$$

The first integral is convergent and easily determined. And using normal distribution symmetry on the second integral, we obtain

$$= \frac{-\sigma}{\sqrt{2\pi}} \lim_{t \to +\infty} \left[e^{\frac{-y^2}{2}} \right]_{-\mu/\sigma}^{t} + \mu \int_{-\infty}^{\mu/\sigma} \frac{e^{\frac{-y^2}{2}}}{\sqrt{2\pi}} dy = \tag{10.17}$$

$$= \frac{\sigma e^{\frac{-\mu^2}{2\sigma^2}}}{\sqrt{2\pi}} + \mu G(\mu/\sigma) = \tag{10.18}$$

$$= \sigma g(\mu/\sigma) + \mu G(\mu/\sigma), \tag{10.19}$$

where $g(x)$ and $G(x)$ are, respectively, the probability and the distribution functions of the $Normal(0, 1)$ distribution. We intend to maximize the expression (10.19).

Maximization of expected utility function

To maximize function (10.19) it is important to know the behavior of its partial derivatives with respect to μ and to σ. Prior to that we shall recall a useful result about the normal distribution. Let $g(x)$ be the probability function of a $Normal(0, 1)$ distribution:

$$g(x) = \frac{e^{\frac{-x^2}{2}}}{\sqrt{2\pi}}. \tag{10.20}$$

Its derivative with respect to x is

$$g'(x) = -x\frac{e^{\frac{-x^2}{2}}}{\sqrt{2\pi}} = -xg(x). \tag{10.21}$$

But if the argument of g is another function $u(\alpha)$, then the partial derivative of g with respect to α will be

$$\frac{\partial g(u(\alpha))}{\partial \alpha} = -u(\alpha)g(u(\alpha))\frac{\partial u}{\partial \alpha}. \tag{10.22}$$

This result will be useful in the following. Deriving (10.19) in order to μ, we obtain

$$\frac{\partial E(U(X))}{\partial \mu} = \sigma\frac{\partial g(\mu/\sigma))}{\partial \mu} + G(\mu/\sigma) + \mu\frac{g(\mu/\sigma)}{\sigma} = \tag{10.23}$$

Applying (10.22),

$$= \sigma(-\mu/\sigma)g(\mu/\sigma)(1/\sigma) + G(\mu/\sigma) + \mu\frac{g(\mu/\sigma)}{\sigma} = \tag{10.24}$$

$$= \frac{-\mu}{\sigma}g(\mu/\sigma) + G(\mu/\sigma) + \frac{\mu}{\sigma}g(\mu/\sigma) = G(\mu/\sigma). \tag{10.25}$$

But $G(x)$ is always nonnegative, so $E(U(X))$ grows with μ.

Deriving (10.19) in order to σ, we have

$$\frac{\partial E(U(X))}{\partial \sigma} = g(\mu/\sigma) + \sigma\frac{\partial g(\mu/\sigma)}{\partial \sigma} + \mu g(\mu/\sigma)\frac{-\mu}{\sigma^2} = \tag{10.26}$$

Applying (10.22) again,

$$= g(\mu/\sigma) + \sigma\frac{\mu^2}{\sigma^3}g(\mu/\sigma) - \frac{\mu^2}{\sigma^2}g(\mu/\sigma) = g(\mu/\sigma). \tag{10.27}$$

The function $g(x)$ is always positive, and we can conclude that $E(U(X))$ grows with σ too. It is possible to build a dominating relation between two paths p_1 and p_2, with $Normal(\mu_1, \sigma_1)$ and $Normal(\mu_2, \sigma_2)$ distributions, respectively:

$p_1 \prec p_2$ if and only if $\mu_1 \leq \mu_2 \wedge, \sigma_1 \leq \sigma_2$
with at least one of the inequalities being strict.

Next we determinate nondominated paths, with respect to the relation defined above. A labeling algorithm could be used Rasteiro and Anjo (2004) and Martins et al. (1999).

Due to problem characteristics, we want to find paths with a certain length, so in each step of the labeling algorithm the path length should also be added to the nodes' labels. Paths bigger than allowed length should be eliminated. We recall that the path length is defined by the squad chief.

The optimal path is the one that maximizes (10.19), chosen from nondominated paths. The value for $G(x)$ can be approximated by polynomial interpolation.

After removing used arcs, we repeat the procedure until the best k are obtained.

- **Lisbon's problem**

1. **Introduction**

This algorithm was applied to Lisbon, specifically to three squads of the Portuguese capital. These squads were chosen because they represented three of the most usual types of Lisbon squads:

a. downtown squads, in the center of the city, with small influence area;
b. periphery squads, with large influence areas; and
c. squads belonging to an urban corona between downtown and periphery, that have areas with intermediate sizes.

For confidential reasons they are identified as *A*, *B*, and *C*. The three zones characteristics are shown in Table 10.1. The number of edges and vertices presented refer to the initial nondirected graph.

"Length sum" is the sum of the lengths of all edges and "Crime pot." is the sum of criminality potential of all edges. Area is measured in km^2.

Table 10.1 Squads characteristics.

Squad	Vertices	Edges	Length sum	Crime pot.	Area
A	159	239	14683.4	28.58	0.72
B	344	467	51222.8	78.75	3.73
C	686	1012	95808.6	61.67	7.15

It should be emphasized that *B* has a higher criminal potential than *C*, but has a significantly smaller influence area.

We intend to find k beats, with approximately 1000 meters of length. The value of k depends on the squad being studied. A tolerance of 10% is assumed, i.e., the only feasible paths have length between 900 and 1100 meters.

The time unit considered is the month, meaning that the criminality index for each arc is the monthly average number of crimes declared to police that occurred in that arc.

It was not possible to compare the results obtained with patrolling schemes presently used by police, so for each squad a thousand solutions with similar characteristics (i.e., with the number of beats equal to the solution obtained by the ORCA algorithm) were randomly generated.

With these solutions we could compare the criminality covered by the algorithm solution with the expected criminality covered by the random ones.

2. **Squad *A***

Having one of the smallest influence areas of Lisbon, this squad has a small number of agents available for patrolling. We settled for solutions with just three beats. With this number of beats it is possible to patrol something around 20% of streets. This zone is next to the Tagus River, and hundreds of persons pass through it every day, because there is an important transport interface in the area.

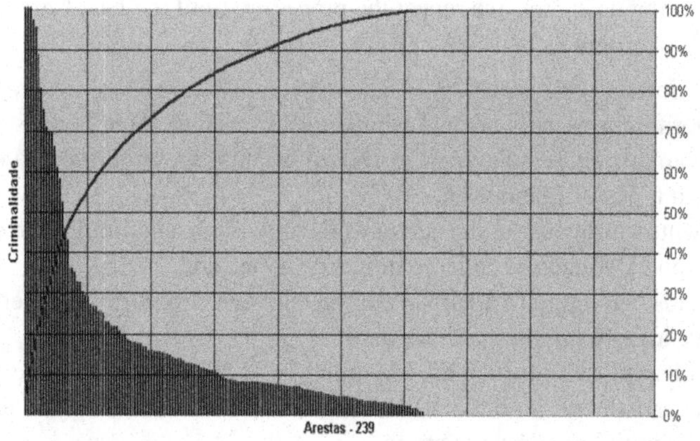

Figure 10.5 Squad *A* – criminality distribution.

The criminal distribution in this zone can be observed in Fig. 10.5. The concentration of crime occurrences on a small number of streets is obvious.

The three beats obtained by the algorithm are presented in Table 10.2.

Table 10.2 Beats in the solution obtained by the algorithm.

Beat	Crime pot.	Length
1	10.34	1098
2	4.12	1082
3	3.85	1085
Total	18.31	3265

The solution was generated in less than two seconds. As can be confirmed by Tables 10.1 and 10.2, a patrol that covers approximately 22% of total extension of streets manages to patrol around 64% of expected criminality.
The average criminality patrolled by the randomly generated solutions is approximately 8.10 crimes per month, i.e., around 28% of monthly squad expected criminality.

Figure 10.6 Squad *A* – beats in the solution obtained by the algorithm.

Fig. 10.6 shows the map of squad *A*'s area of influence with the three obtained beats.
Arcs belonging to this solution represent streets that connect different terminals of the transport or streets that cross one of the "red light" zones of Lisbon.
From the 25 arcs with the highest criminal potential, 22 were chosen by the algorithm, belonging to the solution obtained.

3. **Squad *B***
This squad represents the group of squads that have areas of influence with intermediate size. It is also one with the highest crime rate. There is a football stadium in this zone, which can be a reason for such levels of criminality. The squad has a higher number of human resources so it was decided to generate solutions with six beats. With six beats it is possible to patrol around 11.71% of street extensions.
Again, there is a strong concentration of crimes in a small number of streets, as shown in Fig. 10.7, but this concentration is less felt than in the other two squads. Comparing Fig. 10.7 with Figs. 10.5 and 10.9 it is possible to observe this difference.
The six beats obtained by the algorithm are presented in Table 10.3.
This solution was generated in less than three seconds, and is graphically represented in Fig. 10.8. Approximately 12.6% of the total street extensions is patrolled, and 31% of the squad's expected criminality is covered by these beats.
Only 13 from the 25 "critical" arcs belong to the solution, but the remaining 12 are spread all over the squad's area of influence, significantly distant from the

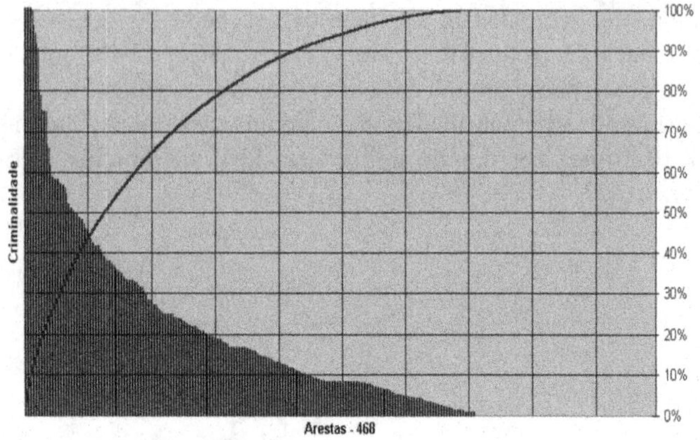

Figure 10.7 Squad B – criminality distribution.

Table 10.3 Beats in the solution obtained by the algorithm.

Beat	Crime pot.	Length
1	5.06	1063
2	4.59	1064
3	4.11	1068
4	3.90	1090
5	3.50	1086
6	3.29	1080
Total	24.45	6451

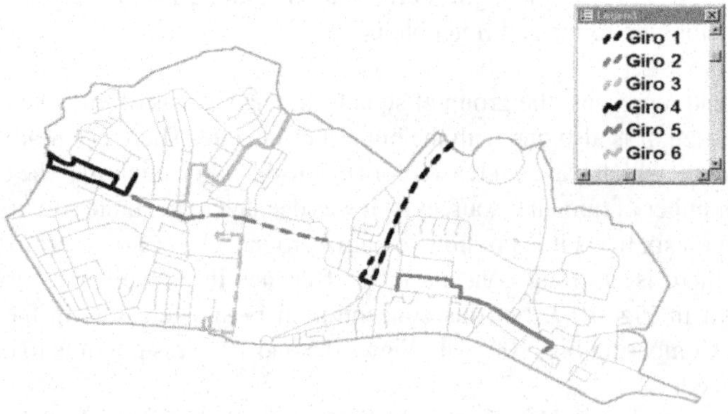

Figure 10.8 Squad B – beats in the solution obtained by the algorithm.

generated beats. If a new beat were created with one, or more, of the "rejected" critical arcs, it would include a high number of arcs with a low criminal potential.

The main reason for the low quality of this particular case is the higher dispersion of criminality, in comparison with other studied squads.
The average criminality of the random solutions is approximately 11.1 occurrences a month, i.e., 14% of the squad's expected criminality.

4. **Squad *C***
This squad was chosen because it is one of the largest of Lisbon. It has more resources than the other two, so it was decided to generate nine beats. It is possible to cover about 9.40% of total street extension. This zone lays by the Tagus too, and is often visited by tourists, due to numerous monuments that can be found here.

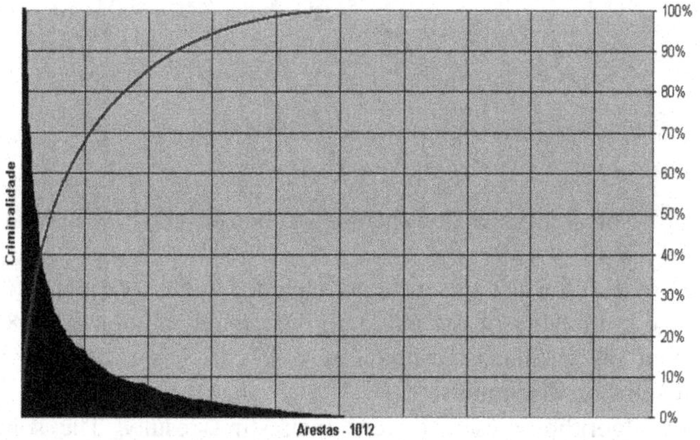

Figure 10.9 Squad *C* – criminality distribution.

In squad *C* the concentration of criminality on a small number of streets is particulary strong. This situation is shown in Fig. 10.9.
This solution was obtained in less than eight seconds, and it can be observed in Fig. 10.10 and Table 10.4.

Table 10.4 Beats in the solution obtained by the algorithm.

Beat	Crime pot.	Length
1	11.60	1093
2	4.44	1019
3	2.67	1017
4	2.64	1086
5	2.47	1036
6	2.35	1088
7	2.19	1052
8	1.91	1085
9	1.73	1049
Total	31.98	9526

Figure 10.10 Squad *C* – beats in the solution obtained by the algorithm.

With just 10% of street extensions being patrolled around 52% of total criminality can be covered.

Only three of the 25 worst arcs were not included in the solution.

The average criminality of the randomly generated solution is 18.8 crimes per month, about 19% of squad *C* criminality.

5. **Proposed solution discussion**

   The tested algorithm presented very interesting results. The solutions were reached vary rapidly, even for the largest squad, where a higher number of beats were generated.

   In the squads with a higher concentration of criminality the solutions obtained have a high preventive capacity, covering almost every "critical" arc. In the case of squad *B*, which has a larger dispersion of criminality than the other squads, the solution was not so good, but, on average, it is much better than solutions generated randomly.

   It is a very fast and versatile algorithm, which allows the testing of several scenarios, by changing the number of beats and/or their "ideal" length.

   It is also possible for the squad chief to ask for a solution with more beats than agents available, and then select the ones really needed, allowing him some degree of control over the streets he assigns to his agents. Another advantage of this practice is the possibility of rotation of patrolled streets, thus assuring that all "critical" streets are periodically patrolled, even for the squads with the highest criminal dispersion.

References

Adlakha, V.G, 1986. An improved conditional Monte-Carlo technique for stochastic shortest route problem. Management Science 32 (10), 1360–1367.

Aiello, W., Lu, L.F.C., 2000. A random graph model for massive graphs. In: Proceedings of the 32nd Annual ACM Symposium on Theory of Computing.

Amaral, L.A.N., Scala, A., Barthélémy, M., Stanley, H.E., 2002. Classes of behavior of small-world networks. arXiv:cond-mat/0001458.

Andreatta, G., Ricaldone, F., Romeo, L., 1985. Exploring stochastic shortest paths problems. Technical report, ATTI Giornale di Lavaro, Bologna, Italy.

Bard, J.F., Bennett, J.E., 1991. Arc reduction and path preference in stochastic acyclic networks. Management Science 31 (7), 198–215.

Bollobás, B., 2001. Random Graphs. Cambridge University Press, United Kingdom.

Chibeles-Martins N., Tavares, L.V., 1999. Análise estatística da criminalidade na cidade de Lisboa. 1995–1998, Reservado, Escola Superior de Polícia.

Erdős, P., Rényi, A., 1959. On random graphs. Publicationes Mathematicae 6, 290–297.

Erdős, P., Rényi, A., 1960. On the evolution of random graphs. Publications of the Mathematical Institute of the Hungarian Academy of Sciences 5, 17–61.

Erdős, P., Rényi, A., 1961. On the strength of connectedness of a random graph. Acta Mathematica Scientia Hungary 12, 261–267.

Fishburn, P.C., 1970. Utility Theory for Decision Making. Wiley, New York.

Frank, H., 1969. Shortest paths in probabilistic graphs. Operations Research 17, 583–599.

Henig, M.I., 1990. Risk criteria in stochastic knapsack problem. Operations Research 38 (5), 820–825.

Jaillet, P., 1992. Shortest paths problems with node failures. Networks 22, 589–605.

Kamburowski, J., 1985. A note on the stochastic shortest route problem. Operations Research 33 (6), 696–698.

Loui, R., 1983. Optimal paths in graphs with stochastic or multidimensional weights. Communications of the ACM 26, 670–676.

Luce, R.D., Raiffa, H., 1957. Games and Decisions. Wiley, New York.

Martins, E., Pascoal, M., Rasteiro, D.D., Santos, J., 1999. The optimal path problem. Investigação Operacional 19 (1), 43–60.

Mirchandini, P.B., Soroush, H., 1985. Optimal paths in probabilistic networks: a case with temporary preferences. Computers and Operations Research 12, 365–383.

Murthy, I., Sarkar, S., 1996. A relaxation-based pruning technique for a class of stochastic shortest path problems. Transportation Science 30 (3), 220–236.

Murthy, I., Sarkar, S., 1997. Exact algorithms for the stochastic shortest path problem with a decreasing deadline utility function. European Journal of Operational Research 103, 209–229.

Murthy, I., Sarkar, S., 1998. Stochastic shortest path problems with piecewise linear concave utility functions. Management Science 44 (11), 125–136.

Newman, M.E.J., Strogatz, S.H., Watts, D.J., 2001. Random graphs with arbitrary degree distribution and their applications.

Pratt, J.W., Raiffa, H., Schlaifer, R.O., 1965. Introduction to Statistical Decision Theory. McGraw-Hill, New York.

Rasteiro, D.M.L.D., Anjo, A.J.B., 2004. Optimal paths in probabilistic networks. Journal of Mathematical Sciences 120 (1), 974–987.

Santos, J., 1997. O problema do trajecto óptimo multiobjectivo.

Savage, L.J., 1954. The Foundations of Statistics. Wiley, New York.

Sigal, C., Pritsker, A.A.B., Solberg, J., 1980. The stochastic shortest route problem. Operations Research 28 (5), 1122–1129.

von Neumann, J., Mogenstern, O., 1947. Theory of Games and Economic Behavior. Princeton University Press, Princeton, N.J.

Snails, snakes, and first-order ordinary differential equations

11

Alberto Alonso Izquierdo, Ascensión H. Encinas, Miguel Ángel González León, Ángel Martín del Rey, Jesús Martín-Vaquero, Araceli Queiruga-Dios, Gerardo Rodríguez
University of Salamanca, Salamanca, Spain

11.1 Background

First-order differential equations and systems of first-order differential equations constitute a powerful modeling tool in different scenarios of science, engineering, and technology. We introduce here, in a very summarized way, the definitions and essential results to make it possible to follow the contents of the chapter, especially referring to the most elementary notions.

Definition 11.1 (First-order ordinary differential equation). A first-order ordinary differential equation (ODE) is a relationship as the following: $f\left(x, y(x), y'(x)\right) = 0$.

Definition 11.2 (System of normalized first-order ordinary differential equations). A system of ordinary differential equations (SODE) is of first order if it only contains first-order derivatives as follows:

$$\begin{cases} f_1\left(x, y_1, y_2, \ldots, y_n, y_1', y_2', \ldots, y_n'\right) & = 0, \\ f_2\left(x, y_1, y_2, \ldots, y_n, y_1', y_2', \ldots, y_n'\right) & = 0, \\ \cdots\cdots\cdots\cdots\cdots\cdots & \\ f_n\left(x, y_1, y_2, \ldots, y_n, y_1', y_2', \ldots, y_n'\right) & = 0. \end{cases}$$

Both differential equations and systems of differential equations have infinite solutions. In most applications, the number of solutions is limited by imposing initial or boundary conditions in what is generally known as initial value problems.

Under fairly general conditions, Picard's theorem establishes the existence and uniqueness of the solution for initial value problems.

This result is fundamental within the theory although, unfortunately, it does not provide a general computable mechanism for obtaining the exact solution of the differential equation. In general, however, it is necessary to develop appropriate numerical methods to obtain an approximate solution to the posed problem.

However, there are some types of first-order differential equations where the calculation of the exact solution is possible. A nonexhaustive list of the types of first-order ODE that can be solved in an exact way includes: differential equations of separable variables, homogeneous and reducible to homogeneous, linear, Bernoulli, exact, with integral factor equations, etc.

Calculus for Engineering Students. https://doi.org/10.1016/B978-0-12-817210-0.00018-7

For example, the general solution of $(y^2 + xy^2)y' + x^2 - yx^2 = 0$, a separable variables' ODE, is $(x + y)(x - y - 2) + 2\log\left|\frac{x+1}{y-1}\right| = C$.

The general solution of the homogeneous ODE $xy' = \sqrt{y^2 + x^2}$ could be calculated from the expression

$$\frac{u^2}{2} + \frac{1}{2}u\sqrt{u^2+1} + \frac{1}{2}\log\left(u + \sqrt{u^2+1}\right) = \log|x| + C,$$

simply undoing the change $y = ux$.

The general solution of the linear equation $y' - y\,\mathrm{tg}\,x = \cos x$ is $y = \frac{K}{\cos x} + \frac{1}{2}\frac{x}{\cos x} + \frac{1}{2}\sin x$. And the general solution of the Bernoulli's differential equation $3xy' - 2y = \frac{x^3}{y^2}$ is $y^3 = x^3 + Kx^2$.

However, most of the ODEs cannot be solved analytically. The different convergent numerical methods provide an approximation to the solution of initial value problems, and are essential methods to solve many scientific and engineering applications. A very important category of numerical methods are the Runge–Kutta methods.

General ideas about Runge–Kutta methods

The general Runge–Kutta method with s steps is given by the following expression:

$$y_{n+1} = y_n + h\sum_{i=1}^{s} b_i k_i,$$

where h is the step size, $k_i = f\left(t_n + c_i h, y_n + h\sum_{j=1}^{s} a_{ij}k_j\right)$ for $i = 1, \ldots, s$, with $a_{ij}, b_i, c_i \in \mathbb{R}$, and such that $c_i = \sum_{j=1}^{s} a_{ij}$, for $i = 1, \ldots, s$.

The Runge–Kutta method is said to be of order p if the truncation local error is $e_n = o(h^{p+1})$.

Runge–Kutta methods convergency

The necessary and sufficient condition for an s-step Runge–Kutta method to be convergent is that $b_1 + b_2 + \cdots + b_s = 1$.

The most widely used fourth-order Runge–Kutta method is similar to the following algorithm:

$$y_{n+1} = y_n + \frac{h}{6}\left(k_1 + 2k_2 + 2k_3 + k_4\right),$$

$$\begin{cases} k_1 = f(t_n, y_n), \\ k_2 = f\left(t_n + \frac{h}{2}, y_n + \frac{h}{2}k_1\right), \\ k_3 = f\left(t_n + \frac{h}{2}, y_n + \frac{h}{2}k_2\right), \\ k_4 = f\left(t_n + h, y_n + hk_3\right). \end{cases}$$

These numerical methods can be applied also to solve SODEs. In general, these types of problems are more difficult to solve analytically, except linear SODEs. A linear first-order SODE could be written as a matrix form $X' = AX + B$, where

$$X = \begin{pmatrix} y_1 \\ y_2 \\ \vdots \\ y_n \end{pmatrix}, \; X' = \begin{pmatrix} y_1' \\ y_2' \\ \vdots \\ y_n' \end{pmatrix}, \; A = \begin{pmatrix} a_{11}(x) & \cdots & a_{1n}(x) \\ a_{21}(x) & \cdots & a_{2n}(x) \\ \vdots & \ddots & \vdots \\ a_{n1}(x) & \cdots & a_{nn}(x) \end{pmatrix}, \text{ and}$$

$$B = \begin{pmatrix} b_1(x) \\ \vdots \\ b_n(x) \end{pmatrix}.$$

If functions $a_{ij}(x)$, with $1 \le i, j \le n$, are constant, complex in general, the linear SODE is called linear SODE with constant coefficients.

Theorem (General solution of a complete linear SODE). *Let us consider the SODE $X' = A(x)X + B(x)$, with $a_{ij}(x)$, $b_i(x)$, $i, j = 1, 2, \ldots, n$, continuous functions in $[a, b] \subset \mathbb{R}$. If $X_1(x), X_2(x), \ldots, X_n(x)$ is a fundamental system of solutions of the homogeneous SODE $X' = A(x)X$ and $X_P(x)$ is a particular solution of the complete linear SODE, then $X(x) = X_P(x) + \sum_{i=1}^{n} C_i X_i(x)$, with $C_1, C_2, \ldots, C_n$ arbitrary constants, is the general solution of the complete linear SODE.*

A particular solution of the complete system is $X_P(x) = F(x) \int F^{-1}(x)B(x)\,dx$, where $F(x)$ is a fundamental matrix of the homogeneous system.

In the case of linear systems with constant coefficients, the homogeneous system solution is expressed through the exponential matrix of the system $e^{Ax} = \sum_{n=0}^{\infty} \frac{A^n x^n}{n!}$. In this case, the system's general solution is $X(x) = e^{Ax} C$, where C is a constant column matrix, while the particular solution is given by $X_P(x) = e^{Ax} \int e^{-Ax} B(x)\,dt$.

Then, for example, the general solution of the system $\begin{cases} x' = x + y, \\ y' = x + y + t \end{cases}$ is

$$\begin{cases} x = C_1 + C_2 e^{2t} - \frac{1}{4}t^2 - \frac{1}{4}t - \frac{1}{8} + \frac{1}{8}e^{2t}, \\ y = -C_1 + C_2 e^{2t} + \frac{1}{4}t^2 - \frac{1}{4}t - \frac{1}{8} + \frac{1}{8}e^{2t}. \end{cases}$$

Below we present some basic results of the theory of the stability of systems of differential equations, which will help us in the examples developed in Section 11.3 of this chapter.

Definition 11.3 (Orbits of an autonomous two-equation SODE. Phase diagram. Critical points. Periodic paths or cycles). Given the autonomous SODE $\left\{\begin{array}{l} x' = F(x, y), \\ y' = G(x, y), \end{array}\right.$ we call orbit or path such a system of a plane curve of parametric equations $\left\{\begin{array}{l} x = x(t), \\ y = y(t), \end{array}\right.$ where $(x(t), y(t))$ is a SODE's solution.

The set of paths of an autonomous system where the path sense is indicated for increasing values of the independent variable is called phase diagram.

A constant solution of an autonomous SODE is called critical point or equilibrium solution or stationary state.

Periodic paths or two-equation autonomous SODE cycles are all paths $(x(t), y(t))$ such that $x(t) = x(t+T)$, $y(t) = y(t+T)$ for any $T > 0$. The associated orbit is therefore a closed curve.

Definition 11.4 (Critical point stability. Basin of attraction). A critical point (x_0, y_0) from the autonomous system $\left\{\begin{array}{l} x' = F(x, y), \\ y' = G(x, y) \end{array}\right.$ is stable if for all $\varepsilon > 0$ there exists a $\delta > 0$ such that all paths $\left\{\begin{array}{l} x = x(t), \\ y = y(t) \end{array}\right.$ that meet

$$\sqrt{(x(t_0) - x_0)^2 + (y(t_0) - y_0)^2} < \delta \tag{11.1}$$

verify that $\sqrt{(x(t) - x_0)^2 + (y(t) - y_0)^2} < \varepsilon$ for all $t \geq t_0$.

The critical stable point (x_0, y_0) is said to be asymptotically stable if there exists a $\delta > 0$ such that, for all the solutions of a system $\left\{\begin{array}{l} x = x(t), \\ y = y(t) \end{array}\right.$ that verify Eq. (11.1), we have $\lim_{t\to\infty} x(t) = x_0$ and $\lim_{t\to\infty} y(t) = y_0$.

The basin of attraction (x_0, y_0) of an asymptotically stable critical point is the set of the initial points $(x(t_0), y(t_0))$ such that the SODE's solution with the initial condition $(x(t_0), y(t_0))$ converges to the critical point, i.e., $\lim_{t\to\infty} x(t) = x_0$, $\lim_{t\to\infty} y(t) = y_0$.

Linear autonomous systems' stability

Let us consider the SODE $\begin{pmatrix} x' \\ y' \end{pmatrix} = \begin{pmatrix} a_1 & a_2 \\ b_1 & b_2 \end{pmatrix} \begin{pmatrix} x \\ y \end{pmatrix}$, written in matrix form as $X' = AX$.

The $(0, 0)$ point is a critical SODE point. Moreover, as we suppose that $|A| \neq 0$, it is the unique critical point.

Let us analyze the behavior of the paths according to matrix A eigenvalues.

1.- The A eigenvalues, λ_1 and λ_2, are real and different. The following situations may arise:

a) $\lambda_1, \lambda_2 < 0$. Then $\lim_{t\to\infty} \Phi(t) = (0,0)^T$, and thus $(0,0)$ is an asymptotically stable critical point. In this case $(0,0)$ is said to be an attractor or sink node.

b) $\lambda_1, \lambda_2 > 0$. In this case the paths are moving away from the critical point, although they started from points close to the origin. Hence $(0,0)$ is a critical unstable point. The point $(0,0)$ is said to be a strong node.

c) λ_1, λ_2 have different sign. The solution moves in the direction given by the eigenvector associated with the negative eigenvalue and approximates the critical point. For the rest of cases, solutions are moving away from the critical point. In this case, $(0,0)$ is an unstable critical point and it is called saddle point.

In Fig. 11.1, an illustration with three examples is shown.

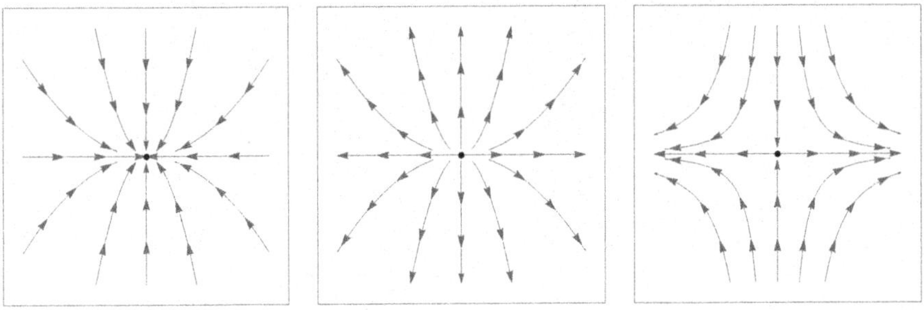

Figure 11.1 Situations when A eigenvalues are real and different. (A) Attractor. (B) Strong node. (C) Saddle point.

2.- Matrix A has a double real eigenvalue λ. The following situations arise:

a) Matrix A is diagonalizable. We can have:

a1) $\lambda < 0$. Then $\lim_{t\to\infty} \Phi(t) = (0,0)^T$ and hence $(0,0)$ is an asymptotically stable critical point and it is called stable degenerate node.

a2) $\lambda > 0$. In this case the paths are moving away from the critical point, although they started from some points close to the origin. Thus, $(0,0)$ is an unstable critical point and it is called unstable degenerate node.

b) Matrix A is not diagonalizable. The following situations may arise:

b1) $\lambda < 0$. Then $\lim_{t\to\infty} \Phi(t) = (0,0)^T$ and the paths are getting close to the critical point $(0,0)$ with the eigenvector direction. Thus $(0,0)$ is an asymptotically stable critical point and it is called stable degenerate node.

b2) $\lambda > 0$. In this case the paths move away from the critical point, although they started in some points close to the origin. So $(0,0)$ is an unstable critical point and it is called unstable degenerate node.

In Fig. 11.2, the reader can find plots of these four examples.

3.- Matrix A eigenvalues are the conjugate complex numbers $\lambda = a + bi$ and $\overline{\lambda} = a - bi$. The following situations may arise:

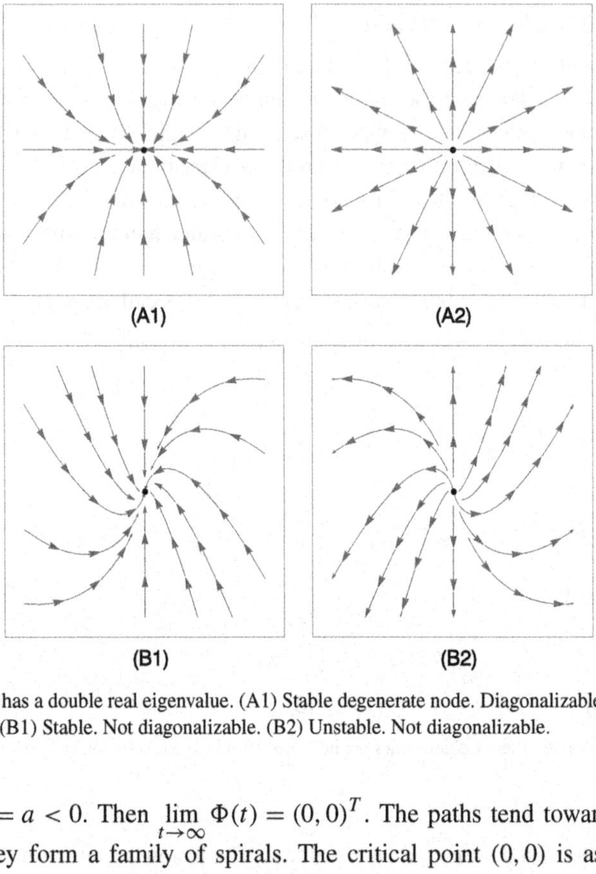

Figure 11.2 A has a double real eigenvalue. (A1) Stable degenerate node. Diagonalizable. (A2) Unstable. Diagonalizable. (B1) Stable. Not diagonalizable. (B2) Unstable. Not diagonalizable.

a) $\text{Re}(\lambda) = a < 0$. Then $\lim_{t\to\infty} \Phi(t) = (0,0)^T$. The paths tend toward the critical point and they form a family of spirals. The critical point $(0,0)$ is asymptotically stable and it is called stable attractor spiral focus.

b) $\text{Re}(\lambda) = a > 0$. Paths are also spiral, but in this case they move away from the critical point. So, $(0,0)$ is an unstable critical point and it is called unstable spiral focus.

c) $\text{Re}(\lambda) = a = 0$. Then the solution is

$$\begin{cases} x(t) = C_1 \cos(bt) + C_2 \sin(bt), \\ y(t) = D_1 \cos(bt) + D_2 \sin(bt). \end{cases}$$

The paths are closed orbits that are not moving closer neither further away from the critical point. So $(0,0)$ is a stable critical point, but not an asymptotically stable point, and it is called stable center.

In Fig. 11.3, we show plots of the new three examples.

n-Equation autonomous linear systems stability

The following result is, somehow, a generalization of the stability and asymptotic stability study of the SODE $X' = AX$ of n equations, where the matrix coefficient, A, is a nonsingular matrix.

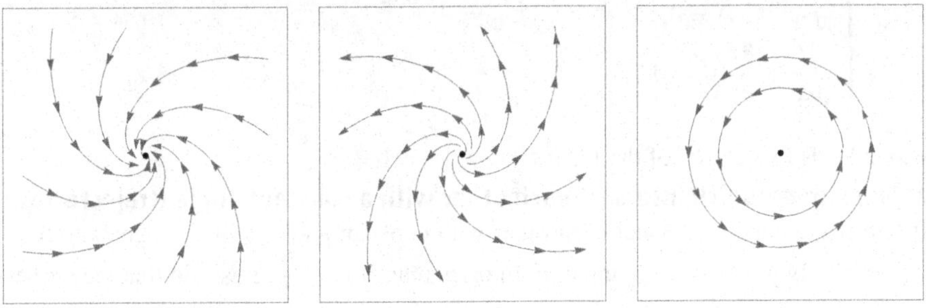

Figure 11.3 Eigenvalues are conjugate complex numbers. (A) Stable attractor spiral focus. (B) Unstable spiral focus. (C) Stable center.

If λ is an A eigenvalue, we will call r_λ the multiplicity of such an eigenvalue, and s_λ the eigensubspace dimension or invariant associated with it.

The following is verified:

i) The origin is asymptotically stable if and only if all A eigenvalues have their real part negative.

ii) The origin is stable, but not asymptotically stable, if and only if:
  - **a)** All A eigenvalues have their real part nonpositive.
  - **b)** At least one eigenvalue has its real part equal to zero.
  - **c)** For all eigenvalue λ_0 with real part equal to zero, $r_{\lambda_0} = s_{\lambda_0}$ is verified.

iii) The origin is unstable if and only if it verifies any of the following conditions:
  - **a)** There exists an eigenvalue with positive real part.
  - **b)** There exists an eigenvalue λ_0 with null real part and $s_{\lambda_0} < r_{\lambda_0}$.

11.2 Description of general problems and areas where they are very common

Differential equations can be applied in many different branches of daily life and also in scientific studies. In this book we are using only two chapters to explain some of these areas. Hence it is not possible to explain all the cases where ODEs can be used, and it is not our intention to cite all types of examples where ODEs appear. In this section we are giving just a few examples and show only a few basic applications.

A) Geometric applications

In general, the problem of finding curves or families of curves with conditions on tangents, normals, etc., involves solving an ODE which, on certain occasions, will be of first order.

1) ODE associated with a uniparameter family of curves: The uniparametric family of curves $f(x, y, a) = 0$ (a being a real parameter) is the general solution of the first-order ODE that can be obtained by eliminating the parameter a in the system

$$\begin{cases} f(x,y,a)=0, \\ \dfrac{\partial f}{\partial x}+\dfrac{\partial f}{\partial y}y'=0, \end{cases}$$

which leads to an ODE of the form $F(x,y,y')=0$.

2) Flat curves which intersect each other with a constant angle (trajectories): If two plane curves intersect each other with constant angle w then tangents to both curves at any point form an angle w. In particular, if $w=\frac{\pi}{2}$, it is said that the curves intersect orthogonally.

1) Let $F(x,y,y')=0$ be the ODE associated with the uniparametric family of curves $f(x,y,a)=0$. The ODE of orthogonal trajectories (curves which intersect orthogonally) is $F\left(x,y,\frac{-1}{y'}\right)=0$.
   The ordinary differential equation of a family of curves that intersect $f(x,y,a)=0$ under a constant angle w, with $0\le w<\frac{\pi}{2}$, is $F\left(x,y,\frac{y'-\operatorname{tg} w}{y'\operatorname{tg} w+1}\right)=0$.
2) Let $F\left(r,\theta,\frac{dr}{d\theta}\right)=0$ be the ODE associated with a family of curves $f(r,\theta,a)=0$, expressed in polar coordinates.
   The ODE of orthogonal trajectories to $f(r,\theta,a)=0$ is $F\left(r,\theta,-\frac{r^2\,d\theta}{dr}\right)=0$.
   The ODE of the family of curves that cut $f(r,\theta,a)=0$ with constant angle w is

$$F\left(r,\theta,\frac{dr+r\operatorname{tg} w\,d\theta}{d\theta-\frac{1}{r}\operatorname{tg} w\,dr}\right).$$

Remark: *Many problems that arise in the study of physical phenomena are formulated using the language of orthogonal trajectories. For example, in electrostatic and gravitational problems, trajectories that are orthogonal to a family of equipotential curves are called lines of force.*

In an electric field the lines with the same power (equipotential lines) are the trajectories that are orthogonal to the flow lines. Similarly, in thermodynamics, the family of curves with the same temperature (isotherms) have orthogonal trajectories to heat flow lines.

B) Population growth

Different models about population growth can be constructed with first-order ODEs:

1) One species. Let $x(t)$ denote the number of individuals in a population at the time t. The Malthus model (1798) assumes that the growth rate $\frac{dx}{dt}$ is proportional to the population, or in other words, the "per capita" growth rate $\frac{1}{x}\frac{dx}{dt}$ is constant. Solving the initial value problem $\frac{dx}{dt}=rx,\ x(0)=x_0$ (note that the ODE is separable), the solution is $x(t)=x_0e^{rt}$, i.e., exponential behavior. The constant r

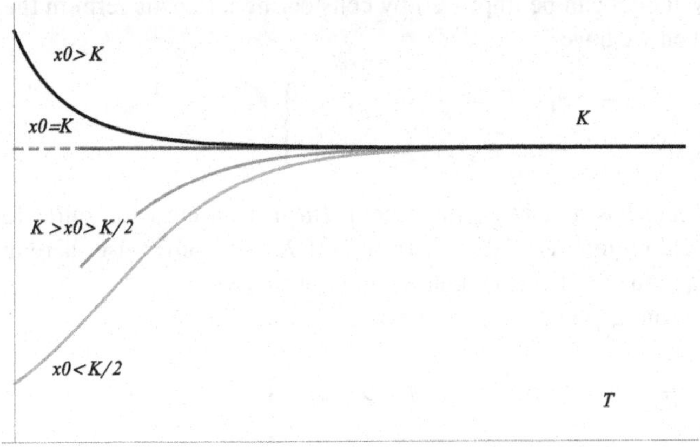

Figure 11.4 Graphs of the solutions of the logistic model for the different relative values of the constants x_0 and K.

represents the intrinsic growth rate for the species, and it would be positive if the birth rate is greater than the death rate (in the absence of migrations).
In 1838 Verhulst introduced the logistic model where a self-limiting growth is considered, motivated by the intraspecific competition between the individuals of the species. The initial value problem $\frac{dx}{dt} = rx\left(1 - \frac{x}{K}\right)$, $x(0) = x_0$, also a separable ODE, has the solution $x(t) = \frac{x_0 K}{x_0+(K-x_0)e^{-rt}}$, where K is the carrying capacity, which represents the maximum population size that the ecosystem can sustain indefinitely. Three well-defined regimes are found according to the value of x_0 with respect to K and $\frac{K}{2}$ (see Fig. 11.4).

2) Two species in coexistence. The Lotka–Volterra (LV) model was proposed between 1925 and 1927 (Lotka, 1910, 1925; Volterra, 1927) and is described as follows:

$$\left.\begin{aligned} \frac{dx_1}{dt} &= r_1 x_1 - \alpha x_1 x_2, \\ \frac{dx_2}{dt} &= -s x_2 + \beta x_1 x_2. \end{aligned}\right\}$$

It is the simplest model describing the coexistence between a prey species x_1 and its predator x_2. The model is composed by two Malthusian terms (with intrinsic rates r_1 and $-s$ for prey and predator, respectively) and two interaction terms $\alpha x_1 x_2$ and $\beta x_1 x_2$, obviously negative for the prey growth and positive for the predator one.
The LV model is an autonomous SODE that presents a stationary point at $P \equiv \left(\frac{s}{\beta}, \frac{r_1}{\alpha}\right)$. The linearization of the LV system around P allows us to classify this point as a stable center, explaining the cyclical (periodical) behavior of the population sizes that can be observed in nature for several real predator–prey coexistence situations.

The LV model can be improved by considering a logistic term in the prey equation. Then we have

$$\left.\begin{aligned}\frac{dx_1}{dt} &= r_1x_1\left(1-\frac{x_1}{K}\right)-\alpha x_1x_2,\\ \frac{dx_2}{dt} &= -sx_2+\beta x_1x_2.\end{aligned}\right\}$$

This model has a stable critical point (in fact an attractor spiral focus) in the ecological region ($x_1 \geq 0, x_2 \geq 0$) only if $K > \frac{s}{\beta}$; otherwise there are no stable points and the predator population will not survive.
Another interesting model for two species is the following:

$$\left.\begin{aligned}\frac{dx_1}{dt} &= r_1x_1\left(1-\frac{x_1}{K_1}\right)-\alpha x_1x_2,\\ \frac{dx_2}{dt} &= r_2x_2\left(1-\frac{x_2}{K_2}\right)-\beta x_1x_2.\end{aligned}\right\}$$

This model describes two species in mutual competition with respective intrinsic growth rates r_1 and r_2 and assumes that the common ecosystem can support carrying capacities K_1 and K_2 for each species, respectively. The interspecific competition is taken into account by the terms αx_1x_2 and βx_1x_2. This system presents a critical point in the ecological region that can be a saddle or a sink node depending on the relative size of the interspecific and intraspecific competition parameters.

C) Trajectories of falling bodies and other movement problems

Newton's law of free fall of a body is $\dfrac{dv}{dt} = g$, with v being the speed of fall.

If we assume that air, or any other external environment, exerts a resistance force opposite to the movement and proportional to the speed of the falling body, the differential equation of this movement is $\dfrac{dv}{dt} = g - \kappa v$.

D) Radioactive decay

We can use ODEs to model radioactive decay phenomena. The speed at which the nuclei of a substance A disintegrate is proportional to the number of such nuclei, i.e.,

$$\frac{dA}{dt} = -kA. \tag{11.2}$$

Example: The ^{14}C carbon extracted from an ancient skull contained only one sixth of that extracted from a bone of modern times. What is the antiquity of the skull? (Note: the k decay constant of C^{14} carbon is approximately equal to 0.000126 year^{-1}.)

Solution: $\frac{x}{x_0} = e^{-kt}$ and therefore $t = \frac{\ln(1/6)}{k} = -14220.3$ years.

E) Newton's law of cooling

The speed with which the temperature T of a body changes is proportional to the difference between the temperature of the body and the temperature of the medium around it, T_m, i.e.,

$$\frac{dT}{dt} = k(T_m - T). \tag{11.3}$$

Example: Suppose that the thermometer of the living room of a house says 35°C and another thermometer is placed on the roof of the house that is above it at a constant temperature of 20°C. One hour later, the thermometer inside the room reads 30°C. Which is the temperature in the living room after two hours if we assume an ideal behavior?

Solution: $T = T_m - \lambda e^{-kt}$, where $T_m = 20$ and $k = \ln(3/2)$. Hence the temperature after two hours in the living room is $\frac{80}{3} = 26.66667$°C.

F) Epidemiology

Some diseases, such as flu and legionella, spread in a community through contact between people. If we call $x(t)$ the people with the contagious disease and $y(t)$ those who do not have it yet, the equation that applies is

$$\frac{dx}{dt} = kxy.$$

G) Chemistry

In chemical reactions we can determine the speed of combination of some compounds or we can study chemical solutions with ODEs such as

$$\frac{dx}{dt} = k(a - x)(b - y), \text{ for compounds } x \text{ and } y, \text{ or}$$
$$\frac{dx}{dt} = v_{x,inside} - v_{x,outside}, \tag{11.4}$$

where $v_{x,inside}$ is the entry speed of x and $v_{x,outside}$ is the exit speed of x.

Example: A deposit contains 100 kilograms of brine that has 200 g of salt in solution. In the tank flows water containing 1 g of salt per kilogram at a rate of 3 kg/minute and the mixture, which remains uniform by agitation, escapes from the tank at the same rate. Calculate the amount of salt after 1 hour and a half.

Solution: Eq. (11.4) gives us $\frac{dx}{dt} = 3 - 0.03x$ because $v_{x,inside} = 1 \times 3$ g/minute, and $v_{x,outside} = \frac{x}{100} \times 3$ g/minute. Hence, $\frac{\ln(0.03x-3)}{0.03} = -t + C$, and $x(0) = 200$, thus $C = \frac{\ln(3)}{0.03}$. Therefore $x(90) = 100 + 100e^{-2.7} = 106.72$ g.

H) Physics. Serial circuits

The second law of Kirchhoff determines that the potential differences in a circuit with a condenser C, an inductance L, and a resistance R are related by the following expression, where $I(t)$ is the intensity that passes through the circuit and is the derivative of the load, q, of the circuit with respect to time:

$$L\frac{dI}{dt} + RI + \frac{1}{C}q = E(t).$$

l) Financial mathematics

The quantitative variations that occur in financial capital over a long time are studied with ODEs, partial differential equations (PDEs), stochastic ODEs, and stochastic PDEs (see Kloeden and Platen (2011), Lee and Shi (2010)).

11.3 Real problems

11.3.1 The epidemiological, and also malware propagation, model of Kermack and McKendrick

In 1978, in a school with 763 students, there was an influenza outbreak. We want to know at every minute the number of infected students with a dynamical system.

Thus, numerically, it was calculated that the associated parameters for a susceptible–infected–recovered (SIR) model (without vital dynamics) were the following:

(a) The infection coefficient (also called transmission coefficient, calculated from the effective contacts with infected individuals by time unit, and the probability that one effective contact becomes infected) was $a = 0.00218$, and the recovery coefficient (the inverse of the duration of the infectious period) was $b = 0.4404$.

(b) Perform a similar analysis, but with $a = 0.00055$ and $b = 0.4404$.

Considering that there is one infectious patient at the beginning (and the rest are healthy), study if there was a proper epidemic outbreak, and calculate the functions of susceptible, infectious, and recovered populations.

One of the most important milestones in mathematical epidemiology occurred in 1927 when William Ogilvy Kermack and Anderson Gray McKendrick proposed a mathematical model (Kermack and McKendrick, 1927) whose objective was to study the dynamics of propagation of the bubonic plague that occurred in England between 1665 and 1666. This epidemic, known as the Great Plague and caused by the bacteria *Yersinia pestis* (transmitted through the fleas of rats), caused the death of between 70,000 and 100,000 people throughout England and more than one-fifth of the population of London. The Kermack–McKendrick model was strongly influenced by two previous results: the *law of mass action* of the epidemiology postulated by Hammer, which states that the rate at which an epidemic spreads is proportional to the product of the number of susceptible individuals and the number of infectious individuals (Hammer, 1906), and the mathematical model of Ross to simulate the spread of malaria.

The Kermack–McKendrick model is a deterministic model of compartmental nature in which the population, which remains constant, is divided into three large groups: susceptible or healthy individuals, those who can acquire the disease; infected, those who have acquired the disease and are capable of transmitting it; and recovered individuals, those who were infected but who currently cannot be infected for various

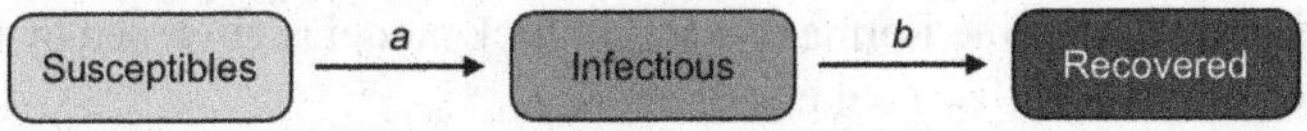

Figure 11.5 Flow diagram representing the dynamics of the SIR model.

reasons (they have been isolated, they have recovered by acquiring immunity, they are dead, etc.). It is assumed that the transmission of the biological agent occurs directly, that is, through effective contact between a infectious individual and a susceptible individual. It is a SIR-type model where susceptible individuals become infected, and these might recover (see Fig. 11.5). Therefore, the dynamic is governed by a nonlinear autonomous system of ordinary differential equations in which two parameters appear: the infection coefficient and the recovery coefficient. The population dynamics does not change in this model (we are assuming a model without vital dynamics): neither births, nor deaths, nor migrations are taken into account, with which the population is supposed to remain constant throughout the simulation period.

The great scientific contribution of this model is the so-called *threshold theorem,* with the so-called *basic reproductive number,* whose numerical value determines if a certain outbreak becomes epidemic or not.

As previously mentioned, this has been the model that has inspired most of the work in mathematical epidemiology in the last decades. From it, new models have been built and analyzed in which new compartments are included (exposed, carriers, quarantine, or other types of individuals), population dynamics, new connection topologies, etc. (Karyotis and Khouzani, 2016). Also, there are other models which take into account the change of total population.

Consequently, three time-dependent variables are considered in the mathematical model:

(1) number of susceptible individuals, $S(t) \geq 0$;
(2) number of infectious, $I(t) \geq 0$;
(3) number of recovered individuals, $R(t) \geq 0$.

Note that $S(t) \in \mathbb{R}$, $I(t) \in \mathbb{R}$, and $R(t) \in \mathbb{R}$. Additionally, since it is assumed that the population remains constant, we have

$$S(t) + I(t) + R(t) = N > 0, \forall t \geq 0, \quad (11.5)$$

where N is the initial population.

In addition, two parameters are taken into account in the model: the transmission coefficient $a > 0$ and the recovery coefficient $b > 0$.

As indicated above, it is assumed that contagion occurs by direct contact between an infectious individual and a susceptible individual. In this way, taking into account the law of mass action of Hammer, we can say that the incidence (number of new infectious per unit of time) is given by the following expression: $a \cdot S(t) \cdot I(t)$. On the other hand, the number of new individuals recovering per unit of time is considered proportional to the number of infected individuals, i.e., $b \cdot I(t)$.

Thus, the dynamics of the Kermack–McKendrick model is governed by the following system of ordinary differential equations:

$$\frac{dS}{dt} = -a \cdot S(t) \cdot I(t), \tag{11.6}$$

$$\frac{dI}{dt} = a \cdot S(t) \cdot I(t) - b \cdot I(t), \tag{11.7}$$

$$\frac{dR}{dt} = b \cdot I(t). \tag{11.8}$$

In general, the initial conditions imposed in the previous system are the following:

$$S(0) > 0, \;\; I(0) > 0, \;\; R(0) = 0, \tag{11.9}$$

that is, it is assumed that initially there are no recovered individuals, but only susceptible and infectious ones. If the time $t = 0$ corresponds to the initial moment of the outbreak, then it would be reasonable to assume that $S(0) \approx N$. Thereby, the Kermack–McKendrick equations verify the following.

- The variation of the number of susceptible individuals is given by Eq. (11.6). It can be seen how the number of new infectious individuals per unit of time (incidence) is proportional to the product between the number of susceptible individuals $S(t)$ and the number of infectious individuals $I(t)$; the proportionality constant is the transmission coefficient a. Given that it is a SIR model in which, therefore, individuals acquire permanent immunity, the number of susceptible individuals must decrease with the passage of time.
- In Eq. (11.7) the variation of the number of infectious individuals is reflected. Thus, this will be given by the difference between the incidence and the recovery, that is, between the new infectious ones that appeared and the old infectious ones that have been recovered.
- Finally, Eq. (11.8) represents the variation of the recovered individuals, which is proportional to the number of infectious individuals. b (the coefficient of recovery) is the proportionality constant. Since the recovered population acquires permanent immunity, the number of recovered individuals cannot decrease.

However, we can observe that, taking into account equation (11.5), the given system of three differential equations can be reduced to the following one, formed by two ordinary differential equations:

$$\frac{dS}{dt} = -a \cdot S(t) \cdot I(t),$$

$$\frac{dI}{dt} = a \cdot S(t) \cdot I(t) - b \cdot I(t).$$

Solving the problem

From Eq. (11.6) we obtain that $S'(t) \leq 0$, and therefore $S(t)$ is a decreasing function. Since we are assuming that $S(t) \geq 0$, it is bounded inferiorly and, therefore, its limit exists, which we will denote by $S(\infty) = \lim_{t\to\infty} S(t)$.

On the other hand, using Eqs. (11.6) and (11.8) we also know that

$$\frac{dS}{dt} = -\frac{a}{b} \cdot S(t) \cdot \frac{dR}{dt}, \tag{11.10}$$

and therefore

$$S(t) = S(0) \cdot e^{-\frac{a}{b} \cdot R(t)}. \tag{11.11}$$

As $R(t) \leq N, \forall t$, we have $S(t) \geq S(0) \cdot e^{-\frac{a}{b} \cdot N} > 0, \forall t$, and then

$$0 < S(0) \cdot e^{-\frac{a}{b} \cdot N} \leq S(\infty) \leq S(t), \forall t. \tag{11.12}$$

Strictly speaking, this does not mean that there are always susceptible individuals, since it can happen that $0 < S(t) < 1$ from a certain instant of time t. To ensure the existence of at least one susceptible individual, it should be imposed as a condition that $S(0) \cdot e^{-\frac{a}{b} \cdot N} > 1$, which occurs when $N < \frac{a}{b} \cdot \log(S(0))$. Note that from this condition we can deduce that knowing the number of individuals of the population and the number of initially susceptible individuals, it is possible to determine if there are always susceptible individuals in this population.

From Eq. (11.8) it follows that $\frac{dR}{dt} \geq 0$ since $b > 0$ and $I \geq 0$ and, consequently, $R(t)$ is a growing function. On the other hand, given that the number of individuals remains constant and in addition $I(t), S(t), R(t) \geq 0$, we know $R(t)$ is bounded superiorly, $R(t) \leq N$, and therefore its limit exists, $R(\infty) = \lim_{t \to \infty} R(t)$. In addition, from Eq. (11.11) we obtain

$$R(t) = -\frac{b}{a} \log\left(\frac{S(t)}{S(0)}\right). \tag{11.13}$$

Now, a simple calculation in Eqs. (11.6) and (11.7) gives us

$$I(t) = N - \frac{b}{a} \log\left(\frac{S(t)}{S(0)}\right) - S(t). \tag{11.14}$$

And, on the other hand, solving Eq. (11.7) and using the fundamental theorem of calculus (seen in Chapter 6 in this book), we obtain the following expression:

$$I(t) = I(0) \cdot e^{a \int_0^t S(x)dx - bt}. \tag{11.15}$$

Now, let us see how the function $I(t)$ behaves. Suppose that $I(t) > 0$. Then from Eq. (11.7) it follows that:

- If $a \cdot S(t) - b > 0$, then $\frac{dI}{dt} > 0$ and function $I(t)$ is monotone increasing.
- If $a \cdot S(t) - b < 0$, then $\frac{dI}{dt} < 0$ and function $I(t)$ is monotone decreasing.
- If $a \cdot S(t) - b = 0$, then $\frac{dI}{dt} = 0$ and there is a critical point.

Let us use this result to recreate the behavior of the function $I(t)$ from the initial instant $t = 0$. A simple calculation shows us the following:

(1) If $\frac{a \cdot S(0)}{b} > 1$, then $I(t)$ grows to a certain maximum,

$$I_{max} = S(0) + I(0) - \frac{b}{a} + \frac{b}{a} \log\left(\frac{a \cdot S(0)}{b}\right), \tag{11.16}$$

which is reached when $\frac{dI}{dt} = 0$, that is, when $S(t) = \frac{b}{a}$, which happens when $t = t_{max}$. Later, it decreases since $a \cdot S(t) - b < 0, \forall t > t_{max}$.

(2) If $\frac{a \cdot S(0)}{b} < 1$, then $I(t)$ decreases since $a \cdot S(t) - b < 0, \forall t > 0$.

Since $I(t)$ is monotone decreasing from a certain t and it is bounded below, $I(t) \geq 0$, it has a limit. Actually, $I(\infty) = \lim_{t\to\infty} I(t) = 0$.

As a consequence of this analysis, the following theorem is corroborated.

Theorem 11.1 (Threshold theorem). *Let us call* $R_0 = \frac{a \cdot S(0)}{b}$ *the basic reproductive number. Then we obtain:*

1. *If* $R_0 \leq 1$, *then* $I(t)$ *is a decreasing monotone function such that* $\lim_{t\to\infty} I(t) = 0$.
2. *If* $R_0 > 1$, *then* $I(t)$ *initially grows to reach its maximum value* $I_{max} = I(0) + S(0) - \frac{b}{a} - \frac{b}{a} \log\left(\frac{a \cdot S(0)}{b}\right)$, *and later decreases so that* $I(\infty) = \lim_{t\to\infty} I(t) = 0$.
3. $S(t)$ *is a decreasing monotone function so that* $S(0) \cdot e^{-\frac{a}{b}N} \leq \lim_{t\to\infty} S(t) = S(\infty)$ *and* $S(\infty)$ *is the only root of the equation* $S(\infty) = S(0) \cdot e^{-\frac{a}{b}(N - S(\infty))}$.
4. $R(t)$ *is a growing monotone function such that* $R(\infty) = \lim_{t\to\infty} R(t) < N$ *and* $R(t) = -\frac{b}{a} \log\left(\frac{S(t)}{S(0)}\right)$.

In our problem, in case (a)

$$R_0 = \frac{aS(0)}{b} = 3.77193 > 1.$$

Hence, there will be a proper epidemic outbreak with an increase of the number of the infectious.

On the contrary, in (b)

$$R_0 = \frac{aS(0)}{b} = 0.951635 < 1,$$

and there is no epidemic outbreak.

Some numerical studies in the past established basic reproductive numbers (for SIR models) for some epidemics, such as Ebola (2014) (in Guinea $R_0 = 1.51$, in Sierra Leone $R_0 = 2.53$, in Liberia $R_0 = 1.59$) and SARS (2002–2003) (reproductive number was $2 < R_0 < 5$); for Spanish flu $2 < R_0 < 3$, for measles $12 < R_0 < 18$, and for AIDS $2 < R_0 < 5$ (for more information see `https://en.wikipedia.org/wiki/Basic_reproduction_number`).

However, compartmental models, such as SIR, have not been used only in medicine or biology. They are also frequently considered by engineers, for example to study the propagation of malware.

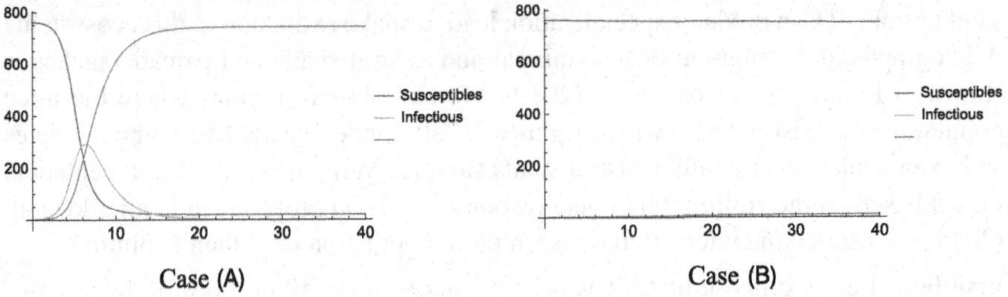

Figure 11.6 Numerical approximations of the number of susceptible, infectious, and recovered individuals in cases (A) and (B) of our problem.

Hence, for the functions $S(t)$, $I(t)$, and $R(t)$ usually numerical solvers are employed. In Fig. 11.6, we show the numerical results obtained with command NDSolve of Mathematica. More explanations about this and other codes for SIR models are developed in the Software Supplement.

11.3.2 Coevolution and chirality: a story of snails and snakes

Interactions between two species can give rise to a chain of biological adaptations. The evolution of one species causes adaptations guided by natural selection in a second species, which, in turn, cause subsequent biological adaptations in the first species. This phenomenon is called coevolution. A fascinating example of this scenario is given by the *Pareas* snake/*Satsuma* snail interacting system (You et al., 2015). *Satsuma* snails live in Japanese and Taiwanese archipelagos and have predominantly dextral (clockwise coiled) shells. *Pareas* is a genus of snail eating specialist snakes. Almost all the 14 different pareatid species involve distinct degrees of mandibular tooth asymmetry (Gotz, 2002). This fact probably facilitates feeding on dextral snails (Hoso et al., 2007; Danaisawadi et al., 2015, 2016). The snake bites an exposed part of the snail from behind at a determined leftward angle and extracts the soft body by alternately protracting and retracting the asymmetric right and left mandibles. The extreme case of this adaptation corresponds to the *Pareas iwasakii* snakes, which have approximately 17 teeth on the left side and 25 teeth on the right side (see Figure 1b in Hoso et al. (2010)). This species is endemic to the Yaeyama Islands in the Southern Japanese Ryukyu Islands.

But this is only half the story. Snail gene mutations can give rise to a sinistral (counterclockwise coiled) snail population. Both types of snails (dextral and sinistral) are identical in nature, but the different chirality leads to a reproductive incompatibility, and thus dextral/sinistral snails can be interpreted as different populations (Asami et al., 1998; Ueshima and Asami, 2003; Robertson, 1993). The left-handed snail variant initially involves a small population but it has an important advantage over the right-handed snails. The chirality shift counteracts the effect of the predator specialization (Gittenberger et al., 2012; Schilthuizen and Davison, 2005; Hoso et al., 2010). For example, lab experiments have determined 87.5% of sinistral *Satsuma* snails survive *Pareas iwasakii* predation. This raises interesting questions: Can coevolution change snail chirality? Can predator specialization lead to snake extinction in this ecosystem? A geographical distribution of the sinistral and dextral snails and pareatic snakes is shown in Figure 5a in Hoso et al. (2010). We can observe regions where the three populations coexist, islands where right- and left-handed snails live without snakes and even islands where only sinistral snails survive (Vermeij, 1975). A mathematical model based on the Holling functional response has been proposed in Izquierdo et al. (2019) to describe the interaction between these populations and their evolution.

Problem: Let us suppose that it has been estimated that 450 millions of dextral *Satsuma* snails and 0.9 millions of *Pareas iwasakii* snakes coexist on one of the Ryukyu Islands. Let us also suppose that the ecological parameters of these populations have been approximately assessed. By using units of million for population and years as the time unit, the intrinsic snail growth rate is approximately equal to 0.025, the snail carrying capacity of the island is assessed as 850, the intrinsic snake mortality rate is 0.01, the per capita predation coefficient per time unit is 0.02, and snake consumption efficiency per time unit is approximately 10^{-4}.

(a) Construct a mathematical model for the interactions between the dextral/sinistral *Satsuma* snails and *Pareas iwasakii* snakes by using Lotka–Voterra-type terms.

(b) Find the stationary points of the model. Are there constant solutions where the three populations coexist?

(c) Describe the model in the absence of the predator population. Obtain analytically the solutions in this competition model.

(d) Describe qualitatively the model in the absence of the mutant snail population. Is there any stationary point? Determine the population evolution on the island by means of a numerical method in this case.

(e) If a small population of 100 left-handed snails emerges on the island, how does the ecological system evolve? Use a numerical method to answer this question. Compare the results with the previous point.

(f) If the intrinsic snake mortality rate doubles, how do the snail and snake populations evolve from the initial conditions employed in the previous points?

Solution:

(a) Let $x_1 = x_1(t)$ and $x_2 = x_2(t)$ denote the populations of dextral and sinistral *Satsuma* snails, respectively, whereas $y = y(t)$ will be used to represent the *Pareas Iwasakii* snake population. We will construct a Lotka–Volterra-type mathematical model where logistic growth will be assumed for the whole snail population. Taking into account that dextral/sinistral snail variants only differ in body chirality, the same intrinsic growth rate r and carrying capacity K is conjectured for the two snail populations. As usual in this type of models, a decreasing Malthusian growth rate s is considered for the predator population $y(t)$ in the absence of preys. The number of snails consumed by the snake is assumed to be proportional to the number of prey/predator interactions. The predation rate on dextral/sinistral snails will be denoted by α_1 and α_2, respectively, whereas β_1 and β_2 will stand for consumption efficiencies increasing the predator growth rate. Note that $\alpha_2 = 0.125\alpha_1$ and $\beta_2 = 0.125\beta_1$ for *Pareas iwasakii* snakes since the previously mentioned lab experiments show that the ratio between hunting efficiencies on sinistral/dextral snails is approximately 0.125. By using the previous notation, the interactions between these three populations can be described by the two-prey one-predator Lotka–Volterra model

$$\begin{aligned}\frac{dx_1}{dt} &= rx_1\Big(1-\frac{x_1+x_2}{K}\Big)-\alpha_1 x_1 y,\\ \frac{dx_2}{dt} &= rx_2\Big(1-\frac{x_1+x_2}{K}\Big)-\alpha_2 x_1 y,\\ \frac{dy}{dt} &= -s\,y+\beta_1 x_1 y+\beta_2 x_2 y.\end{aligned} \tag{11.17}$$

Plugging the values of the ecological parameters $r = 0.025$, $K = 850$, $s = 0.01$, $\alpha_1 = 0.02$, $\beta_1 = 10^{-4}$, $\alpha_2 = 0.125\alpha_1$, and $\beta_2 = 0.125\beta_1$ of our ecosystem into (11.17) leads to the following system of differential equations:

$$\begin{aligned}\frac{dx_1}{dt} &= 0.025x_1\Big(1-\frac{x_1+x_2}{850}\Big)-0.02\,x_1 y,\\ \frac{dx_2}{dt} &= 0.025x_2\Big(1-\frac{x_1+x_2}{850}\Big)-0.0025\,x_1 y,\\ \frac{dy}{dt} &= -0.01\,y+10^{-4}\,x_1 y+1.25\,10^{-5}\,x_2 y.\end{aligned} \tag{11.18}$$

(b) The stationary points are the constant solutions $(\overline{x}_1, \overline{x}_2, \overline{y})$ of the model. They are obtained by solving the system (11.17) when the left-hand sides of these equations vanish. The absence of snails and snakes determines the trivial stationary point $P_0 \equiv (0, 0, 0)$, which lacks ecological interest. A second steady solution of (11.17) is given by

$$P_1 = \Big(\frac{s}{\beta_1}, 0, \frac{r(\beta_1 K - s)}{\alpha_1 \beta_1 K}\Big) \approx (100, 0, 1.103),$$

where the dextral *Satsuma* snail and *Pareas iwasakii* snake populations coexist. Here, the sinistral snail population is absent. In our ecosystem the ratio between the snail and snake populations is approximately 90.66 for this constant solution.

Likewise, there exists a steady population where sinistral snails and snakes coexist without the presence of the dextral variant, which is expressed as

$$P_2 = \Big(\frac{s}{\beta_2}, 0, \frac{r(\beta_2 K - s)}{\alpha_2 \beta_2 K}\Big) \approx (0, 800, 0.588).$$

Now, the ratio between the predator and prey populations is 1360.5, which remarks the difficulty of the snakes on feeding on sinistral snails.

Finally, a one-parametric family of stationary points

$$P^{(\mu)} = (\mu, K-\mu, 0) = (\mu, 850-\mu, 0) \quad \text{with} \quad \mu \in [0, 1]$$

arises in the x_1–x_2-coordinate plane. These steady solutions characterize the situation where the total snail population has reached the carrying capacity where the predator population is not present. Note that there exist no stationary points where the three populations coexist.

Here we shall study the stability of the stationary points P_1 and P_2 for our ecosystem. The linear stability matrix in the first case is given as

$$L[P_1] = \begin{pmatrix} -0.00294118 & -0.00294118 & -2 \\ 0 & 0.0193015 & 0 \\ 0.000110294 & 0.0000137868 & 0 \end{pmatrix}.$$

This matrix involves the positive real eigenvalue 0.0193015 together with the complex conjugate eigenvalues $-0.00147059 \pm 0.0147792i$. This indicates that P_1 is a spiral sink restricted to the x_1–y-plane but it is unstable against mutant population fluctuation. This is an important fact to be considered in the next points.

On the other hand, the linear stability matrix for the stationary point P_2 reads

$$L[P_2] = \begin{pmatrix} -0.0102941 & 0 & 0 \\ -0.0235294 & -0.0235294 & -2 \\ 0.0000588235 & 7.35294 \times 10^{-6} & 0 \end{pmatrix}.$$

Now, all the eigenvalues of $L[P_2]$ are negative. In particular, their numeric values are -0.0228869, -0.0102941, and -0.000642547. The stationary point P_2 is a stable node.

(c) When the predator population is removed from our study, the population evolution is ruled by the following system of differential equations:

$$\begin{aligned} \frac{dx_1}{dt} &= rx_1\left(1 - \frac{x_1 + x_2}{K}\right), \\ \frac{dx_2}{dt} &= rx_2\left(1 - \frac{x_1 + x_2}{K}\right). \end{aligned} \tag{11.19}$$

The symmetry of the previous expressions allows us to analytically find the solutions in this case. In the phase portrait the system (11.19) reads

$$\frac{dx_1}{dx_2} = \frac{x_1}{x_2},$$

which leads to the relation $x_1 = Cx_2$, where C is an integration constant. Indeed, $C = x_{10}/x_{20}$ by employing the initial conditions. On the other hand, if we sum the two equations in (11.19) the global snail populations comply with the logistic differential equation

$$\frac{dx}{dt} = rx\left(1 - \frac{x}{K}\right), \quad \text{where} \quad x = x_1 + x_2.$$

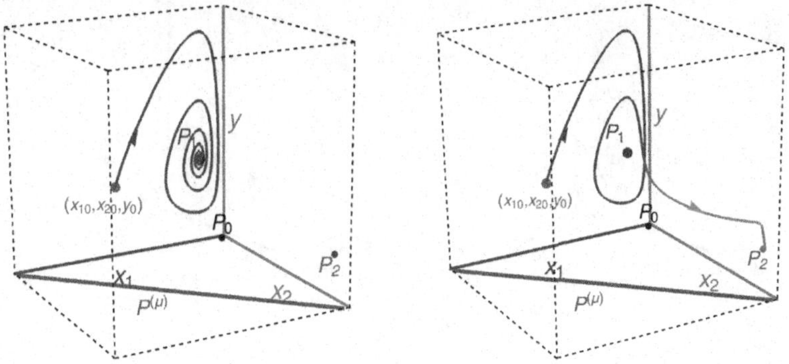

Figure 11.7 Evolution of the dextral/sinistral *Satsuma* snail and *Pareas iwasakii* snake populations with $s = 0.01$ and initial conditions $x_{10} = 450$, $x_{20} = 0$, and $y_0 = 0.9$ (left) and $x_{10} = 450$, $x_{20} = 10^{-4}$, and $y_0 = 0.9$ (right). Note that the introduction of a small group of left-handed snails changes the final result of the evolution.

All the previous facts lead to the analytical solutions

$$x_i(t) = \frac{x_{i0} K}{x_{10} + x_{20} + (K - x_{10} - x_{20})e^{-rt}} \quad \text{with} \quad i = 1, 2.$$

(d) In the absence of left-handed snails, the system (11.17) reduces to

$$\frac{dx_1}{dt} = r x_1 \left(1 - \frac{x_1}{K}\right) - \alpha_1 x_1 y = 0.025 x_1 \left(1 - \frac{x_1}{850}\right) - 0.02 x_1 y,$$
$$\frac{dy}{dt} = -s y + \beta_1 x_1 y = -0.01 y + 10^{-4} x_1 y,$$

which corresponds to a prey–predator Lotka–Volterra model, where the prey population follows a logistic growth. The evolution of the system of differential equations with the initial conditions $x_{10} = 450$ and $y_0 = 0.9$ in the phase space has been depicted in Fig. 11.7 (left). We can observe that the snail and snake populations oscillate while approaching the spiral sink P_1 in the x_1–y-plane.

(e) If a small sinistral snail population is introduced in the ecosystem the behavior of the distinct populations is completely different. In Fig. 11.7 (right) the evolution of the model with the initial conditions $x_{10} = 450$, $x_{20} = 10^{-4}$, and $y_0 = 0.9$ is shown. We can see that the behavior is initially similar to the previous case, although the sinistral snail population is slowly thriving. When this mutant population is large enough, a drastic change takes place; the dextral snail population decreases and is progressively substituted by the sinistral variant. Finally, the populations tend to the stable stationary point P_2. This new situation is more propitious for the snake population than the previous one.

(f) If the intrinsic snake mortality rate increases to $s = 0.02$ in our ecosystem the stationary point P_2 loses its ecological meaning because its third component becomes negative. The numerical simulations for the two initial conditions previously employed in these new circumstances have been depicted in Fig. 11.8. If there is no

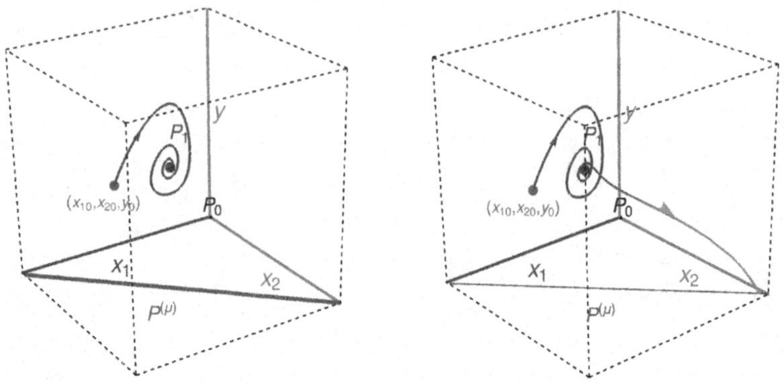

Figure 11.8 Evolution of the dextral/sinistral *Satsuma* snail and *Pareas iwasakii* snake populations with $s = 0.02$ and initial conditions $x_{10} = 450$, $x_{20} = 0$, and $y_0 = 0.9$ (left) and $x_{10} = 450$, $x_{20} = 10^{-4}$, and $y_0 = 0.9$ (right). Note that the introduction of a small group of left-handed snails changes the final result of the evolution.

sinistral snail population initially the dextral snail and snake populations oscillate and approach to the stationary point P_1 (see Fig. 11.8 (left)). However, if $x_{20} = 10^{-4}$ we find a completely different result because, although initially the evolution is similar to the previous case, progressively the sinistral variant is replacing the dextral variant, whereas the snake population becomes extinct. The final result is given by the coexistence of a large sinistral snail population with a small dextral snail population, corresponding to one of the stationary points located on the set $P^{(\mu)}$.

References

Asami, T., Cowie, R., Ohbayashi, K., 1998. Evolution of mirror images by sexually asymmetric mating behavior in hermaphroditic snails. The American Naturalist 152 (2), 225–236.

Danaisawadi, P., Asami, T., Ota, H., Sutcharit, C., Panha, S., 2015. Subtle asymmetries in the snail-eating snake *Pareas carinatus* (Reptilia: Pareatidae). Journal of Ethology 33, 243–246.

Danaisawadi, P., Asami, T., Ota, H., Sutcharit, C., Panha, S., 2016. A snail-eating snake recognizes prey handedness. Scientific Reports 6, 23832.

Gittenberger, E., Hamann, T.D., Asami, T., 2012. Chiral speciation in terrestrial pulmonate snails. PLoS ONE 7 (4), e34005.

Gotz, M., 2002. The feeding behavior of the snail-eating snake Pareas carinatus. Amphibia-Reptilia 23, 487–493.

Hammer, W., 1906. Epidemic disease in England. Lancet, 733–754.

Hoso, M., Asami, T., Hori, M., 2007. Right-handed snakes: convergent evolution of asymmetry for functional specialization. Biology Letters 3 (2), 169–173.

Hoso, M., Kameda, Y., Wu, S., Asami, T., Kato, M., Hori, M., 2010. A speciation gene for left-right reversal in snails results in anti-predator adaptation. Nature Communications 1, 133.

Izquierdo, A.A., León, M.G., de la Torre Mayado, M., 2019. A generalized Holling type ii model for the interaction between dextral-sinistral snails and Pareas snakes. Applied Mathematical Modelling 73, 459–472.

Karyotis, V., Khouzani, M.H.R., 2016. Malware Diffusion Models for Modern Complex Networks: Theory and Applications, 1st edn. Morgan Kaufmann Publishers Inc., San Francisco, CA, USA.

Kermack, W.O., McKendrick, A.G., 1927. A contribution to the mathematical theory of epidemics. Proceedings of the Royal Society of London A: Mathematical, Physical and Engineering Sciences 115 (772), 700–721.

Kloeden, P., Platen, E., 2011. Numerical Solution of Stochastic Differential Equations. Stochastic Modelling and Applied Probability. Springer, Berlin, Heidelberg.

Lee, C.-F., Shi, J., 2010. Application of Alternative ODE in Finance and Economics Research. Springer US, Boston, MA, pp. 1293–1300.

Lotka, A.J., 1910. Contribution to the theory of periodic reactions. The Journal of Physical Chemistry 14 (3), 271–274.

Lotka, A., 1925. Elements of Physical Biology. Williams and Wilkins, Baltimore, Md.

Robertson, R., 1993. Snail handedness. National Geographic Research and Exploration 9, 104–119.

Schilthuizen, M., Davison, A., 2005. The convoluted evolution of snail chirality. Naturwissenschaften, 504–515.

Ueshima, R., Asami, T., 2003. Single-gene speciation by left-right reversal. Nature 425, 679.

Vermeij, G.J., 1975. Evolution and distribution of left-handed and planispiral coiling in snails. Nature 254, 419–420.

Volterra, V., 1927. Variazioni e fluttuazioni del numero d'individui in specie animali conviventi, vol. 6. C. Ferrari.

You, C.-W., Poyarkov, N.A., Lin, S.-M., 2015. Diversity of the snail-eating snakes Pareas (Serpentes, Pareatidae) from Taiwan. Zoologica Scripta 44 (4), 349–361.

Oscillations in higher-order differential equations and systems of differential equations

12

Lucian Niţă[a], Ion Mierluş-Mazilu[a], Leonard Dăuş[a], Daniela Richtáriková[b], Manuel Rodríguez-Martín[c], Pablo Rodríguez-Gonzálvez[d]
[a]Technical University of Civil Engineering, Bucharest, Romania, [b]Slovak University of Technology in Bratislava, Bratislava, Slovakia, [c]University of Salamanca, Salamanca, Spain, [d]University of León, León, Spain

12.1 Basic theory background

Many physical, chemical, and biological processes, and also processes from other scientific domains, are modeled by mathematical equations in which the unknown is a function, representing the physical characteristic of the process which describes the respective action. In most of the cases, in those equation also appear derivatives of various orders of the unknown function. Such equations are called *differential equations.*

Definition 12.1. Let $I \subset R$ be an interval, $V \subset \mathbb{R}^{n+1}$, and let $F = F(x, y, y', \ldots, y^{(n)})$ be a function defined on $I \times V$ having as argument variable $x \in I$ and the real function y and its derivatives $y', y'', \ldots, y^{(n)}$. The relation $F(x, y, y', \ldots, y^{(n)}) = 0$ is called the *nth-order differential equation.* Solving a differential equation means to find the function $y = f(x)$, $f : I \to R$, which has all the derivatives, including its nth-order derivative, defined on I such that $F(x, f(x), f'(x), \ldots, f^{(n)}(x)) = 0$, $\forall x \in I$.

Definition 12.2. (a) We say that the function $\phi(x, C_1, C_2, \ldots, C_n)$ is the general solution of the differential equation

$$F\big(x, y, y', \ldots, y^{(n)}\big) = 0 \tag{12.1}$$

if it satisfies the equation and we obtain every solution of the equation by a convenient choice of the constants $C_1, C_2, \ldots, C_n$.

(b) A function $y = \varphi_0(x)$, obtained from the general solution giving particular values to constants $C_1, C_2, \ldots, C_n$, is called the *particular solution* of the differential equation (12.1).

Example 12.1. The equation $y'' - 4y = 4x$ is a differential equation of order 2. The function $y(x) = C_1 e^{2x} + C_2 e^{-2x} - x$ (C_1, C_2 are arbitrary constants) represents a

Calculus for Engineering Students. https://doi.org/10.1016/B978-0-12-817210-0.00019-9

family of solutions of the given equation. Giving particular values to the constants, we obtain particular solutions. For example, when $C_1 = 1$, $C_2 = 0$, we obtain the solution $y = e^{2x} - x$.

Example 12.2. The differential equation of order 3 $y''' - y'' + y' - y = 0$ admits the solutions $y_1 = e^x$, $y_2 = \sin x$, $y_3 = \cos x$, but also admits the solution $y = C_1 e^x + C_2 \sin x + C_3 \cos x$.

12.1.1 Particular types of differential equations of order $n \geq 2$

12.1.1.1 Linear equations

The general form is $a_0(x)y^{(n)} + a_1(x)y^{(n-1)} + \cdots a_{n-1}(x)y' + a_n(x)y = f(x)$.

The linear equation is called *homogeneous* if $f \equiv 0$ and *nonhomogeneous* in a contrary case.

(a) Homogeneous linear equations

Definition 12.3. Let $y_1, y_2, \ldots, y_n : I \rightarrow \mathbb{R}$ (I interval) be a function. We say that the n functions are *linearly independent* on I if from the equality $\lambda_1 y_1(x) + \lambda_2 y_2(x) + \cdots + \lambda_n y_n(x) = 0$, $\forall x \in I$, it results that $\lambda_1 = \lambda_2 = \cdots = \lambda_n = 0$.

In a contrary case, we will say that the functions are *linearly dependent*.

Example 12.3. The functions $1, x, e^x$ are linearly independent on $\mathbb{R}$. Indeed, considering the relation $\lambda_1 \cdot 1 + \lambda_2 x + \lambda_3 e^x = 0$, for $x = 0$, $x = 1$, and $x = -1$ we obtain the following system:

$$\begin{cases} \lambda_1 + \lambda_3 = 0, \\ \lambda_1 + \lambda_2 + e\lambda_3 = 0, \\ \lambda_1 - \lambda_3 + \frac{1}{e}\lambda_3 = 0, \end{cases}$$

with solution $\lambda_1 = \lambda_2 = \lambda_3 = 0$.

Definition 12.4. Let $y_1, y_2, \ldots, y_n : I \rightarrow \mathbb{R}$ be a function having the derivatives continuous until the order -1. The determinant

$$W(y_1, y_2, \ldots, y_n) = \begin{vmatrix} y_1 & y_2 & \cdots & y_n \\ y_1' & y_2' & \cdots & y_n' \\ \cdots & \cdots & \cdots & \cdots \\ y_1^{(n-1)} & y_2^{(n-2)} & \cdots & y_n^{(n-1)} \end{vmatrix}$$

is called the Wronskian of functions $y_1, y_2, \ldots, y_n$ (or the determinant of Wronski).

Theorem 12.1. *Let*

$$y^{(n)} + a_1(x)y^{(n-1)} + \cdots + a_{n-1}(x)y' + a_n(x)y = 0 \tag{12.2}$$

be the homogeneous linear equation of order n with $a_1, a_2, \ldots, a_n$ continuous functions on interval $I \subset R$. Let $y_1, y_2, \ldots, y_n$ be the solutions of the given equation on I.

If the Wronskian of the functions $y_1, y_2, \dots, y_n$ is not identically null on I, then any solution of the equation is of the form

$$y = C_1 y_1 + C_2 y_2 + \cdots + C_n y_n \tag{12.3}$$

with $C_1, C_2, \dots, C_n$ arbitrary constants. (The function y given by (12.3) represents the general solution of the equation.)

Remark. It can be proved that $W(y_1, y_2, \dots, y_n) \neq 0, \forall x \in I$.

Definition 12.5. Functions $y_1, y_2, \dots, y_n$ from Theorem 12.1 form a *fundamental system of solutions* (FSS) for the nth-order equation (12.2).

Remark. Let $y_1, y_2, \dots, y_n : I \to R$ be functions, with $W(y_1, y_2, \dots, y_n) \neq 0, x \in I$. The equation which has the FSS given by $y_1, y_2, \dots, y_n$ is

$$\begin{vmatrix} y & y' & \dots & y^{(n)} \\ y_1 & y_1' & \dots & y_1^{(n)} \\ \dots & \dots & \dots & \dots \\ y_n & y_n' & \dots & y_n^{(n)} \end{vmatrix} = 0$$

(according to the properties of determinants).

Example 12.4. Equation $x^2 y'' - 2xy' + 2y = 0$ admits solutions $y_1(x) = x$, $y_2(x) = x^2$. Wronskian $W(y_1, y_2) = \begin{vmatrix} x & x^2 \\ 1 & 2x \end{vmatrix} = x^2 \neq 0, \forall x \in \mathbb{R}\setminus\{0\}$, which proves that y_1, y_2 form an FSS for the given equation. The general solution of the equation will be $y(x) = C_1 x + C_2 x^2$ for any interval included in $\mathbb{R}\setminus\{0\}$.

Example 12.5. Functions $y_1(x) = x^2$, $y_1(x) = x^4$ form an FSS on any interval which does not contain 0. Indeed, $W(y_1, y_2) = \begin{vmatrix} x^2 & x^4 \\ 2x & 4x^3 \end{vmatrix} = 2x^5 \neq 0$, $\forall x \neq 0$. The equation which admits the particular solutions y_1, y_2 is

$$\begin{vmatrix} y & y' & y'' \\ x^2 & 2x & 2 \\ x^4 & 4x^3 & 12x^2 \end{vmatrix} = 0, \text{ i.e., } x^2 y'' - 2xy' + 8y = 0.$$

Theorem 12.2 (The solution of the Cauchy problem). *Let*

$$y^{(n)} + a_1(x) y^{(n-1)} + \cdots + a_{n-1}(x) y' + a_n(x) y = 0 \tag{12.4}$$

be a homogeneous linear differential equation of nth order, having FSS $y_1, y_2, \dots, y_n : I \to \mathbb{R}$. Then there is a unique solution $y : I \to \mathbb{R}$, which for any $x_0 \in I$ fixed satisfies the following conditions: $y(x_0) = y_0$, $y'(x_0) = y_{01}, \dots, y^{(n-1)}(x_0) = y_{0n-1}$ ($y_0, y_{01}, \dots, y_{0n-1}$ are arbitrary, fixed, real numbers).

Definition 12.6. Eq. (12.4), with the conditions given in Theorem 12.2, is called the *Cauchy problem.*

(b) Nonhomogeneous linear equations

Theorem 12.3 (The variation of constants methods). *Let*

$$a_0(x)y^{(n)} + a_1(x)y^{(n-1)} + \cdots + a_{n-1}(x)y' + a_n(x)y = f(x) \tag{12.5}$$

be a nonhomogeneous linear differential equation of nth order with $a_1, a_2, \ldots, a_n, f : I \to \mathbb{R}$, continuous functions, $a_0(x) \neq 0$, $\forall x \in I$. Let $y_1, y_2, \ldots, y_n$ be an FSS for the attached homogeneous equation $a_0(x)y^{(n)} + a_1(x)y^{(n-1)} + \cdots + a_n(x)y = 0$. A particular solution of the nonhomogeneous equation will be $y_p(x) = y_1(x)\int C_1'(x)dx + y_2(x)\int C_2'(x)dx + \cdots + y_n(x)\int C_n'(x)dx$, where $C_1', C_2', \ldots, C_n'$ represent the solution of the system

$$\begin{cases} y_1C_1' + y_2C_2' + \cdots + y_nC_n' = 0, \\ y_1'C_1' + y_2'C_2' + \cdots + y_n'C_n' = 0, \\ \ldots \quad \ldots \quad \ldots \quad \ldots \\ y_1^{(n-1)}C_1' + y_2^{(n-1)}C_2' + \cdots + y_n^{(n-1)}C_n' = \dfrac{f(x)}{a_0(x)}. \end{cases}$$

We note $\varphi_1(x) = A_1 + \int C_1'(x)dx, \ldots, \varphi_n(x) = A_n + \int C_n'(x)dx$ ($A_1, \ldots, A_n$ being arbitrary constants). Then the function $y(x) = \varphi_1(x)y_1(x) + \cdots + \varphi_n(x)y_n(x)$ represents the general solution of the nonhomogeneous equation from Theorem 12.3.

(c) Linear differential equations with constant coefficients

(i) Homogeneous We have

$$a_0y^{(n)} + a_1y^{(n-1)} + \cdots + a_{n-1}y' + a_ny = 0,\ a_k \in \mathbb{R},\ k \in \{0, 1, \ldots, n\},\ a_0 \neq 0.$$

We are looking for solutions of the form $y = e^{rx}$. We obtain

$$a_0r^ne^{rx} + a_1r^{n-1}e^{rx} + \cdots + a_ne^{rx} = 0 \quad \text{or} \quad a_0r^n + a_1r^{n-1} + \cdots + a_n = 0$$

called *characteristic equation* associated to the differential equation. According to the type of roots of the characteristic equation, an FSS for the linear differential equation with constant coefficients will be:

(1) $e^{r_1x}, e^{r_2x}, \ldots, e^{r_nx}$, if the characteristic equation has the distinct, real roots $r_1, r_2, \ldots, r_n$.

(2) $e^{rx}, xe^{rx}, \ldots, x^{k-1}e^{rx}$, if the characteristic equation admits the root $r \in \mathbb{R}$, with k multiplicity order $1 < k \leq n$.

(3) $e^{\alpha x}\cos\beta x$, $e^{\alpha x}\sin\beta x$, if the characteristic equation admits complex, conjugated roots $\alpha \pm i\beta$ of order 1.

(4) $e^{\alpha x}\cos\beta x, e^{\alpha x}\sin\beta x,\ x^{\alpha x}\cos\beta x, x^{\alpha x}\sin\beta x, \ldots, x^{k-1}e^{\alpha x}\cos\beta x, x^{k-1}e^{\alpha x} \times \sin\beta x$, if the characteristic equation admits the complex, conjugated roots $\alpha \pm i\beta$ of order k, $1 < k$.

(ii) Nonhomogeneous We have

$$a_0 y^{(n)} + a_1 y^{(n-1)} + \cdots + a_{n-1} y' + a_n y = f(x), \ f \not\equiv 0.$$

One of the solving methods is the variation of constant coefficients illustrated in Theorem 12.3. Another method, called the *undetermined coefficients* method, consists in passing the following steps:

1) finding the general solution y_0 of the homogeneous attached equation;
2) finding a particular solution y_p of the nonhomogeneous equation;
3) the general solution of the nonhomogeneous equation will be $y = y_0 + y_p$.

How to find y_p (α) If $f(x) = P_n(x)$ (nth-degree polynomial), we choose $y_p(x) = Q_n(x)$ (nth-degree polynomial whose coefficients are determined from the condition that y_p satisfies the nonhomogeneous equation), and if the characteristic equation admits the order k root $x = 0$, we choose $y_p(x) = x^k Q_n(x)$.

(β) If $f(x) = P_n(x)e^{\alpha x}$, we choose $y_p(x) = Q_n(x)e^{\alpha x}$ if α is not a root of the characteristic equation, and $y_p(x) = x^k e^{\alpha x} Q_n(x)$ if α is a root of order k of the characteristic equation.

(γ) If $f(x) = e^{\alpha x}(M \cos \beta x + N \sin \beta x)$, we choose $y_p(x) = e^{\alpha x}(A \cos \beta x + B \sin \beta x)$ if $\alpha \pm i\beta$ are not roots of the characteristic equation, and $y_p(x) = x^k e^{\alpha x}(A \cos \beta x + B \sin \beta x)$ if $\alpha \pm i\beta$ are roots of order k of the characteristic equation.

(δ) If $f(x) = f_1(x) + f_2(x)$, where f_1 and f_2 are of type α, β, γ, or δ different from each other, $y_p = y_{p_1} + y_{p_2}$, where y_{p_1} corresponds to f_1 and y_{p_2} corresponds to f_2.

Example 12.6. Solve the differential equation $y^{(7)} - 2y^{(4)} + y' = x^2 + 4$.

Step 1. The homogeneous associated equation is $y^{(7)} - 2y^{(4)} + y' = 0$ and admits the characteristic equation $r^7 - 2r^4 + r = 0$, which can be written equivalently as $r(r^6 - 2r^3 + 1) = 0$ or $r(r^3 - 1)^2 = 0$, i.e., $r(r-1)^2(r^2 + 2 + 1)^2 = 0$. Its roots will be $r_1 = 0$, $r_2 = r_3 = 1$, $r_4 = r_5 = -\frac{1}{2} + i\frac{\sqrt{3}}{2}$, $r_6 = r_7 = -\frac{1}{2} - i\frac{\sqrt{3}}{2}$. Functions $y_1(x) = e^{0 \cdot x} = 1$, $y_2(x) = e^x$, $y_3(x) = xe^x$, $y_4(x) = e^{-\frac{x}{2}} \cos \frac{\sqrt{3}}{2}x$, $y_5(x) = e^{-\frac{x}{2}} \sin \frac{\sqrt{3}}{2}x$, $y_6(x) = xe^{-\frac{x}{2}} \cos \frac{\sqrt{3}}{2}x$, $y_7(x) = xe^{-\frac{x}{2}} \sin \frac{\sqrt{3}}{2}x$ will correspond in FSS to the roots in paragraph (i) (2) and (4). It results that the general solution of the homogeneous equation will be

$$y_0(x) = C_1 + C_2 e^x + C_3 x e^x + C_4 e^{-\frac{x}{2}} \cos \frac{\sqrt{3}}{2}x + C_5 e^{-\frac{x}{2}} \sin \frac{\sqrt{3}}{2}x + \\ + C_6 x e^{-\frac{x}{2}} \cos \frac{\sqrt{3}}{2} + C_7 x e^{-\frac{x}{2}} \sin \frac{\sqrt{3}}{2}x.$$

Step 2. We are searching for y_p, particular solution of the nonhomogeneous equation. The right member is a polynomial of second order. As $x = 0$ is a root of order $K = 1$ in the characteristic equation, we take $y_p(x) = x(ax^2 + bx + c) = ax^3 + bx^2 + cx$. Obviously, $y^{(4)} \equiv 0$, $y^{(7)} \equiv 0$, so we obtain $3ax^2 + 2bx + c = x^2 + 4$, wherefrom the following system results:

$$\begin{cases} 3a = 1, \\ 2b = 0, \\ c = 4, \end{cases} \text{with solution} \begin{cases} a = \frac{1}{3}, \\ b = 0, \\ c = 4. \end{cases} \text{so, } y_p(x) = \frac{x^3}{3} + 4x.$$

Step 3. We have

$$\begin{aligned} y(x) &= y_0(x) + y_p(x) = C_1 + e^x(C_2 + xC_3) + \\ &+ e^{-\frac{x}{2}}\left(C_4 \cos\frac{\sqrt{3}}{2}x + C_5 \sin\frac{\sqrt{3}}{2}x + C_6 x \cos\frac{\sqrt{3}}{2}x + C_7 x \sin\frac{\sqrt{3}}{2}x\right) \\ &+ \frac{x^3}{3} + 4x. \end{aligned}$$

Example 12.7. Solve the differential equation $y''' - 3y'' + 3y' - y = \frac{e^x}{x}, x < 0$.

Firstly, we observe that the undetermined coefficients method does not work here, because the form of the right member does not respect the "pattern" that makes this method applicable. Hence, we will apply the *variation of constants method.*

Step 1. The homogeneous equation is $y''' - 3y'' + 3y' - y = 0$ and it admits the characteristic equation $r^3 - 3r^2 + 3r - 1 = 0$ or $(r-1)^3 = 0$, with the roots $r_1 = r_2 = r_3 = 1$.

Functions $y_1(x) = e^x$, $y_2(x) = xe^x$, $y_3(x) = x^2e^x$, which form an FSS, correspond to these roots, because the characteristic equation does not have any other roots. Hence, $y_0(x) = C_1e^x + C_2xe^x + C_3x^2e^x$ is the general solution of the homogeneous equation.

Step 2. The general solution of the nonhomogeneous equation will be of the form $y(x) = C_1(x)y_1(x) + C_2(x)y_2(x) + C_3(x)y_3(x)$, where functions C_1, C_2, C_3 are deduced from the system

$$\begin{cases} C_1'y_1 + C_2'y_2 + C_3'y_3 = 0, \\ C_1'y_1' + C_2'y_2' + C_3'y_3' = 0, \\ C_1'y_1'' + C_2'y_2'' + C_3'y_3'' = \frac{f}{a_0} \end{cases} \quad \text{(from Theorem 12.3).}$$

In our case, it results in the following system:

$$\begin{cases} C_1'e^x + C_2'xe^x + C_3'x^2e^x = 0, \\ C_1'e^x + C_2'(x+1)e^x + C_3'(x^2+2x)e^x = 0, \\ C_1'e^x + C_2'(x+2)e^x + C_3'(x^2+4x+2)e^x = \frac{e^x}{x}. \end{cases}$$

By subtracting the first two equations and dividing by e^x we find $C_2' = -2xC_3'$; by replacing C_2' with the expression $-2xC_3$ in the first equation, we obtain $C_1' = x^2C_3'$. Hence, from the last equation it results that $x^2C_3' - (2x^2+4x)C_3' + (x^2+4x+2)C_3' = \frac{1}{x}$.

We deduce successively $C_3' = \frac{1}{2x}$, $C_2' = -1$, $C_1' = \frac{x}{2}$ and by integrating these equalities, we will have $C_3(x) = \frac{1}{2}ln(-x) + K_3$, $C_2(x) = -x + K_2$, $C_1(x) = \frac{x^2}{4} + K_1$. The required function will be

$$y(x) = \left(\frac{x^2}{4} + K_1\right)e^x + (-x + K_2)xe^x + \left(\frac{1}{2}ln(-x) + K_3\right)x^2e^x.$$

Example 12.8. We consider the equation $(x^2 + 1)y'' - 2xy' + \lambda y = 0,\ \lambda > 0$. Find the value of λ for which the equation admits a polynomial as solution. Then solve the equation by searching a particular solution of the form of a polynomial of first degree.

We solve this as follows.

Let $y_P(x) = a_0x^n + a_1x^{n-1} + \cdots + a_n$ a polynomial of nth degree which satisfies the equation. It results that $(x^2+1)[n(n-1)a_0x^{n-2} + (n-1)(n-2)a_1x^{n-3} + \cdots +] - 2x[na_0x^{n-1} + (n-1)a_1x^{n-2} + \cdots] + \lambda(a_0x^n + a_1x^{n-1} + \cdots + a_n) = 0$.

By identifying the coefficients of the powers of x, we obtain

$$a_0\big[n(n-1) - 2n + \lambda\big] = 0, a_1\big[(n-1)(n-3) - 2(n-1) + \lambda\big] = 0, \ldots.$$

We have $n^2 - 3n + \lambda = 0$. As $n \in \mathbb{N}, n > 0$, we have

$$\Delta = 9 - 4\lambda \overset{not}{\to} k^2 \overset{\lambda \neq 0}{\Longrightarrow} k = 1, \lambda = 2.$$

Now, let $y_P(x) = ax + b$. By substituting in the equation, it results that $b = 0$, and hence by choosing $a = 1$, we have $y_P(x) = x$. We make the change of function $y = zx$.

We obtain successively $y' = z + z'x$, $y'' = 2z' + xz''$,

$$(x^2+1)(2z' + xz'') - 2x(z + z'x) + 2zx = 0,\ z''(x^3 + x) = -2z'.$$

We note $z' = u$, and hence $z'' = u'$ and the equation becomes $u'(x^3 + x) = -2u$, or $-\frac{1}{2u}du = \frac{1}{x(x^2+1)}dx$, wherefrom, by integration, it results that $u(x) = C(1 + \frac{1}{x^2}) = z'(x)$. We then obtain $z(x) = C(x - \frac{1}{x}) + D$ and finally $y(x) = xz(x) = Cx^2 + Dx - C$ (we observe that for $C = 0$, we obtain the solution $y(x) = Dx$, i.e., of the form of y_P.

12.1.1.2 *Euler equations*

The general form is $a_0x^ny^{(n)} + a_1x^{n-1}y^{(n-1)} + \cdots + a_{n-1}xy' + a_ny = f(x)$ with $a_k \in \mathbb{R}$, $k \in \{0, \ldots, n\}$, $a_0 \neq 0$.

It is solved as follows:

(α) The variable change is executed. For $x > 0$, $x = e^t$; for $x < 0$, $x = -e^t$; and for $x > 0$, $y(x) = y(e^t) \overset{not}{=} \bar{y}(t)$.

(β) From $y'(x) = \frac{dy}{dx} = \frac{d\bar{y}}{dt} : \frac{dx}{dt} = e^{-t}\frac{d\bar{y}}{dt}$ we obtain the differentiation operator $\frac{d}{dx} = e^{-t}\frac{d}{dt}$.

(γ) Then $y''(x) = \frac{d^2y}{dx^2} = \frac{d}{dx}(\frac{dy}{dt}) = e^{-t}\frac{d}{dt}(e^{-t}\frac{d\bar{y}}{dt}) = -e^{-2t}\dot{\bar{y}} + e^{-2t}\ddot{\bar{y}}$, where $\ddot{\bar{y}} = \frac{d^2\bar{y}}{dt^2}$, etc.

With this type of change of variable, the equation of Euler type reduces to a linear differential equation with constant coefficients.

Example 12.9. Solve the differential equation of Euler type

$$x^3y''' + 2xy' - 2y = xln^2x, x > 0.$$

Denoting $x = e^t$, $y(x) \overset{not}{=} \bar{y}(t)$, $y' = \frac{d\bar{y}}{dt} \cdot \frac{dt}{dx} = e^{-t}\dot{\bar{y}}$ (where $\dot{\bar{y}} = \frac{d\bar{y}}{dt}$), we obtain the derivation operator $\frac{d}{dx} = e^{-t}\frac{d}{dt}$. Further, $y'' = \frac{d}{dx}(y') = e^{-t}\frac{d}{dt}(e^{-t}\dot{\bar{y}}) = e^{-2t}(\ddot{\bar{y}} - \dot{\bar{y}})$ $y''' = \frac{d}{dx}(y'') = e^{-t}\frac{d}{dt}[e^{-2t}(\ddot{\bar{y}} - \dot{\bar{y}})] = e^{-3t}(\dddot{\bar{y}} - 3\ddot{\bar{y}} + 2\dot{\bar{y}})$. By replacing these equations in the initial equation, we obtain $e^{3t} \cdot e^{-3t}(\dddot{\bar{y}} - 3\ddot{\bar{y}} + 2\dot{\bar{y}}) + 2e^{2t}e^{-t}\dot{\bar{y}} - 2\bar{y} = t^2e^t$ (because $lnx = lne^t = t$). The problem reduces to the solution of a linear equation with constant coefficients

$$\dddot{\bar{y}} - 3\ddot{\bar{y}} + 3\dot{\bar{y}} - 2\bar{y} = t^2e^t.$$

For the attached homogeneous equation, we obtain the characteristic equation $r^3 - 3r^2 + 4r - 2 = 0$ with roots $r_1 = 1$ and $r_{2,3} = 1 \pm i$.

The function e^t will correspond to r_1 and functions $e^t \cos t$ and $e^t \sin t$ correspond to roots $r_{2,3}$. In consequence, the general solution of the homogeneous equation will be $\bar{y}_0(t) = C_1e^t + C_2e^t \cos t + C_3e^t \sin t$.

Now, we are looking for a particular solution ($\bar{y}_p$) for the nonhomogeneous equation. The right-hand side member is of the form $P_2(t)e^t$ (with $P_2(t) = t^2$). Because $\alpha = 1$ (t's coefficient from the exponent) is a first-order root of the characteristic equation, we will choose $\bar{y}_p$ of the form $tQ_2(t)e^t$, i.e., $\bar{y}_p(t) = t(at^2 + bt + c)e^t = (at^3 + bt^2 + ct)e^t$. We have successively

$$\begin{aligned}
\dot{\bar{y}}_p(t) &= \left[at^3 + (3a + b)t^2 + (2b + c)t + c\right],\\
\ddot{\bar{y}}_p(t) &= \left[at^3 + (6a + b)t^2 + (6a + 4b + c)t + c\right],\\
\dddot{\bar{y}}_p(t) &= \left[at^3 + (9a + b)t^2 + (18a + 6b + c)t + c\right].
\end{aligned}$$

By replacing these expressions in the equation, we obtain

$$\begin{aligned}
&t^3(a - 3a + 4a - 2a) + t^2(9a + b - 18a - 3b + 12a + 4b - 2b)\\
&+ t(18a + 6b + c - 18a - 12b - 3c + 8b + 4c - 2c)+\\
&+ (c - 3c + 4c) = t^2
\end{aligned}$$

It results in the system

$$\begin{cases} 3a = 1, \\ 2b = 0, \\ 2c = 0, \end{cases} \text{so } \begin{cases} a = \frac{1}{3}, \\ b = c = 0, \end{cases} \text{ and } \bar{y}_p(t) = \frac{t^3}{3}e^t.$$

Then $\bar{y}(t) = \bar{y}_0(t) + \bar{y}_p(t) = e^t(C_1 + C_2 \cos t + C_3 \sin t + \frac{t^3}{3})$ and going back from t to variable x, we have $y(x) = x(C_1 + C_2 \cos ln x + C_3 \sin ln x + \frac{ln^3 x}{3})$.

12.1.1.3 Nonlinear equations

There are situations when the differential equations of superior degree are linear; in the cases of those that are not linear, we do not have the theoretical framework used previously anymore and we are forced to apply other methods for finding the solution. Let us illustrate this statement by some examples.

Example 12.10. Solve the equation $y''y^{(4)} - 2(y''')^2 = 0$.

We solve this as follows. We note $y'' = z$, and hence $y''' = z'$, $y^{(4)} = z''$. We have $zz' - 2z'^2 = 0$, or $\frac{z''}{z'} = 2\frac{z'}{z}$.

By integration, it results in $\ln|z'| = 2ln|z| + lnC$, and hence $z' = Cz^2$, which means $\frac{dz}{z^2} = Cdx$. By integrating again, it results in $-\frac{1}{z} = Cx + D$, and hence $z(x) = -\frac{1}{Cx+D} = y''(x)$. We deduce successively

$$\begin{aligned} y'(x) &= -\frac{1}{C}ln|Cx + D| + \frac{1}{C}lnE = -\frac{1}{C}ln|K_1x + K_2|, y(x) \\ &= -\frac{1}{C}\int ln|K_1x + K_2|dx \\ &= -\frac{1}{C}\left[xln|K_1x + K_2| - x + \frac{K_2}{K_1}ln|K_1x + K_2|\right] + \frac{K}{3}. \end{aligned}$$

Example 12.11. Let us solve the Cauchy problem

$$y'' \cos y + y'^2 \sin y = y';\ y(-1) = \frac{\pi}{6};\ y'(-1) = 2.$$

We solve this as follows. We note $y' = p$, and hence $y'' = p' = \frac{dp}{dx} = \frac{dp}{dy}\frac{dy}{dx} = p\frac{dp}{dy}$.

We obtain $p\frac{dp}{dy}\cos y + p^2 \sin y = p$, i.e., $p(\cos y\frac{dp}{dy} + p \sin y - 1) = 0$.

a) If $p = 0 \Rightarrow y' = 0, y(x) = C$ (constant), the function that does not satisfy the initial conditions.

b) $\cos y\frac{dp}{dy} + p \sin y = 1$, which is a linear equation of first order, with the solution $p(y) = \sin y + \sqrt{3}\cos y$.

We have $y' = \sin y + \sqrt{3}\cos y$, i.e., $\frac{dy}{\sin y + \sqrt{3}\cos y} = dx$, wherefrom, by integration, $\frac{1}{2}ln|\frac{\tan\frac{1}{2}(y-\frac{\pi}{6})+1}{\tan\frac{1}{2}(y-\frac{\pi}{6})-1}| = x + k$. From the condition $y(-1) = \frac{\pi}{6}$, we obtain

$$x + 1 = \frac{1}{2}ln\left|\frac{\tan\frac{1}{2}(y - \frac{\pi}{6}) + 1}{\tan\frac{1}{2}(y - \frac{\pi}{6}) - 1}\right|.$$

For determining p we took into account the condition $y'(-1) = 2$, which means $p(\frac{\pi}{6}) = 2$ (because if $x = -1, y = \frac{\pi}{6}$).

12.1.2 Systems of differential equations

If, describing the process, there are more physical characteristics which interact with each other, then the mathematical modeling of the respective process involves at least two differential equations, where each magnitude occurring in the process appears as an unknown function, i.e.,

$$\begin{cases} F_1(x, y_1, y_1', \dots, y_1^{(n)}, \dots, y_k, y_k', \dots, y_k^{(n)} = 0, \\ \dots \; \dots \; \dots \; \dots \; \dots \; \dots \; \dots \; \dots \\ F_k(x, y_1, y_1', \dots, y_1^{(n)}, \dots, y_k, y_k', \dots, y_k^{(n)} = 0, \end{cases} \tag{12.6}$$

where the unknown functions $y_1, \dots, y_k$ are defined on an interval $I \subset \mathbb{R}$. For solving systems of differential equations of order $n \geq 2$, we recommend the following substitution: replace $k-1$ unknown functions, let us say, $y_1, \dots, y_{k-1}$, from $k-1$ equations in terms of y_k and solve the remaining equation with the unknown y_k. After finding y_k we also find $y_1, \dots, y_{k-1}$ from the relations established above.

Remark. For supplementary theoretical details the reader can consult Roşculeţ (1984) or Toma et al. (2014).

Example 12.12. Solve the differential equations system $\begin{cases} z - y'' = x, \\ z'' - y = 3e^{2x}. \end{cases}$

We solve this in the following way. From the first equation, it results that $z = y'' + x$, $z' = y''' + 1$, $z'' = y^{(4)}$, and by substituting in the second equation, it results in $y^{(4)} - y = 3e^{2x}$. The characteristic equation associated with the homogeneous equation $y^{(4)} - y = 0$ is $r^4 - 1 = 0$, with roots $r_1 = 1$, $r_2 = -1$, $r_3 = i$, $r_4 = -i$. Functions e^x, e^{-x}, $\cos x$, and $\sin x$ correspond, in this order, to the roots. We deduce that $y_0(x) = C_1e^x + C_2e^{-x} + C_3\cos x + C_4\sin x$. We are searching for y_p of the form ae^{2x}, since the right-hand side member is of the form $P_0(x)e^{2x}$, and $\alpha = 2$ is not a root of the characteristic equation. It results that

$$y_p'(x) = 2ae^{2x}, \; y_p''(x) = 4ae^{2x}, \; y_p'''(x) = 8ae^{2x}, \; y_p^{(4)}(x) = 16ae^{2x}.$$

The relation $y_p^{(4)} - y_p = 3e^{2x}$ becomes $15a = 3$, so $a = \frac{1}{5}$, $y_p(x) = \frac{1}{5}e^{2x}$,

and $y(x) = C_1e^x + C_2e^{-x} + C_3\cos x + C_4\sin x + \frac{1}{5}e^{2x}$. Finally,

$$z(x) = y''(x) + x = C_1e^x + C_2x^{-x} - C_3\cos x - C_4\sin x + \frac{4}{5}e^{2x} + x.$$

Consider the homogeneous system of first-order linear differential equations with constant coefficients

$$\mathbf{y}' = \mathbf{A}\mathbf{y}, \tag{12.7}$$

where $\mathbf{y} = (y_1, y_2, ..., y_n)^{\mathrm{T}}$ and $\mathbf{A}$ is a matrix of constant coefficients a_{ij}. Solutions of such a system will have the form

$$\mathbf{y} = \mathbf{y}(x) = \mathbf{v}e^{\lambda x}, \tag{12.8}$$

where λ is an eigenvalue of the matrix $\mathbf{A}$ and $\mathbf{v}$ is the associated eigenvector. Eigenvalues are roots of characteristic equations, this time computed as $\mathrm{Det}(\mathbf{A} - \lambda\mathbf{E}) = 0$, where $\mathbf{E}$ is the identity matrix. Composing the FSS for the general solution $\mathbf{y}_G = \sum_{i=1}^{n} c_i \mathbf{y}_i$ we respect the similar rules as in paragraph (*c*) (*i*) (1), (3), and (4). In the case of root $\lambda \in \mathbb{R}$, with multiplicity order k, $1 < k \leq n$, the independent system of solutions $\mathbf{y}_i$ is created from $e^{\lambda x}, \mathbf{P}_1 e^{\lambda x}, \ldots, \mathbf{P}_{k-1} e^{\lambda x}$ (where $\mathbf{P}_i = (P_{1i}, P_{2i}, \ldots P_{ni})^{\mathrm{T}}$ is a vector of polynomials).

12.1.2.1 Numerical solution

However, in most practical problems the analytical solutions are not available. In such situations *numerical methods* are used, taking into consideration that by numerical methods we are looking for approximations of particular solutions fixed by initial conditions or by other constraints within given finite relatively short intervals in specified nodes.

Consider the following system of n initial value problems $\mathbf{y}' = \mathbf{f}(x, y_i, y_2, \ldots y_n)$:

$$\begin{aligned} y_1' &= f_1(x, y_1, y_2, \ldots y_n,) \\ y_2' &= f_2(x, y_1, y_2, \ldots y_n,) \\ &\ldots \\ y_n' &= f_n(x, y_1, y_2, \ldots y_n,) \end{aligned} \tag{12.9}$$

for $x \in \langle a, b \rangle$ and initial values $y_1(a) = \alpha_1$, $y_2(a) = \alpha_2$, $\ldots$, $y_n(a) = \alpha_n$ forming the slope field $\Omega = \{(x, y_1, y_2, \ldots y_n), x \in \langle a, b \rangle, y_i \in \mathrm{M}_i\}$ where $f_i(x, y_1, y_2, \ldots y_n,)$ are defined. The basic assumption for the system (12.9) to be solved by numerical methods is that it has a unique solution for each point $x \in \langle a, b \rangle$.

Theorem 12.4. *If* $f_i(x, y_1, y_2, \ldots y_n,)$ *and its partial derivatives* $\frac{\delta f_i(x, y_1, y_2, \ldots y_n,)}{\delta y_k}$, $k = 1, 2, \ldots, n$, *are continuous on* Ω, f_i *satisfies the Lipschitz condition* ($|f_i - f_j| \leq L \sum_{k=1}^{n} |y_{ik} - y_{jk}|$, $L > 0$), *then the system (12.9) has a unique solution* $\mathbf{y} = (y_1, y_2, \ldots, y_n)^{\mathrm{T}}$ *for* $x \in \langle a, b \rangle$.

To solve the system (12.9) by numerical methods, generalization of methods for single differential equations can be used. In the case of linear differential equations, the system can be rewritten in a matrix form $\mathbf{y}' = \mathbf{A}\mathbf{y} + \mathbf{g}$, and we can follow the matrix modification. The interval $\langle a, b \rangle$ is divided by points x_i, $x_{i+1} = x_i + step$, $x_0 = a$ and initial vector of first entries $\mathbf{y}_0 = (y_{10}, y_{20}, \ldots, y_{n0})^{\mathrm{T}}$ with $y_{10} = \alpha_1$, $y_{20} = \alpha_2$, $\ldots$, $y_{n0} = \alpha_n$. Then vectors $\mathbf{y}_{i+1}$ are computed under the scheme $\mathbf{y}_{i+1} = \mathbf{y}_i + step \cdot \mathbf{k}$, where $\mathbf{k}$ is the vector of slopes computed by some suitable numerical method. For instance, the Runge–Kutta fourth-order method will satisfy the process of computation

$$\begin{aligned} \mathbf{k}_1 &= \mathbf{A} \cdot \mathbf{y}_i + \mathbf{g}(x_i) \\ \mathbf{k}_2 &= \mathbf{A} \cdot \left(\mathbf{y}_i + \frac{step}{2} \mathbf{k}_1 \right) + \mathbf{g}\left(x_i + \frac{step}{2} \right) \\ \mathbf{k}_3 &= \mathbf{A} \cdot \left(\mathbf{y}_i + \frac{step}{2} \mathbf{k}_2 \right) + \mathbf{g}\left(x_i + \frac{step}{2} \right) \end{aligned} \tag{12.10}$$

$$\mathbf{k}_4 = \mathbf{A} \cdot (\mathbf{y}_i + step \cdot \mathbf{k}_3) + \mathbf{g}(x_i + step)$$
$$\mathbf{k} = \frac{\mathbf{k}_1 + 2\mathbf{k}_2 + 2\mathbf{k}_3 + \mathbf{k}_4}{6}$$

The other possibility is to follow the process for each vector item of (12.9) separately (for more detailed information, the reader is advised to consult books on numerical analysis, e.g., Burden and Faires (2011)).

Higher-order differential equations $y^{(n)} = f(x, y, y', ..., y^{(n-1)})$ can be solved in a similar way, when they are expressed by the following substitution:

$$z_1 = y,\ z_2 = y',\ z_3 = y'',\ \ldots,\ z_{n-1} = y^{(n-2)},\ z_n = y^{(n-1)},$$

from which it follows that

$$z_1' = y' = z_2,\ z_2' = y'' = z_3,\ \ldots,\ z_n' = y^{(n)},$$

producing the following system of first-order differential equations:

$$z_1' = f_1(x, z_1, z_2, ..., z_n)$$
$$z_2' = f_2(x, z_1, z_2, ..., z_n)$$
$$\ldots$$
$$z_n' = f_n(x, z_1, z_2, ..., z_n)$$

with transformed initial values

$$z_1(a) = y(a) = \alpha_1,\ z_2(a) = y'(a) = \alpha_2,\ \ldots,\ z_n(a) = y^{(n-1)}(a) = \alpha_n.$$

12.2 Higher-order differential equations in practice

The ability to express and solve relations between a function and its derivatives in one equation or in their system of equations was a very important moment in science. As a result of research in mechanics, going back to the time of Newton and Leibniz, it has been shown that relations between quantity and its rate of change, acceleration, jerk, snap, etc., have many applications from engineering, physical sciences, and bilogical sciences. Although only a few types have closed form solutions, theory on differential equations provides strong scientific tools. To solve more complex models and real-life problems, various numerical methods have been invented; they are inbuilt in solvers of many kinds of technical software. From this point of view we would like to note that it is crucially important to understand the methods and their limitations, to be able to control the processes, design them, and interpret or use the results in subsequent procedures.

The following examples illustrate and give a look into the main areas of higher-order differential equations occurring in engineering, and we continue with challenging problems in the next section.

Remark. The CAS software offers for computation of differential equations the commands DSolve[] and NDSolve[], differing little in syntax and arguments with respect to specific software.

Example 12.13. Consider an *electric circuit* formed by a resistor with resistance R, a coil of inductance L, and a capacitor with capacitance C, connected in series to the terminals of a generator of constant voltage E. The aim is to determine the current $i(t)$ in the transitory regime which appears after closing the switch. Kirchhoff's theorem gives us the expression $E = Ri + Li' + \frac{1}{C}\int_0^t i(\tau)d\tau$. But $i(t) = q'(t)$, q being the charge of the capacitor. Then $i'(t) = q''(t)$, and for q it results in the differential equation $Lq'' + Rq' + \frac{q}{C} = E$.

Let $L = 1$, $R = 4$, $C = \frac{1}{4}$, $E = 5$. Then $q'' + 4q' + 4q = 5$.

The attached homogeneous equation is $q'' + 4q' + 4q = 0$ and the corresponding characteristic equation is $r^2 + 4r + 4 = 0$, with the double root $r_1 = r_2 = -2$ and the following FSS: $\{e^{-2t}, te^{-2t}\}$. The general solution of the homogeneous equation will be $q_0(t) = C_1e^{-2t} + C_2te^{-2t}$. A particular solution q_p of the nonhomogeneous equation will be of the form of the right-hand side member, $q_p(t) = a$ (constant). It results that $q_p' = q_0'' = 0$ and by replacing q_p in the equation, we obtain $4a = 5$, $a = \frac{5}{4} = q_p(t)$. Then

$$q(t) = q_0(t) + q_p(t) = C_1e^{-2t} + C_2te^{-2t} + \frac{5}{4},$$

$$i(t) = q'(t) = -2C_1e^{-2t} - 2tC_2e^{-2t} + C_2e^{-2t}.$$

Usually, the differential equations, modeling a certain process, have attached certain initial conditions, a fact that transforms them in Cauchy problems (see Theorem 12.2). For example, if we impose the conditions $q(\frac{1}{2}) = 1$, $q'(\frac{1}{2}) = 2$, we will have successively $\frac{C_1}{e} + \frac{C_2}{2e} + \frac{5}{4} = 1$, $\frac{-2C_1}{e} = 2$, from which $C_1 = -e$ and $C_2 = \frac{3}{2}e$. The particular solution

$$q(t) = e^{1-2t}\left(\frac{3}{2}t - 1\right) + \frac{5}{4}, \quad i(t) = e^{1-2t}\left(\frac{7}{2} - 3t\right).$$

Example 12.14 (Linear mechanical oscillator). We consider a spiral spring, of elastic constant K, length l_0, fixed at one end. To stretch the spring of length l, we need an elastic force of magnitude Kl. Let us suppose that we attach to the free end of the spring a body of mass m. This body will stretch the spring with length l_1, given by the relation $mg = Kl_1$. Then stretching the spring by an extra length l_2 and releasing it will cause the spring to oscillate vertically. Let $X(t)$ be the distance between the equilibrium position and the position of the free end at moment t. Let us suppose that the movement is made in an environment which resists a resistance proportional to the speed X' of the body, i.e., $\alpha X'(t)$. The motion equation will be $mX'' = -KX - \alpha X'$ or: $mX'' + \alpha X' + KX = 0$, with conditions $X(0) = l_2$, $X'(0) = 0$.

Let $m = 1$, $\alpha = 2$, $K = 2$, K – consistent, $l_2 = 1$. Then we can solve

$$X'' + 2X' + 2X = 0, \text{ with } X(0) = 1 \text{ and } X'(0) = 0.$$

The characteristic equation is $r^2 + 2r + 2 = 0$ with simple complex-conjugated roots $r_{1,2} = -1 \pm i$. The functions $e^{-t} \cos t$ and $e^{-t} \sin t$ corresponding to them form an FSS. It results that $X(t) = e^{-t}(C_1 \cos t + C_2 \sin t)$. Then $X'(t) = e^{-t}(-C_1 \sin t + C_2 \cos t - C_1 \cos t - C_2 \sin t)$. From $X(0) = 1 \Rightarrow C_1 = 1$; $X'(0) = C_2 - C_1 = 0$. Hence, $C_1 = C_2 = 1$, so $X(t) = e^{-t}(\cos t + \sin t)$.

Note. $\lim\limits_{t\to\infty} X(t) = 0$ (because $e^{-t} \to 0$ and $\cos t$, $\sin t$ are bounded functions). This result proves that motion of the arch attenuates over time, the arch being at a certain time in equilibrium position.

Example 12.15. In *material's resistance* one can deduce the differential equation of the inflexion of a beam on an elastic medium as $y^{(4)} + 4\alpha^4 y = \frac{q(x)}{EI}$, where $y = y(x)$ is the beam movement at inflexion, α is a positive material constant, q is a distributed load of the beam at the point x, E is the modulus of elasticity of the beam (constant), and I is the inertia moment of the beam (constant). On an uncharged area of the beam where $q = 0$, the equation becomes a homogeneous equation, $y^{(4)} + 4\alpha^4 y = 0$, with the characteristic equation $r^4 + 4\alpha^4 = 0 \Leftrightarrow r^2 = \pm 2i\alpha^2$.

a) If $r^2 = 2i\alpha^2$, by noting $r = c + id$, we obtain $c^2 - d^2 = 0$, $2cd = 2\alpha^2$, $c^2 + d^2 = |r^2| = 2\alpha^2$, so $c = d = \pm\alpha$, $r_1 = \alpha + \alpha i$, $r_2 = -\alpha - \alpha i$.

b) If $r^2 = -2i\alpha^2$, we deduct analogically the roots $r_3 = \alpha - \alpha i$, $r_4 = -\alpha + \alpha i$.

Functions $e^{\alpha x} \cos \alpha x$ and $e^{\alpha x} \sin \alpha x$ correspond in FSS to roots (simple, complex-conjugated) r_1 and r_3, and functions $e^{-\alpha x} \cos \alpha x$ and $e^{-\alpha x} \sin \alpha x$ correspond in FSS to r_2 and r_4.

Hence, $y(x) = e^{-\alpha x}(A \cos \alpha x + B \sin \alpha x) + e^{\alpha x}(C \cos \alpha x + D \sin \alpha x)$.

By imposing supplementary conditions, we can deduce the constants A, B, C, D. For example, if we impose $\lim\limits_{x\to\infty} y(x) = 0$ (supposing the beam is infinite), it results in $C = D = 0$ (because $\lim\limits_{x\to\infty} e^{\alpha x} = \infty$).

Example 12.16. The *Bessel equation* $x^2 y'' + xy' + (x^2 - \nu)y = 0$ has multiple applications in many physical phenomena, such as heat propagation and mechanical and electromagnetic vibration. Obviously, the solution of such equation depends essentially on the value of the parameter ν. In the general case, when we consider that ν can take any real value, we search the solution y of the form of a power series' sum, which leads to the construction of so-called *Bessel functions* (for more details, the reader is referred to Tikhonov and Samarski (1956)). We will deal with the case $\nu = \frac{1}{2}$, when the equation becomes $x^2 y'' + xy' + (x^2 - \frac{1}{4})y = 0$. We will consider $x > 0$ and we will make the change of variable $y(x) = \frac{z(x)}{\sqrt{x}} = x^{-\frac{1}{2}} z(x)$. We deduce successively $y'(x) = -\frac{1}{2}x^{-\frac{3}{2}} z(x) + x^{-\frac{1}{2}} z'(x)$, $y''(x) = x^{-\frac{1}{2}} z''(x) - x^{-\frac{3}{2}} z'(x) + \frac{3}{4}x^{-\frac{5}{2}} z(x)$. By replacing these expressions in the equation, we obtain

$$x^{\frac{3}{2}} z'' + \left(x^{\frac{1}{2}} - x^{\frac{1}{2}}\right) z'(x) + \left(x^{\frac{3}{2}} - \frac{1}{4}x^{-\frac{1}{2}} - \frac{1}{2}x^{-\frac{1}{2}} + \frac{3}{4}x^{-\frac{1}{2}}\right) z = 0,$$

i.e., we reach the homogeneous equation $z'' + z = 0$, with the characteristic equation $r^2 + 1 = 0$, $r_{1,2} = \pm i$, $z(x) = C_1 \cos x + C_2 \sin x$. Hence, we obtain

$$y(x) = \frac{C_1 \cos x + C_2 \sin x}{\sqrt{x}}.$$

Example 12.17. The wall of thickness δ of an oven has on the interior front temperature T_1, and on the exterior one temperature T_2 ($T_1 > T_2$). Determine the *temperature's variation on the wall's thickness*, knowing the thermal conductivity of the material which the wall is built from, $\lambda = \lambda_0(1 + bT)$, λ_0 and b being material positive constants.

We solve this as follows. From physical considerations, we demonstrate that the wall's temperature satisfies the equation (*) $\lambda \frac{d^2T}{dx^2} + \frac{d\lambda}{dT}(\frac{dT}{dx})^2 = 0$ (resulting from the differentiation with respect to x of the equality $\lambda(T(x))\frac{dT}{dx}(x) = C$ (constant)).

By replacing λ and $\frac{d\lambda}{dT}$ in (*), it results that $(1 + bT)\frac{d^2T}{dx^2} + b(\frac{dT}{dx})^2 = 0$.

We observe that the equation is not linear. We note $\frac{dT}{dx} = p$, and hence $\frac{d^2T}{dx^2} = \frac{dp}{dx} = \frac{dp}{dT}\frac{dT}{dx} = p\frac{dp}{dt}$.

We deduce the equation $p[(1 + bT)\frac{dp}{dt} + bt] = 0$.

i) If $p = 0$, it results that $\frac{dT}{dx} = 0$, i.e., T is a constant, which contradicts the hypothesis $T_1 > T_2$.

ii) $(1 + bT)\frac{dp}{dt} + bt = 0$, wherefrom it results that $\frac{1}{p}dp = -\frac{b}{bT+1}dT$.

By integrating, we obtain $ln|p| = -ln|bT + 1| + lnC$, i.e. $p(1 + bT) = C$.

Hence, $\frac{dT}{dx} = -\frac{C}{bT+1}$, or $(1 + bT)dT = Cdx$.

Integrating again, we will obtain the relation (**) $\frac{bT^2}{2} + T = Cx + D$.

We consider that the exterior front of the oven corresponds to $x = \delta$ and the interior one corresponds to $x = 0$. We obtain the conditions $T(0) = T_1,\ T(\delta) = T_2$.

By replacing these values in (**), we find $D = \frac{bT_1^2}{2} + T_1,\ C = \frac{T_2 - T_1}{2\delta}(bT_1 + bT_2 + 2)$.

The relation (**) becomes $\frac{b}{2}T^2 + T - \frac{T_2 - T_1}{2\delta}(bT_1 + bT_2 + 2)x - (\frac{bT_1^2}{2} + T_1) = 0$.

By solving this equation of second order with respect to T, it results that

$$T_{1,2}(x) = -\frac{1}{b} \pm \sqrt{\left(\frac{1}{b} + T_1\right)^2 - \left[\left(\frac{1}{b} + T_1\right)^2 - \left(\frac{1}{b} + T_2\right)^2\right]\frac{x}{\delta}}.$$

From conditions $T(0) = T_1,\ T(\delta) = T_2$, we will obtain

$$T(x) = -\frac{1}{b} + \sqrt{\left(\frac{1}{b} + T_1\right)^2 - \left[\left(\frac{1}{b} + T_1\right)^2 - \left(\frac{1}{b} + T_2\right)^2\right]\frac{x}{\delta}}.$$

12.3 Challenging problems in applications

Problem 12.1. In seismic event calculations, building blocks are modeled as rigid bodies supported by two springs, one with translational stiffness and one with rotational stiffness (Fig. 12.1). We are interested in a description of building roof point motion under a seismic impulse given by proposed speed of earth surface translation. The mechanical model of building block motion is determined by the homogeneous system of differential equations (modified from Starek, 2009)

$$\begin{aligned} & mx'' - mh\varphi'' + kx = 0 \\ & -mhx'' + \left(mh^2 + I_G\right)\varphi'' - (mgh - k_t)\varphi = 0 \end{aligned} \tag{12.11}$$

where $x = x(t)$ [m] represents a time function of building base translation displacement and $\varphi = \varphi(t)$ [rad] represents a time function of building rotation angle (Fig. 12.1).

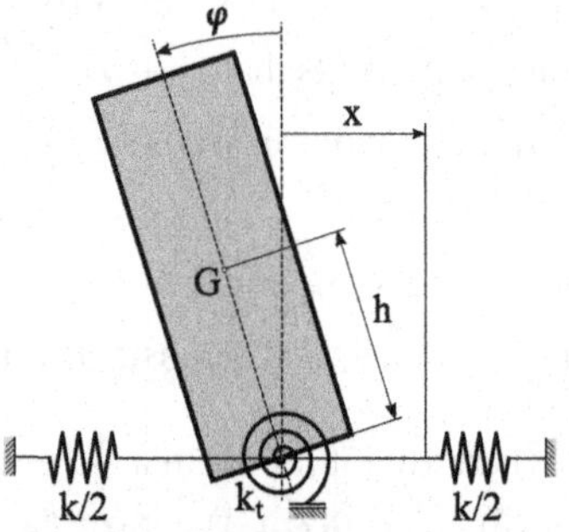

Figure 12.1 The model of building block for seismic event calculations.

We solve this as follows. From (12.11), we explicitly express second derivatives of variables.

$$\begin{aligned} x'' &= -\left(\frac{h^2k}{I_G} + \frac{k}{m}\right)x - \frac{gh^2m + hk_t}{I_G}\varphi, \\ \varphi'' &= -\frac{hk}{I_G}x + \frac{ghm - k_t}{I_G}\varphi. \end{aligned}$$

Transforming into a system of first-order differential equations we obtain

$$\begin{aligned} & z_1 = x \Rightarrow z_1' = z_2 \\ & z_2 = x' \Rightarrow z_2' = -\left(\frac{h^2k}{I_G} + \frac{k}{m}\right)z_1 + \frac{gh^2m + hk_t}{I_G}z_3 \\ & z_3 = \varphi \Rightarrow z_3' = z_4 \\ & z_4 = \varphi' \Rightarrow z_4' = -\frac{hk}{I_G}z_1 + \frac{ghm - k_t}{I_G}z_3 \end{aligned}$$

In matrix form

$$\begin{pmatrix} z_1' \\ z_2' \\ z_3' \\ z_4' \end{pmatrix} = \begin{pmatrix} 0 & 1 & 0 & 0 \\ -(\frac{h^2k}{I_G} + \frac{k}{m}) & 0 & \frac{gh^2m+hk_t}{I_G} & 0 \\ 0 & 0 & 0 & 1 \\ -\frac{hk}{I_G} & 0 & \frac{ghm-k_t}{I_G} & 0 \end{pmatrix} \cdot \begin{pmatrix} z_1 \\ z_2 \\ z_3 \\ z_4 \end{pmatrix}. \tag{12.12}$$

Consider a building where height of a gravity point $h = 21$ m, building mass $m = 5.6 \cdot 10^4$ kg, gravity acceleration $g = 9.81$ m/s^2, inertia moment $I_G = 8.232 \cdot 10^6$ kg m^2, spring translational stiffness $k = 1.5 \cdot 10^6$ N/m, spring rotational stiffness $k_t = 1.35 \cdot 10^7$ N m/rad, and initial speed of earth surface translation $x'(0) = 0.15$ m/s.

Then the system $\mathbf{z}' = \mathbf{A} \cdot \mathbf{z}$ represents the homogeneous system of first-order linear differential equations

$$\begin{pmatrix} z_1' \\ z_2' \\ z_3' \\ z_4' \end{pmatrix} = \begin{pmatrix} 0 & 1 & 0 & 0 \\ -107.143 & 0 & -5.00878 & 0 \\ 0 & 0 & 0 & 1 \\ -3.82653 & 0 & -0.238513 & 0 \end{pmatrix} \cdot \begin{pmatrix} z_1 \\ z_2 \\ z_3 \\ z_4 \end{pmatrix}. \tag{12.13}$$

Eigenvalues of the system $\lambda_n \in \{10.3596\,\mathrm{i}, -10.3596\,\mathrm{i}, 0.243985\,\mathrm{i}, -0.243985\,\mathrm{i}\}$ with corresponding eigenvectors

$$\begin{aligned}\mathbf{v}_n \in \big\{&(0.0960207\,\mathrm{i}, -0.994739, 0.00343122\,\mathrm{i}, 0.0355461)^{\mathrm{T}},\\ &(-0.0960207\,\mathrm{i}, -0.994739, 0.00343122\,\mathrm{i}, 0.0355461)^{\mathrm{T}},\\ &(-0.0453919, -0.011075\,\mathrm{i}, -0.970441, -0.236773\,\mathrm{i})^{\mathrm{T}},\\ &(-0.0453919, +0.011075\,\mathrm{i}, -0.970441, 0.236773\,\mathrm{i})^{\mathrm{T}}\big\}\end{aligned}$$

produce the general solution of (12.13) $\mathbf{z} = \sum_{n=1}^{4} c_n \mathbf{v}_n e^{\lambda_n t}$, $c_n \in R$. Using initial values $z_1(0) = 0$, $z_2(0) = 0.15$, $z_3(0) = 0$, $z_4(0) = 0$, the solution vector $\mathbf{z}$ has entries

$$\begin{aligned}x(t) &= z_1(t) = 0.001026\sin(0.243985t) + 0.014455\sin(10.3596t)\\ x'(t) &= z_2(t) = 0.000250\cos(0.243985t) + 0.149750\cos(10.3596t)\\ \varphi(t) &= z_3(t) = 0.021932\sin(0.243985t) - 0.000517\sin(10.3596t)\\ \varphi'(t) &= z_4(t) = 0.005351\cos(0.243985t) - 0.005351\cos(10.3596t)\end{aligned}$$

which oscillate in the superposition of two vibrations with natural angular frequencies of the system $\omega_1 = 0.243985$ rad/s and $\omega_2 = 10.3596$ rad/s (determined as eigenvalues of $\mathbf{A}$ with plus and minus signs). The maximal values do not exceed the sum of partial signal amplitudes, which can be considered to be their upper bound. We can conclude that the base of the building will vibrate horizontally maximally between $\pm\, 0.015481$ m from the origin position, $x(t) \in \langle -0.015481, 0.015481 \rangle$[m] (Fig. 12.2). The angle of building leaning will change maximally between $\pm\, 0.0224489$ rad (1.44533°), $\varphi(t) \in \langle -0.02245, 0.02245 \rangle$ rad (Fig. 12.3).

The deflection at the top of the building can be expressed as

$$x_{roof}(t) = x(t) + 2h * \sin\varphi(t).$$

Amplitude of $\varphi(t)$ does not exceed 0.0224489 rad (1.44533°). For small angles $\sin\varphi(t) \doteq \varphi(t)$, so the motion at the top of the building can be determined as

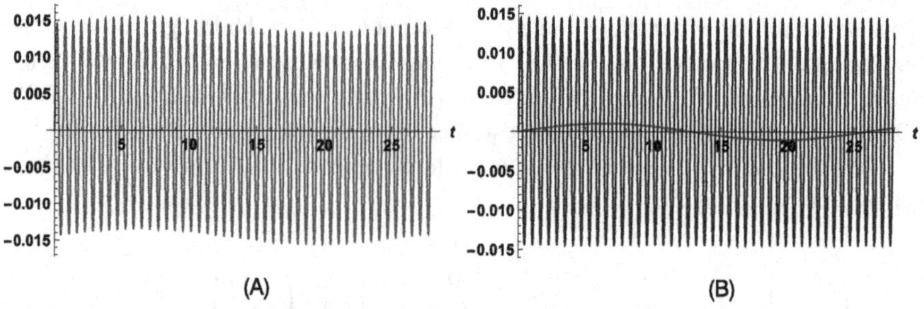

Figure 12.2 Translation of the base of the building block. (A) $x(t)$. (B) Partial signals of $x(t)$: $0.001026\sin(0.243985t)$ and $0.014455\sin(10.3596t)$.

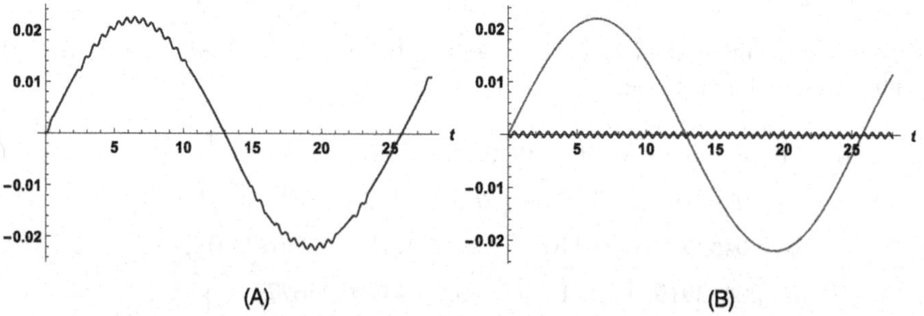

Figure 12.3 Rotation of the building block. (A) $\varphi(t)$. (B) Partial signals of $\varphi(t)$: $0.021932\sin(0.243985t)$ and $-0.000517\sin(10.3596t)$.

$$x_{roof}(t) = x(t) + 2h * \varphi(t) = 0.922185\sin(0.243985t) - 0.007240\sin(10.3596t),$$

not exceeding the maximal possible roof swing $\pm$ 0.929425 m (Fig. 12.4).

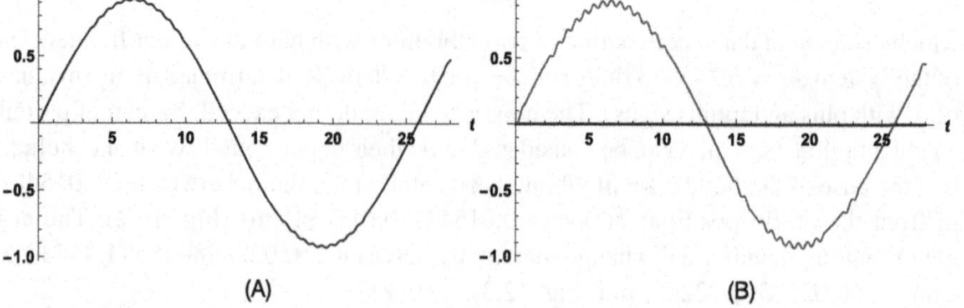

Figure 12.4 Building top point motion. (A) $x_{roof}(t)$. (B) Partial signals of $x_{roof}(t)$: $1.22958 \times \sin(0.243985t)$ and $-0.009653\sin(10.3596t)$.

Problem 12.2. Let us consider the problem from Chapter 5 on a simplified manipulator forced by a polygonal periodic chain (Fig. 5.14C in Chapter 5), described by the second-order differential equation (5.20) with particular values

$$6.97729\ddot{x}(t) + 47.0837\dot{x}(t) + 7943.19x(t) = F(t),$$

where $F(t)$ (Fig. 5.5 in Chapter 5) is given by (5.30), in particular (5.32). In contrast to Chapter 5, where $F(t)$ was decomposed into Fourier series of harmonics and the solution was obtained as a superposition of its responses, now we will proceed using methods for differential equations.

Having given the right-hand side of the equation as the piecewise determined function, numerical methods have to be used for solution. Therefore the differential equation (5.20) has to be transformed into a system of first-order differential equations,

$$z_1 = x \Rightarrow z_1' = z_2$$
$$z_2 = \dot{x} \Rightarrow z_2' = -\frac{k}{m}z_1 - \frac{b}{m}z_2 + F(t)$$

in matrix form $\mathbf{z}' = \mathbf{A} \cdot \mathbf{z} + \mathbf{g}$,

$$\begin{pmatrix} z_1' \\ z_2' \end{pmatrix} = \begin{pmatrix} 0 & 1 \\ -\frac{k}{m} & -\frac{b}{m} \end{pmatrix} \begin{pmatrix} z_1 \\ z_2 \end{pmatrix} + \begin{pmatrix} 0 \\ F(t) \end{pmatrix}$$

where the matrix

$$\mathbf{A} = \begin{pmatrix} 0 & 1 \\ -\frac{k}{m} & -\frac{b}{m} \end{pmatrix} = \begin{pmatrix} 0 & 1 \\ -1138.43 & -6.74814 \end{pmatrix}$$

Then numerical algorithms for single differential equations can be applied in matrix form. In the case the method is executable, numerically stable, and convergent to a desired solution, the vector of solution, containing the function of position $x(t)$ and its derivative, velocity of the motion $\dot{x}(t)$, is determined by point functions $z_1(t)$ and $z_2(t)$. Here the solution is calculated for $t \in \langle a, b \rangle$, $a = 0$, $b = 2T \doteq 1.256$ s with step $h = 0.001$ s. Using the Runge–Kutta fourth-order method (see the computation scheme (12.10) and the code in Chapter 15) we receive the coordinates of a point function vector $\mathbf{z}$ for 1257 points. In Table 12.1, coordinates of each 157th point are displayed. The graphs of vibration displacement $z_1(t) = x(t)$ and velocity $z_2(t) = \dot{x}(t)$ are shown in Fig. 12.5. The comparison with the solution computed in Chapter 5 is shown in Fig. 12.6.

Problem 12.3. A sports equipment manufacturing company is designing a new line of bungee cords. In this type of cords it is critical to know the maximum traction length they can reach in order to avoid accidents (e.g., falling to the ground).

a) The company's engineering department wants to calculate whether it is safe to use a bungee cord with an elastic constant of 300 N/m and 70 m of length to jump from an 85-meter high bridge considering a person of 80 kg. They want to obtain the graph which shows the position of the jumper with respect to time to

Table 12.1 Table of selected values from numerical solution $z_1(t) = x(t)$, $z_2(t) = \dot{x}(t)$.

i	t	$z_1(t)$	$z_2(t)$
0	0	0	0
157	0.157	0.003718	−0.085272
314	0.314	−0.004573	0.158848
471	0.471	−0.000340	−0.032326
628	0.628	0.000392	−0.017457
785	0.785	0.004094	−0.085303
942	0.942	−0.004545	0.165500
1099	1.099	−0.000431	−0.029439
1256	1.256	0.000325	−0.017962

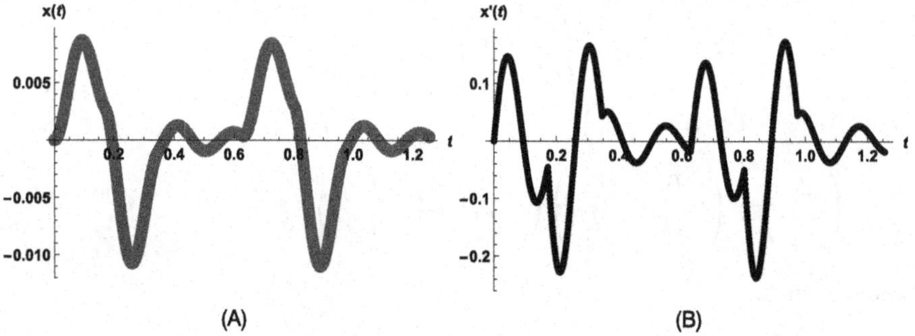

Figure 12.5 (A) Displacement $z_1(t) = x(t)$ m. (B) Velocity $z_2(t) = \dot{x}(t)$ m/s.

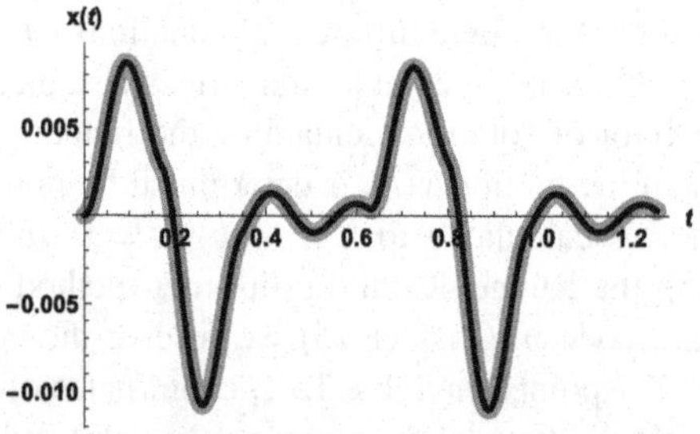

Figure 12.6 Comparison with results from Chapter 5 (see also Fig. 5.26A).

know if the cord satisfies the safety regulations and if, at any time, the person could hit the ground.

b) They require an expression that allows to know the maximum length that the cord will reach depending on the mass, the coefficient of elasticity, and the length of the bungee cord. From this expression they want to graph the values of maximum safe fall as a function of the mass for cords with elastic constant K = 300, 400, and 500 N/m.

To solve the problem, firstly it is necessary to obtain a mathematical model that allows one to predict the elastic behavior of the bungee jumping cord when a person jumps.

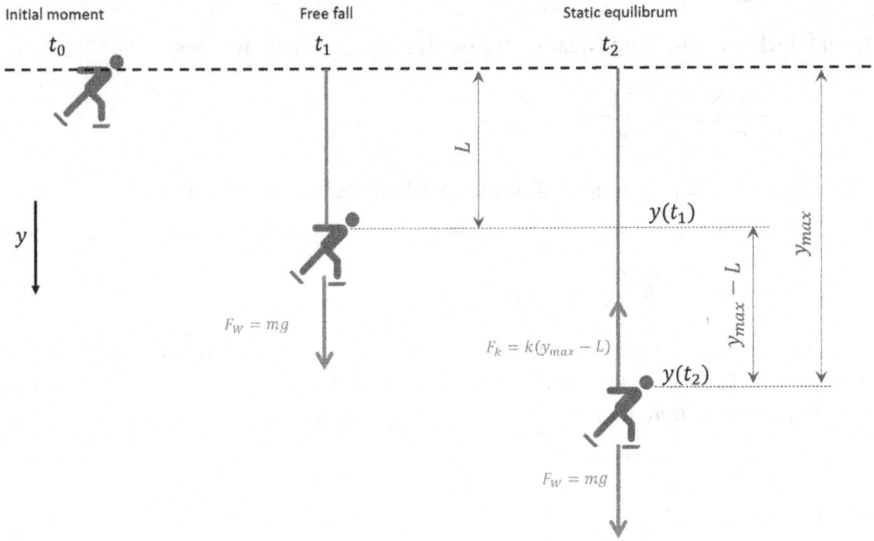

Figure 12.7 Scheme of the jump from the initial moment.

The model must allow one to obtain the height of the fall as a function of time ($y(t)$). The scheme of the jump from the initial moment is given in Fig. 12.7. From a physical point of view, we can divide the jumping process in two different parts:

1. Free fall phase: the person jumps in a free fall. The elastic phenomenon in the cord has no influence because the deformation of the cord is null. This first phase occurs during $0 < t < t_1$. The person falls just by effect of the gravity acceleration, $\sum F_y = ma \to mg = ma \to g = y''(t)$. Velocity and position can be obtained from the acceleration expression

$$y''(t) = g \to y'(t) = \int gdt = gt \to y(t) = \int gtdt = \frac{gt^2}{2}.$$

The phase is finished at time t_1, which can be obtained from $y(t_1) = L \to L = \frac{gt_1^2}{2} \to t_1 = \sqrt{\frac{2L}{g}}$. Now, the cord is fully extended ($y(t_1) = L \to y(\sqrt{\frac{2L}{g}}) = L$) and it will begin to deform due to the weight of the person. The speed of the fall in (t_1) can be calculated considering the previous equation obtained for the velocity: $y'(t_1) = gt_1 \to y'(\sqrt{\frac{2L}{g}}) = g\sqrt{\frac{2L}{g}} = \sqrt{2Lg} \to y'(\sqrt{\frac{2L}{g}}) = \sqrt{2Lg}$.

2. Oscillation phase: it occurs at $t > t_1$; the cord is deformed due to the mechanical energy of the person so the total length of the cord will oscillate in elongation and contraction cycles. While in the free fall phase, the person is only subject to the effect of the gravity acceleration, in this case the elastic force F_k is affecting the

movement, where $F_k = k(y(t) - L)$ and

$$\sum F_y = my''(t) \rightarrow mg - k\big(y(t) - L\big) = my''(t),$$

from which the following equation for oscillating movement follows:

$$y'' + \frac{k}{m}y = g + \frac{kL}{m}.$$

The characteristic polynomial of associated homogeneous equation $y'' + \frac{k}{m}y = 0$ is $r^2 + \frac{k}{m} = 0$ with roots $r = \pm\sqrt{k/m}\,i$. Then

$$y_h = c_1\cos(t\sqrt{k/mt}) + c_2\sin(t\sqrt{k/m}).$$

The right-hand side function is a constant and it can be considered as a zero-degree polynomial. The particular solution will have the form $y_p = x^0 A_0 = A_0$. Considering this issue in the differential equation,

$$\frac{k}{m}A_0 = g + \frac{kL}{m} \rightarrow A_0 = \frac{gm}{k} + L.$$

The solution of the differential equation for modeling the position will be

$$y(t) = c_1\cos(t\sqrt{k/m}) + c_2\sin(t\sqrt{k/m}) + \frac{gm}{k} + L. \tag{12.14}$$

(a) Applying the numerical values of the parameters and initial conditions calculated at the beginning ($y'(\sqrt{\frac{2L}{g}}) = \sqrt{2Lg}$ and $y(\sqrt{\frac{2L}{g}}) = L$), the constants c_1 and c_2 are computed as follows:

$$y(t) = c_1\cos(1.936t) + c_2\sin(1.936t) + 72.6,$$

$$y\left(\sqrt{\frac{2L}{g}}\right) = L \rightarrow 0.512c_1 + 0.85c_2 + 72.6 = 70,$$

$$y'(t) = -1.936c_1\sin(1.936t) + 1.936c_2\cos(1.936t),$$

$$y'\left(\sqrt{\frac{2L}{g}}\right) = \sqrt{2Lg} \rightarrow -1.663c_1 + 0.992c_2 = 37.06.$$

Solving the linear system equations, we obtain c_1 and c_2:

$$y(t) = -17.732\cos(1.936t) + 7.622\sin(1.936t) + 72.6.$$

Looking at the plot of the function solution (Fig. 12.8), we can see the fall (y) is higher than 85 m (the exact value of the maximum height can be easily calculated using $y'(t_{max}) = 0$ and $y(t_{max}) = y_{max}$) but it is enough for the graph to demonstrate that it is not safe to use that rope because the person would hit the ground.

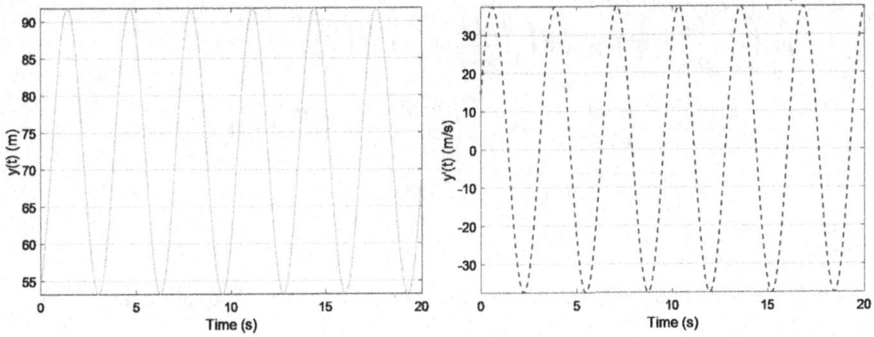

Figure 12.8 (left) Position ($y(t)$) as a function of time. The reader can see that the maximum fall is higher than 85 m. (right) Velocity ($y'(t)$) as a function of time. During the oscillation, when the position is maximum/minimum, the velocity is zero.

(b) We want to know the y_{max} for the other parameters, $y_{max} = f(k, L, m)$. The starting point is the solution of the differential equation calculated in the previous paragraphs but in this case it is convenient to express the homogeneous solution in the amplitude-phase mode, i.e., $y_h = c_1 \cos(t\sqrt{k/m t}) + c_2 \sin(t\sqrt{k/m}) = R\cos(t\sqrt{k/m} + \varphi)$, where R is the oscillation amplitude value and φ is the phase angle (for more information see Chapter 5), so we have

$$y(t) = R\cos(t\sqrt{k/m} + \varphi) + \frac{gm}{k} + L.$$

Applying the initial values we get

$$y\left(\sqrt{\frac{2L}{g}}\right) = L \rightarrow R\cos\left(\sqrt{\frac{2L}{g}}\sqrt{\frac{k}{m}} + \varphi\right) + \frac{gm}{k} + L = L,$$

$$y'\left(\sqrt{\frac{2L}{g}}\right) = \sqrt{2Lg} \rightarrow -R\sqrt{\frac{k}{m}}\sin\sqrt{\left(\sqrt{\frac{2L}{g}}\sqrt{\frac{k}{m}} + \varphi\right)} = \sqrt{2Lg},$$

and we express trigonometric functions

$$\cos\left(\sqrt{\frac{2Lk}{gm}} + \varphi\right) = -\frac{gm}{kR},$$

$$\sin\left(\sqrt{\frac{2Lk}{gm}} + \varphi\right) = -\frac{\sqrt{2Lg}}{R\sqrt{\frac{k}{m}}} = \frac{\sqrt{2Lgm}}{R\sqrt{k}}.$$

Then we can apply the trigonometric equation $\sin^2(x) + \cos^2(x) = 1$ to obtain an expression which allows one to calculate R as a function of the rest of the parameters, i.e.,

$$\sin^2\left(\sqrt{\frac{2Lk}{gm}}+\varphi\right)+\cos^2\left(\sqrt{\frac{2Lk}{gm}}+\varphi\right)=1$$

$$\to \frac{g^2m^2}{k^2R^2}+\frac{2Lgm}{R^2k}=1 \to \frac{1}{R^2}\left(\frac{g^2m^2}{k^2}+\frac{2Lgm}{k}\right)=1$$

$$\to R=\sqrt{\frac{g^2m^2}{k^2}+\frac{2gmL}{k}}.$$

Once R has been obtained, we can include it in the differential equation, i.e.,

$$y_p(t)=\sqrt{\frac{g^2m^2}{k^2}+\frac{2gmL}{k}}\cos(t\sqrt{k/m}t+\varphi)+\frac{gm}{k}+L.$$

We are interested in the maximum value of $y_p(t)$ as it cannot exceed the value of the ground distance. The function reaches the maximum value when $\cos(t\sqrt{k/m}t+\varphi)=1$. This occurs when $t\sqrt{k/m}t+\varphi=n\frac{\pi}{2}$, $n=1, 2, 3\ldots$, and $y_p(t_2)=y_{max}$. To calculate y_{max} for different values of L, m, and k we express

$$y_{max}(m,k,L)=\sqrt{\frac{9.81^2m^2}{k^2}+\frac{2(9.81)mL}{k}}+\frac{9.81m}{k}+L.$$

$$\text{For } k=300,\ y_{max}(m)=\sqrt{96.24\frac{m^2}{300^2}+1373.40\frac{m}{300}}+9.81\frac{m}{300}+70$$
$$=\sqrt{1.069.10^{-3}m^2+4.580m}+0.0330m+70.$$

$$\text{For } k=400,\ y_{max}(m)=\sqrt{96.24\frac{m^2}{400^2}+1373,40\frac{m}{400}}+9,81\frac{m}{400}+70$$
$$=\sqrt{6,02.10^{-4}m^2+3.460m}+0.0245m+70.$$

$$\text{For } k=500,\ y_{max}(m)=\sqrt{96.24\frac{m^2}{500^2}+1373,40\frac{m}{500}}+9,81\frac{m}{500}+70$$
$$=\sqrt{3.85.10^{-4}m^2+2.747m}+0.0196m+70.$$

Graphs of $y_{\max}$ for calculated mass values k are shown in Fig. 12.9.

Problem 12.4. If $y: I\to \mathbb{R}$ (I interval) is a function of class C^n on I, the expression $I(y)=\int_a^b F(x,y,y',\ldots,y^{(n)})dx$ is called *functional*. In many optimization problems, the minimization of such a functional appears to be necessary, i.e., to find such function y for which $I(y)$ is minimum. We will illustrate this statement by two examples.

i) A material point leaves from point $O(0,0)$ without an initial speed and moves under the action of gravity on an arch of curve $\widehat{OA}$ included in a vertical plan

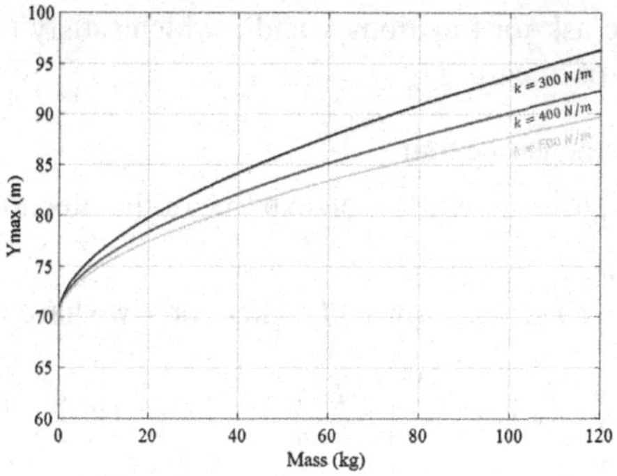

Figure 12.9 Graph of y_{max} for different mass values.

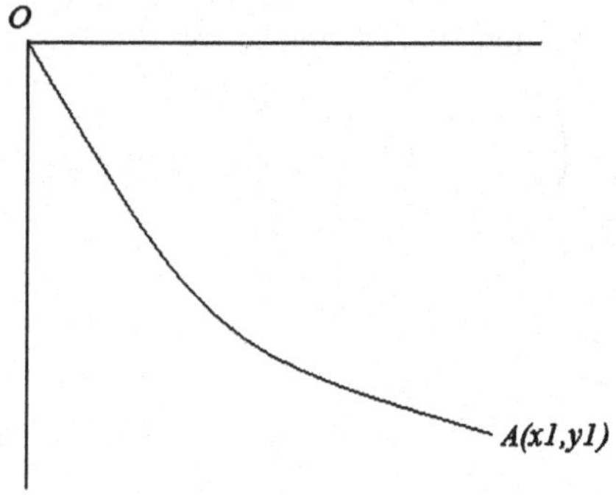

Figure 12.10 The arch of the curve by which the point $A(x_1, y_1)$ is reached from O in the shortest time.

(see Fig. 12.10). Find the arch of the curve by which the material point reaches point $A(x_1, y_1)$ from O in the shortest time.
The velocity in each point of the arch $\widehat{OA}$ $V = \frac{ds}{dt} = \sqrt{2gy}$, g being the gravitational acceleration. The time in which the moving covers the arch of the curve will be given by the curvilinear integral $\int_{\widehat{OA}} \frac{ds}{V} = \int_{\widehat{OA}} \frac{ds}{\sqrt{2gy}}$. By writing the equation of the curve of the form $y = y(x), x \in [0, x_1]$, the integral becomes $T = \int_0^{x_1} \frac{\sqrt{1+y'^2}}{\sqrt{2gy}} dx$.
Hence, we need to determine the function $y : [0, x_1] \to \mathbb{R}$, of class C^1, for which $y(0) = 0$, $y(x_1) = y_1$ and which realizes the minimum value for integral T.

ii) (The problem of the geodesics, met in aeronautics)
From all arches of curve met on a surface given S and which unit 2 points A and B on the surface, determine the arch of minimum length.

We solve this as follows. Let $F(x, y, z) = 0$ be the Cartesian implicit equation of a surface and let $A(x_1, y_1, z_1)$, $B(x_2, y_2, z_2)$ be those two points. An arch of curve marked on the surface will have the equation $F(x, y(x), z(x)) = 0$ (1), where $y(x_1) = y_1, z(x_1) = z_1, y(x_2) = y_2, z(x_2) = z_2$ (2). The length of arch $\widehat{AB}$ will be $L = \int_{x_1}^{x_2} \sqrt{1 + y'^2 + z'^2} dx$.

In other words, we ask for functions y and z which satisfy (1) and conditions (2) and for minimization of integral L.

Remark. One can demonstrate that

A) If function $y : [a,b] \to \mathbb{R}$ realizes an extreme of the functional

$I[y] = \int_a^b F(x, y', \ldots, y^{(n)})dx$, then for y we have the equation

$$\frac{\partial F}{\partial y} - \frac{d}{dx}\left(\frac{\partial F}{\partial y'}\right) + \frac{d^2}{dx^2}\left(\frac{\partial F}{\partial y''}\right) - \cdots + (-1)^n \frac{d^n}{dx^n}\left(\frac{\partial F}{\partial y^{(n)}}\right) = 0.$$

B) If functions $y_1, y_2, \ldots, y_n : [a,b] \to \mathbb{R}$ realize an extreme for the functional $I[y_1, y_2, \ldots, y_n] = \int_a^b F(x, y_1, y_1' \ldots, y_n, y_n')dx$, then the n functions satisfy the following system of differential equations:

$$\begin{cases} \dfrac{\partial F}{\partial y_1} - \dfrac{d}{dx}\left(\dfrac{\partial F}{\partial y_1'}\right) = 0, \\ \dfrac{\partial F}{\partial y_2} - \dfrac{d}{dx}\left(\dfrac{\partial F}{\partial y_2'}\right) = 0, \\ \ldots \ \ldots \ \ldots \ \ldots \ \ldots \\ \dfrac{\partial F}{\partial y_n} - \dfrac{d}{dx}\left(\dfrac{\partial F}{\partial y_n'}\right) = 0. \end{cases}$$

Then the function y from subsection (A) and functions $y_1, \ldots, y_n$ from subsection (B) are called *extremals*.

Example 1°: Determine the extremals of the functional

$$I[y] = (y''^2 - 2y'^2 + y - 2xye^{3x})dx.$$

We solve this as follows. Let $F(x, y, y', y'') = y''^2 - 2y'^2 + y - 2xye^{3x}$. According to (A), an extremal y for functional I will satisfy the equation $\frac{\partial F}{\partial y} - \frac{d}{dx}(\frac{\partial F}{\partial y'}) + \frac{d^2}{dx^2}(\frac{\partial F}{\partial y''}) = 0$ (attention! When we compute the partial derivatives of F, we consider x, y, y', y'' as the independent variables, while when one applies the operators $\frac{d}{dx}$ and $\frac{d^2}{dx^2}$, we treat y, y' and y'' as functions of x). We have

$$\frac{\partial F}{\partial y} = 2y - 2xe^x, \ \frac{\partial F}{\partial y'} = -4y', \ \frac{d}{dx}\left(\frac{\partial F}{\partial y'}\right) = -4y'', \ \frac{\partial F}{\partial y''} = 2y'', \ \frac{d^2}{dx^2}\left(\frac{\partial F}{\partial y''}\right) = 2y^{(4)}.$$

We obtain the equation $y^{(4)} + 2y'' + y = xe^{3x}$.

The associated characteristic equation of the homogeneous equation will be $r^4 + 2r^2 + 1 = 0$ with the roots $r_1 = r_2 = i$, $r_3 = r_4 = -i$ and the fundamental system of solutions $\{\cos x, \ \sin x, \ x \ \cos x, \ x \ \sin x\}$. The general solution of the homogeneous equation will be $y_O(x) = C_1 \cos x + C_2 \sin x + C_3 x \cos x + C_4 x \sin x + (\frac{x}{100} - \frac{3}{250})e^{3x}$.

Remark. If we had given the course of y and y' at the margins of the interval (i.e., some conditions of the form $y(a) = y_1$, $y(b) = y_2$, $y'(a) = \widetilde{y}_1$, $y'(b) = \widetilde{y}_2$) we could have determined also constants C_1, C_2, C_3, C_4. In this case, we would have determined a unique extremal of the functional.

Example 2°: Determine the extremals of the functional

$$I[y,z] = \int_a^b (y'^2 + z'^2 + y' \sin x + z' \cos +2yz)dx.$$

We solve this as follows. Let $F(x, y, y', z, z') = y'^2 + z'^2 + y' \sin x + z' \cos x + 2yz$. According to (B) the extremals y and z of the functional I will satisfy the system

$$\begin{cases} \dfrac{\partial F}{\partial y} - \dfrac{d}{dx}\left(\dfrac{\partial F}{\partial y'}\right) = 0, \\ \dfrac{\partial F}{\partial z} - \dfrac{d}{dx}\left(\dfrac{\partial F}{\partial z'}\right) = 0 \end{cases}$$

(with the same precaution at partial derivatives of F, respectively, at using $\frac{d}{dx}$ as in Example 1°). We obtain successively

$$\frac{\partial F}{\partial y} = 2z, \quad \frac{\partial F}{\partial y'} = 2y' + \sin x, \quad \frac{d}{dx}\left(\frac{\partial F}{\partial y'}\right) = 2y'' + \cos x,$$

$$\frac{\partial F}{\partial z} = 2y, \quad \frac{\partial F}{\partial y'} = 2z' + \cos x, \quad \frac{d}{dx}\left(\frac{\partial F}{\partial z'}\right) = 2z'' - \sin x.$$

The system becomes $\begin{cases} 2z - 2y'' - \cos x = 0, \\ 2y - 2z'' + \sin x = 0. \end{cases}$

From the first equation it results that $z = y'' + \frac{\cos x}{2}$, and hence $z' = y^{(4)} - \frac{\cos x}{2}$.

From the second equation it results that $2y^{(4)} - 2y = \sin x + \cos x$.

By proceeding as in similar exercises, we find $y_O(x) = C_1 e^x + C_2 e^{-x} + C_3 \cos x + C_4 \sin x$ (the solution of the associated homogeneous equation).

The right member can be written $e^{0x}(\cos(1x) + \sin(1x))$. As $0 \pm 1i$ are roots of order $k = 1$ of the characteristic equation, according to subsection c) ii) from theoretical part, we choose y_P of the form $y_P(x) = x(A \cos x + B \sin x)$. Putting the condition that y_P satisfies the nonhomogeneous equation, we find $A = \frac{1}{8}$, $B = -\frac{1}{8}$.

Finally, it results that $y(x) = C_1 e^x + C_2 e^{-x} + C_3 \cos x + C_4 \sin x + \frac{x}{8}(\sin x - \cos x)$.

From the relation $z = y'' + \frac{\cos x}{2}$, we obtain

$$z(x) = C_1 e^x + C_2 e^{-x} - \left(C_3 - \frac{1}{4}\right)\cos x - \left(C_4 + \frac{1}{4}\right)\sin x + \frac{x}{8}(\sin x - \cos x).$$

References

Burden, R.L., Faires, J.D., 2011. Numerical Analysis. Cengage Learning.

Roşculeţ, 1984. Analiză Matematică (Mathematical Analysis). Editura Didactică şi Pedagogică, Bucureşti (in Romanian).
Starek, L., 2009. Kmitanie s riadením (Vibration Control). STU, Bratislava (in Slovak).
Tikhonov, A.N., Samarski, A.A., 1956. Ecuaţiile Fizicii Matematice (Mathematical Physics Equations). Editura Tehnică (in Romanian).
Toma, I., Moşneguţu, V., Constantinescu, Ş., 2014. Analyse Mathématique, Équations Differentielles Ordinaires Calcul Integral (Mathematical Analysis, Ordinary Differential Equations Integral Calculation). Editura Conspress (in Romanian).

Partial differential equations 13

Cormac Breen, Michael Carr
Technological University Dublin, Dublin, Ireland

13.1 Introduction

1. What is a differential equation? In engineering and science ordinary differential equations (ODEs) arise as models for systems where there is one independent variable (often x) and one dependent variable (often y), i.e., the value of y depends on the value of x. Consider, for example, the following equations:

$$\frac{dy}{dx} + 8y = 0, \tag{13.1}$$

$$\frac{d^2y}{dx^2} - 2\frac{dy}{dx} + 11y = 0. \tag{13.2}$$

2. What is a partial differential equation (PDE)? However, in many real problems you may have a variable u (the dependent variable) that is a function of two (or more) independent variables, say, x and t, and we write $u(x,t)$, e.g., the temperature u along a metal bar that is being heated is both a function of where you are along the bar, x, and the time t. Then any derivatives of u will be partial derivatives such as $\frac{\partial u}{\partial x}$ or $\frac{\partial^2 u}{\partial t^2}$ and any differential equation arising will be known as a PDE. The independent variables may be space variables only (x, y, z) or one or more space variables (x, t) or (x, y, t) and time. Mathematical modeling of many real-life situations leads to PDEs (see section 13.2).

The subject of PDEs is a very large one. We shall discuss only a few special PDEs which model a wide range of applied problems.

13.1.1 Some properties of PDEs

Some classifying properties of PDEs are:

- The order of a PDE refers to the highest-order derivative present.
- We say a PDE is linear if:
  1. it is linear in the unknown function u and all of its derivatives;
  2. all coefficients depend only on the independent variables.
- We say a PDE is semilinear if it is linear in the highest-order derivative (i.e., not squared, etc.) and the coefficient of this term does not depend on the function itself.

Calculus for Engineering Students. https://doi.org/10.1016/B978-0-12-817210-0.00020-5

- We say a PDE is quasilinear if it is linear in the highest-order derivative (i.e., not squared, etc.) and the coefficient of this term *does* depend on the function itself.
- An equation which has a power (other than 1) for the highest-order derivative is referred to as nonlinear.
- A PDE for which *all* terms contain the unknown function or its derivatives is called homogeneous. Otherwise, it is referred to as inhomogeneous.
- General solution to a PDE depends on arbitrary functions rather than constants (as is the case for ODEs); hence PDEs have infinite solutions.

In this chapter we will mostly restrict our attention to linear PDEs.

13.1.2 First-order PDEs

We will look at two methods of solving first-order PDEs, the method of characteristics and the method of separation of variables.

1. Method of characteristics: The method of characteristics offers a means to reduce a PDE to ODEs. The principle behind this method is the following. Given a first-order linear PDE

$$a(x, y)\frac{\partial u}{\partial x} + b(x, y)\frac{\partial u}{\partial y} = c(x, y), \tag{13.3}$$

the general solution is obtained using

$$\eta = f(\chi)$$

or equivalently

$$\chi = g(\eta),$$

where f, g are arbitrary functions and η and ξ are the characteristic curves

$$\eta(x, y, u) = c_1, \quad \xi(x, y, u) = c_2.$$

To obtain the characteristic curves, we solve the ODEs

$$\frac{dx}{a} = \frac{dy}{b} = \frac{du}{c} \tag{13.4}$$

or, in the homogeneous case,

$$\frac{dx}{a} = \frac{dy}{b}, \quad du = 0. \tag{13.5}$$

To see why the characteristic curve equations given above lead to solutions of Eq. (13.3), consider the differential

$$du = \frac{\partial u}{\partial x}dx + \frac{\partial u}{\partial y}dy.$$

Using the relationships contained in Eq. (13.4) leads to

$$\frac{c}{b}dy = \frac{\partial u}{\partial x}\frac{a}{b}dy + \frac{\partial u}{\partial y}dy$$

or

$$\frac{\partial u}{\partial y} = \frac{c}{b} - \frac{\partial u}{\partial x}\frac{b}{a}$$

Inserting this in Eq. (13.3) we see that the equation is identically satisfied. Hence the set of ODEs contained in Eq. (13.4) is equivalent to the PDE Eq. (13.3). Consideration of Eq. (13.5) leads to the same conclusion for the homogeneous case ($c = 0$).

Example 13.1. Consider the equation

$$x\frac{\partial u}{\partial x} + y\frac{\partial u}{\partial y} = 2u.$$

The characteristic curves of this equation are

$$\frac{dx}{x} = \frac{dy}{y} = \frac{du}{2u}.$$

Integrating gives

$$\eta = \frac{y}{x} = c_1, \quad \xi = \frac{\sqrt{u}}{x} = c_2.$$

The general solution is

$$u = x^2 g\left(\frac{y}{x}\right).$$

Note that this solution depends on an arbitrary function. To obtain a unique solution we require some initial or boundary conditions.

Example 13.2. Consider the equations

$$\frac{\partial u}{\partial t} + u\frac{\partial u}{\partial x} = x, \quad u(x,0) = 1.$$

Characteristic equations are

$$dt = \frac{dx}{u} = \frac{du}{x},$$

$$dt = \frac{dx}{u} = \frac{du}{x} = \frac{d(x+u)}{x+u}.$$

The solutions are

$$(u+x)\mathrm{e}^{-t} = C_1, \quad u^2 - x^2 = C_2.$$

The general solution is given by

$$u^2 - x^2 = f\big((u+x)\mathrm{e}^{-t}\big).$$

Imposing the initial data

$$1 + x = c_1, \quad 1 - x^2 = c_2 \Longrightarrow c_2 = 2c_1 - c_1^2.$$

Hence

$$\big(u^2 - x^2\big) = 2(u+x)\mathrm{e}^{-t} - (u+x)^2\mathrm{e}^{-2t}.$$

2. Separation of variables: Similarly to the method of characteristics, separation of variables presents a way of reducing a PDE to an ODE problem. The principle is simply the following.
If the general solution $u(x, y)$ to a PDE is in the form of a product

$$u(x, y) = X(x)Y(y)$$

or a sum

$$u(x, y) = X(x) + Y(y),$$

we can obtain the solution by solving ODEs. (Otherwise, it does not work, so the simple strategy is just to try it.)

Example 13.3. Using a trial solution

$$u(x,t) = X(x)T(t) \tag{13.6}$$

and separation of variables, find the solution of the PDE

$$\frac{\partial u}{\partial x} + 4u = \frac{\partial u}{\partial t} \tag{13.7}$$

subject to the initial condition $u(x, 0) = 14e^{-23x}$. We split $U(x,t)$ up into two functions, $X(x)$, which is only a function of x, and $T(t)$, which is only a function of t, in the following manner:

$$U(x,t) = X(x)T(t), \tag{13.8}$$

$$\frac{\partial u}{\partial x} = T(t)\frac{dX(x)}{dx}, \tag{13.9}$$

$$\frac{\partial u}{\partial t} = X(x)\frac{dT(t)}{dt}, \tag{13.10}$$

$$\frac{\partial u}{\partial x} + 4u = \frac{\partial u}{\partial t} \Rightarrow \tag{13.11}$$

$$T(t)\frac{dX(x)}{dx} + 4X(x)T(t) = X(x)\frac{dT(t)}{dt}. \tag{13.12}$$

We can now divide both sides by $X(x)T(t)$, i.e.,

$$\frac{1}{X(x)}\frac{dX(x)}{dx} + 4 = \frac{1}{T(t)}\frac{dT(t)}{dt}. \tag{13.13}$$

This is true for all x and all t.
The left-hand side of this equation has only x terms, whereas the right-hand side has only t terms. This means that no matter what value of x or t we choose, the left-hand side = the right-hand side. This is true if and only if

$$\frac{1}{X(x)}\frac{dX(x)}{dx} + 4 = k, \tag{13.14}$$

$$\frac{1}{T(t)}\frac{dT(t)}{dt} = k. \tag{13.15}$$

We can now solve these two equations by straightforward separation of variables,

$$U(x,t) = X(x)T(t) = Ce^{(k-4)x}e^{kt}, \tag{13.16}$$

$u(x,0) = 14e^{-23x}$.
So

$$U(x,t) = X(x)T(t) = 14e^{-23x}e^{-19t}. \tag{13.17}$$

Exercise 13.1. Give three examples of $U(x,t)$ that can/cannot be represented by a trial solution of the form

$$U(x,t) = X(x)T(t).$$

Exercise 13.2. Using a trial solution

$$u(x,t) = X(x)T(t) \tag{13.18}$$

and separation of variables, find the solution of the PDE

$$\frac{\partial u}{\partial x} + 7u = \frac{\partial u}{\partial t}, \tag{13.19}$$

subject to the initial condition $u(x,0) = 7e^{-13x}$.

In the next section we examine second-order PDEs. We begin by introducing an important PDE known as the wave equation.

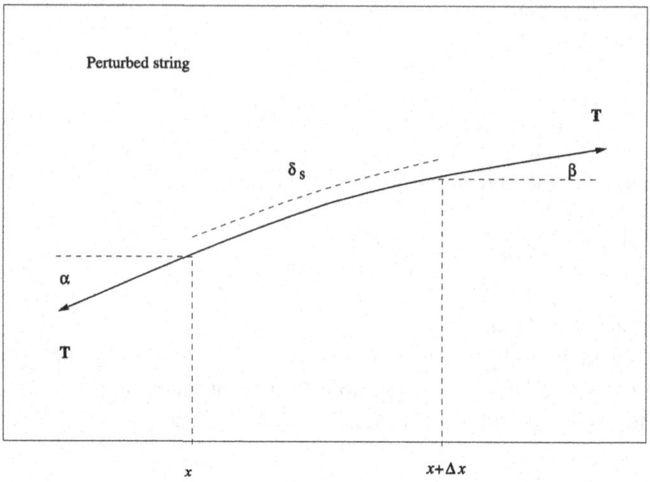

Figure 13.1 Element of a perturbed string.

13.2 Applications of partial differential equations

Wave equation: A very important PDE for applied mathematics is the wave equation, which describes how a disturbance (wave) propagates through a medium. In this section we will derive the 1D version of this equation.

The simplest situation to give rise to the 1D wave equation is the motion of a stretched string – specifically the transverse vibrations of a string such as the string of a musical instrument.

Assume that a string is placed along the x-axis, stretched, and then fixed at ends $x = 0$ and $x = L$. It is then deflected and at some instant, which we call $t = 0$, it is released and allowed to vibrate. The quantity of interest is the deflection u of the string at any point x, $0 \leq x \leq L$ and at any time $t > 0$. We write $u = u(x, t)$. We then consider this system subject to various assumptions:

- damping forces such as air resistance are negligible;
- the weight of the string is negligible;
- the tension T in the string is tangential to the curve of the string at any point;
- the string performs small transverse oscillations, i.e., every particle of the string moves strictly vertically and such that its deflection and slope at every point on the string are small.

Consider an element of the perturbed string of length δs between x and $x + \Delta x$ as shown in Fig. 13.1. Since T is constant, the total tension force acting on the element shown is

$$T(\sin\beta - \sin\alpha) \approx T(\tan\beta - \tan\alpha) = T\left(\frac{\partial u}{\partial x}(x+\Delta x, t) - \frac{\partial u}{\partial x}(x, t)\right) \approx T\frac{\partial^2 u}{\partial x^2}\Delta x. \tag{13.20}$$

If an external force f per unit length ($f\delta s \approx f\Delta x$) is acting on the element, then the total force is given by

$$T\frac{\partial^2 u}{\partial x^2}\Delta x + F\rho\Delta x,$$

where $F = f/\rho$. And by Newtons second law,

$$\text{total force} = \rho\delta s\frac{\partial^2 u}{\partial t^2} \approx \rho\Delta x\frac{\partial^2 u}{\partial x^2}. \tag{13.21}$$

Hence we arrive at the 1D wave equation

$$\frac{\partial^2 u}{\partial t^2} = c^2\frac{\partial^2 u}{\partial x^2} + F, \quad \text{where } c^2 \equiv T/\rho.$$

This is now a second-order PDE, and we find a solution to this using the method of separation of variables.

13.2.1 Second-order PDEs

In most applications of PDEs we are required to solve second-order equations. There are three PDEs that are of particular importance to applied mathematics, i.e., the wave equation (derived in the last section), the heat/diffusion equation, and Laplace's equation.

13.2.2 Wave equation

We will begin this section by finding a solution to the wave equation derived in the last section.

Example. Solve the wave equation subject to the following boundary and initial conditions:

$$\begin{aligned} u(x,0) &= f(x) && \text{initial displacement,} \\ \frac{\partial u}{\partial t}(0,t) &= 0 && \text{initial velocity,} \\ u(0,t) = u(L,t) &= g(x) && \text{boundary conditions.} \end{aligned}$$

We seek a solution of the form

$$u(x,t) = X(x)T(t)$$

leading to the general solutions of the form given in the previous example. We now impose the following boundary conditions to restrict the form of the solution:

$$u(0,t) = X(0)T(t) = 0 \implies X(0) = 0, \tag{13.22}$$

$$u(L,t) = X(L)T(t) = 0 \implies X(L) = 0. \tag{13.23}$$

If we take the separation constant $\lambda = k^2$ to be positive, then

$$X(x) = Ae^{kx} + Be^{-kx}, \tag{13.24}$$

and hence the boundary conditions ($X(0) = X(L) = 0$) imply that $A = 0 = B$ and the only solution is the trivial one $u(x,t) = 0$. Therefore we must take $\lambda = -k^2$ to be negative, giving

$$X(x) = A\cos(kx) + B\sin(kx). \tag{13.25}$$

We impose the boundary conditions,

$$X(0) = 0 \implies A = 0,$$

$$X(L) = 0 \implies B\sin kL = 0 \implies k = \frac{n\pi}{L},$$

where $n = 1, 2, 3\ldots$. Hence

$$X_n(x) = B_n \sin\left(\frac{n\pi x}{L}\right),$$

where the subscript n denotes that we have a solution for each value of n. The constraints on λ outlined above imply that the temporal solution $T(t)$ must take the form

$$T(t) = C\cos\left(\frac{n\pi t}{L}\right)kt + D\sin\left(\frac{n\pi t}{L}\right).$$

So for any particular positive integer n the general form of our solution $u(x,t)$ takes the form

$$u(x,t) = \left[a_n \cos\left(\frac{n\pi t}{L}\right)kt + b_n \sin\left(\frac{n\pi t}{L}\right)\right]\sin\left(\frac{n\pi x}{L}\right),$$

with $a_n = A_n C_n$ and $b_n = A_n D_n$. The sum of any number of solutions to a homogeneous PDE is also a solution to that PDE (this is known as the principle of superposition), and hence we may write the solution to the wave equation which satisfies the boundary conditions given above in the following form:

$$u(x,t) = \sum_{n=1}^{\infty}\left[a_n \cos\left(\frac{n\pi t}{L}\right)kt + b_n \sin\left(\frac{n\pi t}{L}\right)\right]\sin\left(\frac{n\pi x}{L}\right).$$

We impose the initial conditions,

$$u(x,0) = f(x) = \sum_{n=1}^{\infty} a_n \sin\left(\frac{n\pi x}{L}\right), \tag{13.26}$$

$$\frac{\partial u}{\partial t}(0,t)=0=\sum_{n=1}^{\infty}\frac{b_n n\pi x}{L}\sin\left(\frac{n\pi x}{L}\right). \tag{13.27}$$

To proceed we make use of the identity

$$\int_0^L \sin\left(\frac{n\pi x}{L}\right)\sin\left(\frac{m\pi x}{L}\right)=\begin{cases}\frac{L}{2} & \text{if } n=m,\\ 0 & \text{if } n\neq m.\end{cases}$$

Multiplying both sides of the equations in (13.26) by $\sin(\frac{m\pi x}{L})$ and integration from 0 to L yields

$$a_n=\frac{2}{L}\int_0^L f(x)\sin\left(\frac{n\pi x}{L}\right),$$
$$b_n=\frac{2}{n\pi L}\int_0^L g(x)\sin\left(\frac{n\pi x}{L}\right).$$

Therefore for a given initial displacement and velocity, we can determine the coefficients a_n and b_n and hence the solution $u(x,t)$.

Exercise. Determine the coefficients a_n and b_n above, and hence the solution to the wave equation $u(x,t)$, for a string with the following initial displacement and velocity:

$$f(x)=\begin{cases}\frac{hx}{a} & 0\le x\le a,\\ \frac{h(L-x)}{a} & a\le x\le L,\end{cases}$$
$$g(x)=0.$$

This can be thought of as a string that is "plucked" to height h at a point a along its length.

13.2.3 Heat equation

The 3D heat or diffusion equation is a second-order PDE of the form

$$\frac{\partial u}{\partial t}=\kappa\left(\frac{\partial^2 u}{\partial^2 x}+\frac{\partial^2 u}{\partial^2 y}+\frac{\partial^2 u}{\partial^2 z}\right).$$

As in the case of the wave equation, we will focus on solving the 1D heat equation using the separation of variables method, which takes the following form:

$$\frac{\partial u}{\partial t}=\kappa\frac{\partial^2 u}{\partial^2 x}.$$

Solution. We use a trial solution

$$Y(x,t)=X(x)T(t). \tag{13.28}$$

We split $Y(x,t)$ up into two functions, $X(x)$, which is only a function of x, and $T(t)$, which is only a function of t. Following the same method as used for the wave equation leads to the following equations:

$$T'(t) = -\alpha^2 \kappa T(t),$$

$$X''(x) = -\alpha^2 X(t),$$

where α^2 is the separation constant. These equations are readily solved using the theory of ODEs to give firstly

$$T(t) = Ae^{-\alpha^2 \kappa t}.$$

Given that we examine the case of heat diffusion in a body, we would expect that our solution would tend to 0 as $t \to \infty$. This physical requirement constrains α^2 to be positive, which implies that

$$X(x) = B\cos(\alpha x) + C\sin(\alpha x).$$

As in the case of the wave equation, we require some boundary and initial conditions to determine the unknown constants A, B, C. Let us take the following set of conditions:

$$\begin{aligned} u(x,0) = f(x) &\quad \text{initial temperature distribution,} \\ u(0,t) = T = u(L,t) &\quad \text{boundary conditions.} \end{aligned}$$

These conditions correspond to the heat distribution on a rod of length L, where both ends are at kept at temperature T, which for simplicity we will take to be 0.
The boundary conditions imply the following:

$$X(0) = B = 0,$$

$$X(L) = C\sin(\alpha L) = 0.$$

The second equation implies that $\alpha L = n\pi$, where n is an integer and hence for any integer $n \geq 1$ we obtain the following solution for $X(x)$:

$$X(x) = C_n \sin\left(\frac{n\pi x}{L}\right),$$

and hence

$$u_n(x,t) = D_n e^{-\frac{n^2\pi^2 t}{l^2}} \sin\left(\frac{n\pi x}{L}\right),$$

where we have a different solution for each integer value of n. We have

$$u(x,t) = \sum_{n=1}^{\infty} D_n e^{-\frac{n^2\pi^2 t}{L^2}} \sin\left(\frac{n\pi x}{L}\right).$$

We have yet to determine the unknown coefficients D_n in this expression. These are determined by the initial condition in the following way:

$$u(x,0) = f(x) = \sum_{n=1}^{\infty} D_n \sin\left(\frac{n\pi x}{L}\right).$$

Multiplying both sides of the above equation by $\sin(n\pi/L)$, integrating with respect to x between 0 and L and using the orthogonality of $\sin(x)$ we arrive at the following result:

$$D_n = \frac{2}{L}\int_0^L f(x)\sin\left(\frac{n\pi x}{L}\right)dx.$$

We are now in a position to find the solution to the heat equation subject to the boundary conditions given above, for any initial temperature distribution $f(x)$.

Exercise. Find the solution to the heat equation

$$\frac{\partial u}{\partial t} = \kappa \frac{\partial^2 u}{\partial^2 x}$$

subject to the following boundary and initial conditions:

$$\begin{aligned} u(x,0) &= x(L-x) \quad \text{initial temperature distribution,} \\ u(0,t) = 0 &= u(L,t) \quad \text{boundary conditions.} \end{aligned}$$

13.2.4 Laplace's equation

We will close this section by considering Laplace's equation in two dimensions, which is given by

$$\frac{\partial^2 u}{\partial^2 x} + \frac{\partial^2 u}{\partial^2 y} = 0. \tag{13.29}$$

We will seek a solution using the separation of variables method, in other words, we seek a solution of the form

$$u(x,y) = X(x)Y(y),$$

which yields upon substitution the following equations:

$$\begin{aligned} X''(x) &= \lambda X(x), \\ Y''(y) &= -\lambda Y(y), \end{aligned}$$

where $-k^2$ is the separation constant. As in the case of the wave equation we must consider the case where $\lambda = k^2 > 0$ and $\lambda = -k^2 < 0$. Following the arguments made in the wave equation section, we have the following possible solutions:

$$\lambda = -k^2 < 0: \quad u(x, y) = (A \sin kx + B \cos kx)\big(Ce^{ky} + De^{ky}\big), \tag{13.30}$$

$$\lambda = k^2 < 0: \quad u(x, y) = \big(Ae^{kx} + Be^{kx}\big)(C \sin ky + D \cos ky). \tag{13.31}$$

Of course the actual form of the solution will depend on the boundary conditions that we require our solution to satisfy.

As an example, we will consider the following set of boundary conditions:

$$\begin{aligned}
&u(x, 0) = 0 \;\; (0 < x < 1),\\
&u(x, \pi) = 0 \;\; (0 < x < 1),\\
&u(0, y) = 0 \;\; (0 < y < \pi),\\
&u(1, y) = \sin(y) \;\; (0 < y < \pi).
\end{aligned}$$

By consideration of the final boundary condition, we see that our solution must contain $\sin(\mu y)$ terms. Hence we take the solution to have the general form

$$u(x, y) = \big(Ae^{kx} + Be^{kx}\big)(C \sin ky + D \cos ky).$$

The first boundary condition gives

$$\big(Ae^{kx} + Be^{kx}\big)D = 0 \quad (0 < x < 1),$$

and hence $D = 0$. Thus

$$u(x, y) = \big(\hat{A}e^{kx} + \hat{B}e^{kx}\big) \sin(ky),$$

with $\hat{A} = AC$ and $\hat{B} = BC$. The second boundary condition gives

$$\big(\hat{A}e^{kx} + \hat{B}e^{kx}\big) \sin(k) = 0 \quad (0 < x < 1),$$

and hence $k = n$, where n is an integer. We then have

$$u(x, y) = \big(\hat{A}e^{nx} + \hat{B}e^{nx}\big) \sin ny.$$

Imposing the third boundary condition yields

$$(\hat{A} + \hat{B}) \sin ny = 0 \quad (0 < y < \pi),$$

giving $\hat{A} = -\hat{B}$ and so

$$\begin{aligned}
u(x, y) &= \hat{A}\big(e^{nx} - e^{nx}\big) \sin ny\\
&= 2\hat{A} \sinh nx \sin ny
\end{aligned}$$

Finally the last boundary condition requires that

$$u(x, y) = 2\hat{A} \sinh nx \sin ny = \sin(y).$$

This equality holds if we take $n = 1$ and $\hat{A} = 1/2(\sinh 1)$; therefore the solution to Laplace's equation that satisfies the four boundary conditions given above is

$$u(x, y) = \frac{\sinh(x)}{\sinh 1} \sin(y).$$

Exercise. Solve Laplace's equation subject to the following boundary conditions:

$$u(x, 0) = 0 \quad (x > 0),$$
$$u(x, \pi) = 0 \quad (x > 0),$$
$$u(0, y) = a \quad (0 < y < \pi),$$
$$u(x, y) = 0 \quad x \to \infty.$$

13.2.5 Laplace transforms

Thus far we have mostly used the method of separation of variables to solve PDEs; we will now look briefly at another solution method based on the theory of a type of integral transforms known as Laplace transforms. Like the method of separation of variables and the method of characteristics, we will see that using Laplace transforms converts a PDE to an ODE for a transformed function. If we can solve this ODE, then we obtain the solution to the original PDE provided we are able to invert the Laplace transform.

Using the same procedure as used to obtain the Laplace transform of standard derivatives in Chapter 14, we obtain the following:

1. $\mathcal{L}\left\{\frac{\partial u}{\partial x}\right\} = \int_0^\infty e^{-st}\frac{\partial u}{\partial x} = \frac{d}{dx}\int_0^\infty e^{-st}u(x,t) = \frac{d}{dx}U(x,s),$
   where $U(x,s) = \int_0^\infty e^{-st}u(x,t) = \mathcal{L}\{u(x,t)\}$.
2. Repeating this procedure yields
   $\mathcal{L}\left\{\frac{\partial^2 u}{\partial x^2}\right\} = \frac{d^2}{dx^2}U(x,s).$
3. $$\mathcal{L}\left\{\frac{\partial u}{\partial t}\right\} = \int_0^\infty e^{-st}\frac{\partial u}{\partial t} = \left[e^{-st}u(x,t)\right]_0^\infty + s\int_0^\infty e^{-st}u(x,t)$$
   $$= sU(x,s) - u(x,0),$$
   where we have assumed that $\lim_{t\to\infty} e^{-st}u(s,t) = 0$. This sets a condition on the solution $u(x,t)$ in order for the Laplace transform solution method to be applicable.
4. Repeating this procedure leads to
   $\mathcal{L}\left\{\frac{\partial^2 u}{\partial t^2}\right\} = s^2U(x,s) - su(x,0) - \left.\frac{\partial u}{\partial t}\right|_{t=0}.$

We will now revisit the heat equation to demonstrate how the Laplace transform method can be used to solve PDEs.

13.2.6 Heat equation revisited

Consider the heat equation in a semiinfinite 1D rod,

$$\frac{\partial u}{\partial t} = \kappa \frac{\partial^2 u}{\partial x^2},$$

subject to the following boundary and initial conditions:

$$u(0,t) = T_0, \quad \lim_{x\to\infty} u(x,t) = 0, \quad t > 0,$$
$$u(x,0) = 0, \quad x > 0.$$

If we take the Laplace transform of both sides of the heat equation we obtain the following:

$$sU(x,s) - u(x,0) = \frac{d^2}{dx^2}U(x,s)$$

with $U(x,s) = \mathcal{L}\{u(x,t)\}$. Imposing the initial condition $u(x,0) = 0$ results in an ODE for $U(x,s)$, i.e.,

$$\frac{d^2}{dx^2}U(x,s) - \frac{s}{\kappa}U(x,s) = 0$$

subject to the boundary conditions

$$U(0,s) = \mathcal{L}\{T_0\} = \frac{T_0}{s}, \quad \lim_{x\to\infty} U(x,s) = 0,$$

where we have used the property derived in Chapter 14 that $\mathcal{L}\{k\} = k/s$, where k is a constant. The above ODE is readily solved, giving

$$U(x,s) = Ae^{\alpha x} + Be^{-\alpha x},$$

with $\alpha = \sqrt{s/\kappa}$. Imposing the boundary conditions, we see that we must have $A = 0$ and $B = T_0/s$. Hence

$$U(x,s) = \frac{T_0}{s}e^{-\alpha s},$$

and so $u(x,t)$, the solution to the heat equation, is given by

$$u(x,t) = \mathcal{L}^{-1}\left\{\frac{T_0}{s}e^{-\alpha s}\right\},$$

where $\mathcal{L}^{-1}\{f\}$ denotes the inverse Laplace transform of a function f (see Chapter 14 for further details).

To perform the required inverse transformation we note that

$$\mathcal{L}\left\{\text{efrc}\left(\frac{a}{2\sqrt{t}}\right)\right\} = \frac{1}{s}e^{-a\sqrt{s}},$$

where erfc(x) denotes the complementary error function defined by

$$\text{erfc}(x) = \frac{2}{\pi}\int_x^{\infty} e^{-y^2}\,dy.$$

Clearly we must have

$$\mathcal{L}^{-1}\left\{\frac{1}{s}e^{-a\sqrt{s}}\right\} = \text{efrc}\left(\frac{a}{2\sqrt{t}}\right),$$

and therefore we arrive at a solution to the heat equation that satisfies the required boundary and initial conditions detailed above, i.e.,

$$u(x,t) = T_0\ \text{erfc}\left(\frac{x}{2\sqrt{\kappa t}}\right).$$

Exercise. Use the Laplace transform solution method to solve the 1D wave equation

$$\frac{\partial^2 y}{\partial t^2} = \frac{\partial^2 y}{\partial x^2}$$

subject to the following initial and boundary conditions:

$$\begin{aligned} u(x,0) &= 0, \quad \frac{\partial u}{\partial t}(0,t) = 0, \quad x > 0 \ \text{ initial conditions,}\\ u(0,t) &= H_0, \quad \lim_{x\to\infty} u(x,t) = 0, \quad t > 0 \ \text{ boundary conditions.} \end{aligned}$$

13.3 Real engineering problems

1. **Problem 1.** Consider the following equations:

$$\frac{\partial^2 y}{\partial t^2} = \frac{\partial^2 y}{\partial x^2}, \tag{13.32}$$

$$c^2 = 1.$$

This is the equation giving the displacement of a tightly stretched vibrating string. We want to find y in terms of x and t if the motion is started by displacing a stretched string of length 6 into the form $y = 12\sin\frac{\pi x}{6}$ and then releasing it.
Solution: We use a trial solution

$$Y(x,t) = X(x)T(t). \tag{13.33}$$

We split $Y(x,t)$ up into two functions, $X(x)$, which is only a function of x, and $T(t)$, which is only a function of t. Then we have

$$\frac{\partial y}{\partial x} = T(t)\frac{dX(x)}{dx}, \tag{13.34}$$

$$\frac{\partial^2 y}{\partial x^2} = T(t)\frac{d^2X(x)}{dx^2}, \tag{13.35}$$

$$\frac{\partial y}{\partial t} = X(x)\frac{dT(t)}{dt}, \tag{13.36}$$

$$\frac{\partial^2 y}{\partial t^2} = X(x)\frac{d^2T(t)}{dt^2}. \tag{13.37}$$

We now insert these expressions into the original equation,

$$\frac{\partial^2 y}{\partial t^2} = \frac{\partial^2 y}{\partial x^2}, \tag{13.38}$$

$$X(x)\frac{d^2T(t)}{dt^2} = T(t)\frac{d^2X(x)}{dx^2} \tag{13.39}$$

and divide across by $X(x)T(t)$ to give

$$\frac{1}{T(t)}\frac{d^2T(t)}{dt^2} = \frac{1}{X(x)}\frac{d^2X(x)}{dx^2}. \tag{13.40}$$

The left-hand side contains only t terms and the right-hand side contains only x terms. So both sides must be equal to a constant λ.

Exercise. Explain in your own words why both sides are equal to a constant $\lambda = +k^2$ **or** $\lambda = -k^2$**.** We must now decide if we should let $\lambda = -k^2$ or $\lambda = +k^2$. This decision has implications for the nature of our solution.
$\lambda = -k^2$**.** We have

$$\frac{1}{X(x)}\frac{d^2X(x)}{dx^2} = -k^2, \tag{13.41}$$

$$\frac{d^2X(x)}{dx^2} + k^2X(x) = 0. \tag{13.42}$$

We create the auxiliary equation

$$m^2 + k^2 = 0, \tag{13.43}$$

$$m^2 = -k^2, \tag{13.44}$$

$$m = \sqrt{-k^2} = \pm jk. \tag{13.45}$$

where $j = \sqrt{-1}$
$\lambda = -k^2$**.** We now have a complex root

$$m = \sqrt{-k^2} = \pm jk, \tag{13.46}$$

$$X = e^{\alpha x}(A\cos\beta x + B\sin\beta x), \tag{13.47}$$

where

$$m = \alpha \pm j\beta. \tag{13.48}$$

In this case

$$m = \sqrt{-k^2} = \pm jk, \tag{13.49}$$

so $\alpha = 0$ and $\beta = k$,

$$X = e^{0x}(A\cos kx + B\sin kx), \tag{13.50}$$

$$X = (A\cos kx + B\sin kx). \tag{13.51}$$

$\lambda = +k^2$**.** We have

$$\frac{1}{X(x)}\frac{d^2X(x)}{dx^2} = +k^2, \tag{13.52}$$

$$\frac{d^2X(x)}{dx^2} - k^2X(x) = 0. \tag{13.53}$$

We create the auxiliary equation

$$m^2 - k^2 = 0, \tag{13.54}$$

$$m^2 = +k^2, \tag{13.55}$$

$$m = \pm k. \tag{13.56}$$

$\lambda = +k^2$**.** Roots are real and different,

$$X(x) = Ae^{+kx} + Be^{-kx} \tag{13.57}$$

or

$$X(x) = A\cosh kx + B\sinh kx. \tag{13.58}$$

We will use the second form from here on.
Hyperbolic functions arise frequently in the solutions of differential equations

$$\cosh z = \frac{1}{2}\left(e^z + e^{-z}\right),$$

$$\sinh z = \frac{1}{2}\left(e^z - e^{-z}\right).$$

We now have two possible scenarios.
$\lambda = +k^2$**.** Roots are real and different,

$$x = Ae^{+kx} + Be^{-kx}, \tag{13.59}$$

$$X(x) = A\cosh kx + B\sinh kx. \tag{13.60}$$

$\lambda = -k^2$**.** Roots are complex,

$$X(x) = (A\cos kx + B\sin kx). \tag{13.61}$$

To solve this we compare with the boundary conditions *the motion that is started by displacing a stretched string of length 6 into the form* $y = 12\sin\frac{\pi x}{6}$ *and then releasing it.*
We choose the $\lambda = -k^2$ option so that our solution can satisfy the boundary condition.
We solve

$$\frac{\partial^2 y}{\partial t^2} = -k^2, \tag{13.62}$$

$$T = (A\cos kt + B\sin kt). \tag{13.63}$$

Our solution for $Y(x,t)$ now becomes

$$Y(x,t) = X(x)T(t), \tag{13.64}$$

$$Y(x,t) = (A_1\cos kx + B_1\sin kx)(A_2\cos kt + B_2\sin kt). \tag{13.65}$$

We may now use the boundary conditions to determine the unknown constants A_1, A_2, B_1, B_2.

a. $Y = 0$ at $x = 0$ for all $t > 0$.
b. String is released from rest $\frac{\partial y}{\partial t} = 0$ at $t = 0$.
c. At $t = 0$ $Y = 12\sin\frac{\pi x}{6}$.

$y = 0$ **at** $x = 0$ **for all** t**.** We have

$$Y(x,t) = (A_1\cos kx + B_1\sin kx)(A_2\cos kt + B_2\sin kt), \tag{13.66}$$

$$0 = (A_1\cos 0 + B_1\sin 0)(A_2\cos kt + B_2\sin kt), \tag{13.67}$$

$$0 = (A_1)(A_2\cos kt + B_2\sin kt), \tag{13.68}$$

for this to be true for all t $A_1 = 0$.
The string is released from rest $\frac{\partial y}{\partial t} = 0$ **at** $t = 0$**.** We have

$$Y(x,t) = (B_1\sin kx)(A_2\cos kt + B_2\sin kt), \tag{13.69}$$

$$Y(x,t) = (\sin kx)(A\cos kt + B\sin kt), \tag{13.70}$$

$$\frac{\partial y}{\partial t} = (\sin kx)(-kA\sin kt + kB\cos kt), \tag{13.71}$$

where $A = A_2B_1$ and $B = B_1B_2$,

$$0 = (\sin kx)(-kA\sin 0 + kB\cos 0), \tag{13.72}$$

$$0 = (\sin kx)(kB), \tag{13.73}$$

$$0 = B, \tag{13.74}$$

$$Y(x,t) = A(\sin kx)(\cos kt). \tag{13.75}$$

At $t = 0$ $y = 12\sin\frac{\pi x}{6}$. We have

$$y(x,t) = A(\sin kx)(\cos kt), \tag{13.76}$$

$$y(x,t) = A(\sin kx)(\cos k0), \tag{13.77}$$

$$y(x,t) = A(\sin kx)(\cos k0), \tag{13.78}$$

$$y = 12\sin\frac{\pi x}{6}, \tag{13.79}$$

$$A = 12\ k = \frac{\pi}{6}, \tag{13.80}$$

$$y(x,t) = 12\left(\sin\frac{\pi}{6}x\right)\left(\cos\frac{\pi}{6}t\right). \tag{13.81}$$

Exercise. Sketch $y(x,t)$ for $x = 0$ to 5 in steps of 1 for $t = 1$.
In the previous example we arrived at a single solution to the wave equation. However, as can be seen in the next example, for certain boundary conditions we in fact have infinitely many solutions, which we can add together to form a single solution.

2. **Problem 2**
   a. The equation

$$\frac{\partial^2 y}{\partial t^2} = 49\frac{\partial^2 y}{\partial x^2}$$

   is the equation of a tightly stretched guitar string. Find the solution of y in terms of x and t if the motion is started by displacing a string of length 0.25 m into the form $y = 12\sin 4\pi x$ and then releasing it from rest.
   b. Find the displacement of the string at $x = 0.2$ after 3 seconds.
   c. If the tension in the string is increased by one-third, describe how this would effect your solution.

Solution: We have

$$\frac{\partial^2 y}{\partial t^2} = 49\frac{\partial^2 y}{\partial x^2}, \tag{13.82}$$

$$c^2 = 9.$$

This is the equation giving the displacement of a tightly stretched vibrating string. We want to find y in terms of x and t if the motion is started by displacing a stretched string of length 5 into the form $y = 12\sin 4\pi x$ and then releasing it.

We use a trial solution,

$$Y(x,t) = X(x)T(t). \tag{13.83}$$

We split $Y(x,t)$ up into two functions, $X(x)$, which is only a function of x, and $T(t)$, which is only a function of t. We have

$$\frac{\partial y}{\partial x} = T(t)\frac{dX(x)}{dx}, \tag{13.84}$$

$$\frac{\partial y^2}{\partial x^2} = T(t)\frac{d^2X(x)}{dx^2}, \tag{13.85}$$

$$\frac{\partial y}{\partial t} = X(x)\frac{dT(t)}{dt}, \tag{13.86}$$

$$\frac{\partial y^2}{\partial t^2} = X(x)\frac{d^2T(t)}{dt^2}. \tag{13.87}$$

We now substitute these expressions into the original equation,

$$\frac{\partial^2 y}{\partial t^2} = 49\frac{\partial^2 y}{\partial x^2}, \tag{13.88}$$

$$X(x)\frac{d^2T(t)}{dt^2} = 49T(t)\frac{d^2X(x)}{dx^2}. \tag{13.89}$$

We divide across by $X(x)T(t)$ to obtain

$$\frac{1}{T(t)}\frac{d^2T(t)}{dt^2} = \frac{49}{X(x)}\frac{d^2X(x)}{dx^2}. \tag{13.90}$$

The left-hand side contains only t terms and the right-hand side contains only x terms. So both sides must be equal to a constant λ.
We must now decide if we should let $\lambda = -k^2$ or $\lambda = +k^2$. This decision has implications for the nature of our solution.
$\lambda = -k^2$. We have

$$\frac{49}{X(x)}\frac{d^2X(x)}{dx^2} = -k^2, \tag{13.91}$$

$$49\frac{d^2X(x)}{dx^2} + k^2X(x) = 0. \tag{13.92}$$

We create the auxiliary equation

$$49m^2 + k^2 = 0, \tag{13.93}$$

$$49m^2 = -k^2, \tag{13.94}$$

$$m = \sqrt{\frac{-k^2}{7}} = \pm j\frac{k}{7}. \tag{13.95}$$

This is the special case.
$\lambda = -k^2$. We now have a complex root

$$m = \sqrt{\frac{-k^2}{49}} = \pm j\frac{k}{7}, \tag{13.96}$$

$$X = e^{\alpha x}(A\cos\beta x + B\sin\beta x), \tag{13.97}$$

where

$$m = \alpha \pm j\beta. \tag{13.98}$$

In this case

$$m = \sqrt{\frac{-k^2}{49}} = \pm j\frac{k}{7}, \tag{13.99}$$

so $\alpha = 0$ and $\beta = \frac{k}{7}$,

$$X = e^{0x}\left(A\cos\frac{k}{7}x + B\sin\frac{k}{7}x\right), \tag{13.100}$$

$$X = \left(A\cos\frac{k}{3}x + B\sin\frac{k}{3}x\right). \tag{13.101}$$

$\lambda = +k^2$ We have

$$\frac{49}{X(x)}\frac{d^2X(x)}{dx^2} = +k^2, \tag{13.102}$$

$$49\frac{d^2X(x)}{dx^2} - k^2X(x) = 0. \tag{13.103}$$

We create the auxiliary equation

$$49m^2 - k^2 = 0, \tag{13.104}$$

$$49m^2 = +k^2, \tag{13.105}$$

$$m = \pm\frac{k}{7}. \tag{13.106}$$

Roots are real and different,

$$X(x) = Ae^{+\frac{k}{7}x} + Be^{-\frac{k}{7}x}, \tag{13.107}$$

or

$$X(x) = A\cosh\frac{k}{7}x + B\sinh\frac{k}{7}x. \tag{13.108}$$

We now have two possible scenarios.
$\lambda = +k^2$. Roots are real and different,

$$x = Ae^{+\frac{k}{7}x} + Be^{-\frac{k}{7}x}, \tag{13.109}$$

$$X(x) = A\cosh\frac{k}{7}x + B\sinh\frac{k}{7}x. \tag{13.110}$$

$\lambda = -k^2$. Roots are complex,

$$X(x) = \left(A\cos\frac{k}{7}x + B\sin\frac{k}{7}x\right). \tag{13.111}$$

To solve this we compare with the boundary conditions.
We choose the $\lambda = -k^2$ option so as to confirm the boundary conditions.
We solve

$$\frac{1}{T(t)}\frac{d^2T(t)}{dt^2} = -k^2, \tag{13.112}$$

$$T = (A\cos kt + B\sin kt). \tag{13.113}$$

Our solution for y now becomes

$$Y(x,t) = X(x)T(t), \tag{13.114}$$

$$Y(x,t) = \left(A_1\cos\frac{k}{7}x + B_1\sin\frac{k}{7}x\right)(A_2\cos kt + B_2\sin kt). \tag{13.115}$$

We must now use the boundary conditions to solve this.

a. – $y = 0$ at $x = 0$ for all t.
– String is released from rest $\frac{\partial y}{\partial t} = 0$ at $t = 0$.
– At $t = 0$ $y = 12\sin 4\pi x$.

$y = 0$ at $x = 0$ for all t. We have

$$Y(x,t) = \left(A_1\cos\frac{k}{7}x + B_1\sin\frac{k}{7}x\right)(A_2\cos kt + B_2\sin kt), \tag{13.116}$$

$$0 = (A_1\cos 0 + B_1\sin 0)(A_2\cos kt + B_2\sin kt), \tag{13.117}$$

$$0 = (A_1)(A_2 \cos kt + B_2 \sin kt), \tag{13.118}$$

for this to be true for all t $A_1 = 0$.
The string is released from rest $\frac{\partial y}{\partial t} = 0$ at $t = 0$. We have

$$Y(x,t) = \left(B_1 \sin \frac{k}{7} x \right)(A_2 \cos kt + B_2 \sin kt), \tag{13.119}$$

$$Y(x,t) = \left(\sin \frac{k}{7} x \right)(A \cos kt + B \sin kt), \tag{13.120}$$

$$\frac{\partial y}{\partial t} = \left(\sin \frac{k}{3} x \right)(-kA \sin kt + kB \cos kt), \tag{13.121}$$

$$0 = \left(\sin \frac{k}{7} x \right)(-kA \sin 0 + kB \cos 0), \tag{13.122}$$

$$0 = (\sin kx)(kB), \tag{13.123}$$

$$0 = B, \tag{13.124}$$

$$Y(x,t) = A\left(\sin \frac{k}{7} x \right)(\cos kt). \tag{13.125}$$

At $t = 0$ $y = 12 \sin \frac{4\pi x}{1}$. We have

$$y(x,t) = A\left(\sin \frac{k}{7} x \right)(\cos k0), \tag{13.126}$$

$$y = 12 \sin \frac{4\pi x}{1}, \tag{13.127}$$

$$A = 12, \ k = 28\pi, \tag{13.128}$$

$$y(x,t) = 12\left(\sin \frac{4\pi}{1} x \right)(\cos 28\pi t). \tag{13.129}$$

b. We have

$$y(x,t) = 12 \sin\left(\frac{4\pi}{1} 0.2 \right)(\cos 28\pi 3) = 10.17. \tag{13.130}$$

c. We have

$$\frac{\partial^2 y}{\partial t^2} = 9 \frac{\partial^2 y}{\partial x^2}, \tag{13.131}$$

$$c^2 = \frac{T}{\rho}.$$

If T is increased by a third, then the equation changes to

$$\frac{\partial^2 y}{\partial t^2} = 12\frac{\partial^2 y}{\partial x^2}, \tag{13.132}$$

and the general solution looks like

$$Y(x,t) = A\left(\sin\frac{k}{\sqrt{12}}x\right)(\cos kt), \tag{13.133}$$

$$\frac{k}{\sqrt{12}} = 4\pi, \tag{13.134}$$

$$\frac{k}{1} = 4\sqrt{12}\pi, \tag{13.135}$$

$$y(x,t) = 2\left(\sin\frac{4\pi}{1}x\right)(\cos 4\sqrt{12}\pi t), \tag{13.136}$$

making the oscillations in time further apart and thus changing the frequency of the guitar string.

Problem 3. The following equation models the flow of current for a submarine cable:

$$\frac{\partial^2 i}{\partial x^2} = RC\frac{\partial i}{\partial t}.$$

Solve the equation for a cable of length 0.90 m with values of $R = 8\ \Omega$ and $C = 1\ \mu\text{F}$ subject to the boundary conditions at $i = 0$ at $x = 0$ and $x = L$, and the initial condition

$$i(x,t) = 7\sin\frac{\pi x}{0.30} + 11\sin\frac{\pi x}{0.45}$$

at $t = 0$.

Solution: We have $i(x,t) = X(x)T(t)$,

$$\frac{\partial i}{\partial t} = X(x)\frac{dT(t)}{dt},$$

$$\frac{\partial i}{\partial x} = T(t)\frac{dX(x)}{dx},$$

$$\frac{\partial^2 i}{\partial x} = T(t)\frac{d^2X(x)}{dx^2},$$

$$\frac{1}{X(x)}\frac{d^2X(x)}{dx^2} = RC\frac{1}{T(t)}\frac{dT(t)}{dt}.$$

Both sides equal a constant $-k^2$, i.e.,

$$\frac{1}{X(x)}\frac{d^2X(x)}{dx^2} - k^2,$$

$$X(x) = A\cos kx + B\sin kx,$$

$$RC\frac{1}{T(t)}\frac{dT(t)}{dt} = -k^2,$$

$$T = Ae^{-\frac{k^2}{RC}t},$$

$$i(x,t) = (A\cos kx + B\sin kx)e^{-\frac{k^2}{RC}t},$$

$$i = 0 \text{ at } x = 0 \Rightarrow A = 0,$$

$$i(x,t) = B(\sin kx)e^{-\frac{k^2}{RC}t},$$

$$i = 0 \text{ at } x = L \Rightarrow k = \frac{n\pi}{L} = \frac{n\pi}{0.9},$$

$$i(x,t) = \Sigma B_n\left(\sin\frac{n\pi}{L}x\right)e^{-\frac{(\frac{n\pi}{L})^2}{RC}t}.$$

We expand this expression nonzero for $n = 2$ and $n = 3$, i.e.,

$$RC = 8\times 10^{-6},$$

$$i(x,t) = 11\sin\frac{2\pi x}{0.9}e^{-\frac{(\frac{2\pi}{0.9})^2}{RC}t} + 7\sin\frac{3\pi x}{0.9}e^{-\frac{(\frac{3\pi}{0.9})^2}{RC}t}.$$

Problem 4. Find the solution to the high-frequency line equation

$$\frac{\partial^2 i}{\partial x^2} = LC\frac{\partial^2 i}{\partial t^2}.$$

We want to find i in terms of x and t if the initial condition is $i = 5\sin 4\pi x + 9\sin 7\pi x$.

Solution. We have

$$I(x,t) = X(x)T(t). \tag{13.137}$$

We split $Y(x,t)$ up into two functions, $X(x)$, which is only a function of x, and $T(t)$, which is only a function of t. We have

$$\frac{\partial I}{\partial x} = T(t)\frac{dX(x)}{dx}, \tag{13.138}$$

$$\frac{\partial^2 I}{\partial x^2} = T(t)\frac{d^2X(x)}{dx^2}, \tag{13.139}$$

$$\frac{\partial I}{\partial t} = X(x)\frac{dT(t)}{dt}, \tag{13.140}$$

$$\frac{\partial^2 I}{\partial t^2} = X(x)\frac{d^2T(t)}{dt^2}. \tag{13.141}$$

We now substitute these expressions into the original equation,

$$\frac{\partial^2 I}{\partial t^2} = \frac{\partial^2 y}{\partial x^2}, \tag{13.142}$$

$$X(x)\frac{d^2T(t)}{dt^2} = T(t)\frac{d^2X(x)}{dx^2}. \tag{13.143}$$

We divide across by $X(x)T(t)$, i.e.,

$$\frac{1}{T(t)}\frac{d^2T(t)}{dt^2} = \frac{1}{X(x)}\frac{d^2X(x)}{dx^2}. \tag{13.144}$$

The left-hand side contains only t terms and the right-hand side contains only x terms, so $\lambda = +k^2$ or $\lambda = -k^2$. We must now decide if we should let $\lambda = -k^2$ or $\lambda = +k^2$. This decision has implications for the nature of our solution.

$\lambda = -k^2$. We have

$$\frac{1}{X(x)}\frac{d^2X(x)}{dx^2} = -k^2, \tag{13.145}$$

$$\frac{d^2X(x)}{dx^2} + k^2X(x) = 0. \tag{13.146}$$

We create the auxiliary equation

$$m^2 + k^2 = 0, \tag{13.147}$$

$$m^2 = -k^2, \tag{13.148}$$

$$m = \sqrt{-k^2} = \pm jk. \tag{13.149}$$

This is the special case. $\lambda = -k^2$ So we now have a complex root

$$m = \sqrt{-k^2} = \pm jk, \tag{13.150}$$

$$X = e^{\alpha x}(A\cos\beta x + B\sin\beta x), \tag{13.151}$$

where

$$m = \alpha \pm j\beta. \tag{13.152}$$

In this case

$$m = \sqrt{-k^2} = \pm jk, \tag{13.153}$$

so $\alpha = 0$ and $\beta = k$,

$$X = e^{0x}(A\cos kx + B\sin kx), \tag{13.154}$$

$$X = (A\cos kx + B\sin kx). \tag{13.155}$$

$\lambda = +k^2$. We have

$$\frac{1}{X(x)}\frac{d^2X(x)}{dx^2} = +k^2, \tag{13.156}$$

$$\frac{d^2X(x)}{dx^2} - k^2X(x) = 0. \tag{13.157}$$

We create the auxiliary equation

$$m^2 - k^2 = 0, \tag{13.158}$$

$$m^2 = +k^2, \tag{13.159}$$

$$m = \pm k. \tag{13.160}$$

$\lambda = +k^2$. Roots are real and different,

$$X(x) = Ae^{+kx} + Be^{-kx} \tag{13.161}$$

or

$$X(x) = A\cosh kx + B\sinh kx. \tag{13.162}$$

We will use the second form from here on.

We now have two possible scenarios

$\lambda = +k^2$. Roots are real and different,

$$x = Ae^{+kx} + Be^{-kx}, \tag{13.163}$$

$$X(x) = A\cosh kx + B\sinh kx. \quad (13.164)$$

$\lambda = -k^2$. Roots are complex,

$$X(x) = (A\cos kx + B\sin kx). \quad (13.165)$$

To solve this we compare with the boundary conditions $i = 5\sin 4\pi x + 9\sin 7\pi x$.

We choose the $\lambda = -k^2$ option so as to confirm the boundary conditions.

We solve

$$LC\frac{\partial^2 I}{\partial t^2} = -k^2, \quad (13.166)$$

$$T = \left(A\cos\frac{k}{\sqrt{LC}}t + B\sin\frac{k}{\sqrt{LC}}t\right). \quad (13.167)$$

Our solution for I now becomes

$$I(x,t) = X(x)T(t), \quad (13.168)$$

$$I(x,t) = (A_1\cos kx + B_1\sin kx)\left(A_2\cos\frac{k}{\sqrt{LC}}t + B_2\sin\frac{k}{\sqrt{LC}}t\right). \quad (13.169)$$

We must now use the boundary conditions to solve this.

List the boundary conditions.

1. $I = 0$ at $x = 0$ for all t.
2. $i = 0$ at $t = 0$.
3. $\frac{\partial i}{\partial t} = 0$ at $t = 0$.
4. $i = 5\sin 4\pi x + 9\sin 7\pi x$.

$i = 0$ at $x = 0$ for all t. We have

$$i(x,t) = (A_1\cos kx + B_1\sin kx)\left(A_2\cos\frac{k}{\sqrt{LC}}t + B_2\sin\frac{k}{\sqrt{LC}}t\right), \quad (13.170)$$

$$0 = (A_1\cos 0 + B_1\sin 0)\left(A_2\cos\frac{k}{\sqrt{LC}}t + B_2\sin\frac{k}{\sqrt{LC}}t\right), \quad (13.171)$$

$$0 = (A_1)\left(A_2\cos\frac{k}{\sqrt{LC}}t + B_2\sin\frac{k}{\sqrt{LC}}t\right), \quad (13.172)$$

for this to be true for all t $A_1 = 0$.

We have

$$I(x,t) = (B_1\sin kx)\left(A_2\cos\frac{k}{\sqrt{LC}}t + B_2\sin\frac{k}{\sqrt{LC}}t\right), \quad (13.173)$$

$$i(x,t) = (\sin kx)\left(A\cos\frac{k}{\sqrt{LC}}t + B\sin\frac{k}{\sqrt{LC}}t\right). \quad (13.174)$$

$i = 0$ at $x = L$. We have

$$i(x,t) = (\sin kL)\left(A\cos\frac{k}{\sqrt{LC}}t + B\sin\frac{k}{\sqrt{LC}}t\right), \tag{13.175}$$

$$k = \frac{n\pi}{L}, \tag{13.176}$$

$$\frac{\partial i}{\partial t} = 0 \text{ at } t = 0, \tag{13.177}$$

$$i(x,t) = \Sigma\left(\sin\frac{n\pi x}{L}\right)\left(-A\frac{k}{\sqrt{LC}}\sin\frac{k}{\sqrt{LC}}t + B\frac{k}{\sqrt{LC}}\cos\frac{k}{\sqrt{LC}}t\right), \tag{13.178}$$

$$B = o, \tag{13.179}$$

$$i(x,t) = \Sigma A_n\left(\sin\frac{n\pi x}{L}\right)\left(\cos\frac{\frac{n\pi}{L}}{\sqrt{LC}}t\right). \tag{13.180}$$

At $t = 0$ $i = 5\sin 4\pi x + 9\sin 7\pi x$,

$$n = 160 \text{ and } n = 280. \tag{13.181}$$

On comparing only nonzero terms, $n = 160$ and 280 and $A_{160} = 5$ and $A_{280} = 9$,

$$i = 5\sin 4\pi x\cos\frac{4\pi}{\sqrt{LC}}t + 9\sin 7\pi x\cos\frac{7\pi}{\sqrt{LC}}t. \tag{13.182}$$

Finally,

$$L = 40\times 1\times 10^{-6} = 0.04\times 10^{-3}\,HC = 40\times 100\times 10^{-12}\,F, \tag{13.183}$$

$$LC = 1.6\times 10^{-13}. \tag{13.184}$$

Exercises.

1. Name the following equations and write a brief note explaining their application:

   a.

$$\frac{\partial^2 u}{\partial x^2} + \frac{\partial^2 u}{\partial y^2} = 0.$$

   b.

$$\frac{\partial^2 u}{\partial x^2} = \frac{1}{k}\frac{\partial u}{\partial t}.$$

   c.

$$\frac{\partial^2 u}{\partial x^2} = \frac{1}{c^2}\frac{\partial^2 u}{\partial t^2}.$$

d.

$$\frac{\partial^2 i}{\partial x^2} = RC\frac{\partial i}{\partial t}.$$

e.

$$\frac{\partial^2 i}{\partial x^2} = LC\frac{\partial^2 i}{\partial t^2}.$$

2. Using a trial solution $u(x,t) = X(x)T(t)$ and separation of variables, find the solution of the PDE

$$\frac{\partial u}{\partial x} + u = \frac{\partial u}{\partial t},$$

where $u(x,0) = 10e^{-7x}$.

3. Using a trial solution $u(x,t) = X(x)T(t)$ and separation of variables, find the solution of the PDE

$$\frac{\partial u}{\partial x} - 8u = \frac{\partial u}{\partial t},$$

where $u(0,t) = 9e^{-2t}$.

4. Using a trial solution $u(x,t) = X(x)T(t)$ and separation of variables, find the solution of the PDE

$$3\frac{\partial u}{\partial x} - 7u = 5\frac{\partial u}{\partial t},$$

where $u(0,t) = 2e^{-7t}$.

5. Using a trial solution $u(x,t) = X(x)T(t)$ and separation of variables, find the solution of the PDE

$$2\frac{\partial u}{\partial x} - 9u = 4\frac{\partial u}{\partial t},$$

where $u(0,t) = 23e^{-1t}$.

6. The equation

$$\frac{\partial^2 y}{\partial t^2} = 81\frac{\partial^2 y}{\partial x^2}$$

is the equation of a tightly stretched guitar string. We want to find y in terms of x and t if the motion is started by displacing a cable into the form $y = 3\sin\frac{\pi}{2}x$ and then releasing it from rest.

If the string is replaced by a cable with twice the density, describe how this would effect your solution.

Find the displacement of the cable at x=0.01 after 3 seconds.

7. Consider the equation

$$\frac{\partial^2 i}{\partial x^2} = RC\frac{\partial i}{\partial t}.$$

Solve the equation for a cable of length 100 m with values of $R = 2$ kΩ and $C = 4$ μF, subject to the boundary conditions at $i = 0$ at $x = 0$ and $x = L$, and the initial condition

$$i(x,t) = 10\sin\frac{\pi x}{50} + 5\sin\frac{\pi x}{25}$$

at $t = 0$.

Solution 1: Exercise 1.

1. Name the following equations and write a brief note explaining their application.

a.

$$\frac{\partial^2 u}{\partial x^2} + \frac{\partial^2 u}{\partial y^2} = 0$$

Solution. The Laplace equation. This is the time-independent 2D heat equation.

b.

$$\frac{\partial^2 u}{\partial x^2} = \frac{1}{k}\frac{\partial u}{\partial t}$$

Solution. This is the 1D heat equation.

c.

$$\frac{\partial^2 u}{\partial x^2} = \frac{1}{c^2}\frac{\partial^2 u}{\partial t^2}$$

Solution. This is the wave equation.

d.

$$\frac{\partial^2 i}{\partial x^2} = RC\frac{\partial i}{\partial t}$$

Solution. This is the submarine or telegraph equation.

e.

$$\frac{\partial^2 i}{\partial x^2} = LC\frac{\partial^2 i}{\partial t^2} \tag{13.185}$$

Solution. This is the high-frequency line equation.[1]

[1] https://ejpam.com/index.php/ejpam/article/viewFile/418/91.

Laplace transforms

Engineering applications of Laplace transforms

14

Michael Carr[a], *Mark Mc Grath*[b], *Eabhnat Ní Fhloinn*[c]
[a]Technological University Dublin, Dublin, Ireland, [b]School of Mechanical and Design Engineering, TU Dublin, Dublin, Ireland, [c]Dublin City University, Dublin, Ireland

14.1 Introduction to Laplace transforms

Let $f(t)$ be a given function that is defined for all $t \geq 0$, which is integrable with respect to t from 0 to ∞. Let

$$F(s) = \int_0^\infty f(t)e^{-st}dt. \tag{14.1}$$

$F(s)$ is then called the *Laplace transform* of $f(t)$, alternatively denoted $L\{f(t)\}$, i.e.,

$$\boxed{L\{f(t)\} = \int_0^\infty f(t)e^{-st}dt = F(s)}. \tag{14.2}$$

The *inverse Laplace transform* is then denoted L^{-1}, with

$$f(t) = L^{-1}\{F(s)\}. \tag{14.3}$$

Example 14.1. Find the Laplace transform of $f(t) = a$, where a is a constant.
We have

$$\begin{aligned} L\{a\} &= \int_0^\infty ae^{-st}dt \\ &= -\frac{a}{s}\left[e^{-st}\right]_0^\infty \\ &= -\frac{a}{s}(0-1) \\ &= \frac{a}{s} \qquad (s > 0). \end{aligned} \tag{14.4}$$

Calculus for Engineering Students. https://doi.org/10.1016/B978-0-12-817210-0.00021-7

If we then take the specific case of $f(t) = 4$, we have

$$L\{4\} = \int_0^\infty 4e^{-st}\,dt$$
$$= \frac{4}{s} \qquad (s > 0). \tag{14.5}$$

Example 14.2. Find the Laplace transform of $f(t) = e^{at}$, where a is a constant. We have

$$L\{e^{at}\} = \int_0^\infty e^{at}e^{-st}\,dt$$
$$= \int_0^\infty e^{-(s-a)t}\,dt$$
$$= \left[\frac{e^{-(s-a)t}}{-(s-a)}\right]_0^\infty$$
$$= -\frac{1}{s-a}(0-1)$$
$$= \frac{1}{s-a} \qquad (s > a). \tag{14.6}$$

Then, for the specific case where $a = 2$, we have

$$L\{e^{2t}\} = \frac{1}{s-2} \qquad (s > 2), \tag{14.7}$$

and if instead $a = -3$, we have

$$L\{e^{-3t}\} = \frac{1}{s+3} \qquad (s > -3). \tag{14.8}$$

14.1.1 Standard transforms

We now have established two standard transforms:

$$\boxed{L\{a\} = \frac{a}{s} \quad (s > 0)} \tag{14.9}$$

$$\boxed{L\{e^{at}\} = \frac{1}{s-a} \quad (s > a)} \tag{14.10}$$

Exercises. Evaluate each of the following, using either of the two standard transforms shown in (14.9) or (14.10) as appropriate:

1. $L\{11\}$
2. $L\{-7\}$
3. $L\{-123\}$
4. $L\{e^{4t}\}$
5. $L\{e^{-9t}\}$
6. $L\{e^{-87t}\}$
7. $L\{e^{3t}\}$
8. $L\{e^{123t}\}$
9. $L\{e^{-123t}\}$

We now introduce two further transforms:

$$\boxed{L\{t\} = \frac{1}{s^2}} \tag{14.11}$$

$$\boxed{L\{t^n\} = \frac{n!}{s^{n+1}}} \tag{14.12}$$

Example 14.3. Here are some examples using these new transforms we have just met in (14.11) and (14.12):

$$L\{t^2\} = \frac{2!}{s^{2+1}} = \frac{2}{s^3} \tag{14.13}$$

$$L\{t^3\} = \frac{3!}{s^{3+1}} = \frac{6}{s^4} \tag{14.14}$$

$$L\{t^6\} = \frac{6!}{s^{6+1}} = \frac{720}{s^7} \tag{14.15}$$

Exercises. Evaluate each of the following, using the transforms shown in (14.11) or (14.12):

1. $L\{t^4\}$
2. $L\{t^8\}$
3. $L\{t^5\}$

Another useful transform is given by

$$\boxed{L\{t^n e^{at}\} = \frac{n!}{(s-a)^{n+1}}} \tag{14.16}$$

Example 14.4. Here is how to use the transform shown in (14.16) for both a positive and a negative value of a:

$$L\{t^5 e^{2t}\} = \frac{5!}{(s-2)^6} \tag{14.17}$$

$$L\{t^2 e^{-4t}\} = \frac{2!}{(s+4)^3} \tag{14.18}$$

Exercises. Evaluate each of the following, using the transform shown in (14.16):

1. $L\{t^4e^{3t}\}$ **2.** $L\{t^8e^{-7t}\}$ **3.** $L\{t^5e^{-t}\}$

We now introduce transforms involving sin and cos:

$$\boxed{L\{\sin at\} = \frac{a}{s^2+a^2}} \tag{14.19}$$

$$\boxed{L\{\cos at\} = \frac{s}{s^2+a^2}} \tag{14.20}$$

Example 14.5. Now consider how to use these transforms to calculate:

$$L\{\sin 3t\} = \frac{3}{s^2+3^2} = \frac{3}{s^2+9} \tag{14.21}$$

$$L\{\cos 3t\} = \frac{s}{s^2+3^2} = \frac{s}{s^2+9} \tag{14.22}$$

Exercises. Evaluate the following, using the transforms shown in (14.19) or (14.20):

1. $L\{\sin -12t\}$ **2.** $L\{\cos 5t\}$

Next, we learn the transforms for sinh and cosh:

$$\boxed{L\{\sinh at\} = \frac{a}{s^2-a^2}} \tag{14.23}$$

$$\boxed{L\{\cosh at\} = \frac{s}{s^2-a^2}} \tag{14.24}$$

Example 14.6. Now consider how to use these transforms to calculate:

$$L\{\sinh 3t\} = \frac{3}{s^2-3^2} = \frac{3}{s^2-9} \tag{14.25}$$

$$L\{\cosh 3t\} = \frac{s}{s^2-3^2} = \frac{s}{s^2-9} \tag{14.26}$$

Exercises. Evaluate the following, using the transforms shown in (14.23) or (14.24):

1. $L\{\sinh 4t\}$ **2.** $L\{\cosh -9t\}$

Linearity of the Laplace transform

$$\boxed{L\{af(t)+bg(t)+ch(t)\}=aL\{f(t)\}+bL\{g(t)\}+cL\{h(t)\}} \tag{14.27}$$

Example 14.7. We have

$$\begin{aligned} L\{14t+27\sin t+5\cos t\} &= 14L\{t\}+27L\{\sin t\}+5L\{\cos t\} \\ &= 14\left(\frac{1}{s^2}\right)+27\left(\frac{1}{s^2+1}\right)+5\left(\frac{s}{s^2+1}\right) \\ &= \frac{14}{s^2}+\frac{27}{s^2+1}+\frac{5s}{s^2+1} \end{aligned} \tag{14.28}$$

Exercises. Evaluate each of the following, using (14.27):

1. $L\{17+4t^2+19\sin 4t\}$
2. $L\{24t^3+17t+\cos 15t\}$
3. $L\{-7\cos 3t\}$

14.1.2 Inverse Laplace transforms

Inverse Laplace transforms involve the reverse process, whereby we are given a Laplace transform and have to find the function of t to which it belongs. For example, we can use Eq. (14.19) to tell us that

$$L^{-1}\left\{\frac{a}{s^2+a^2}\right\}=\sin at. \tag{14.29}$$

As mentioned earlier, the symbol L^{-1} means inverse transforms.

Example 14.8. Each of the following examples are evaluated by reversing the standard transforms shown in Eqs. (14.9)–(14.24):

$$L^{-1}\left\{\frac{7}{s^2+49}\right\}=\sin 7t \tag{14.30}$$

$$L^{-1}\left\{\frac{s}{s^2+81}\right\}=\cos 9t \tag{14.31}$$

$$L^{-1}\left\{\frac{1}{s-17}\right\}=e^{17t} \tag{14.32}$$

$$L^{-1}\left\{\frac{s}{s^2-49}\right\}=\cosh 7t \tag{14.33}$$

$$L^{-1}\left\{\frac{9}{s^2-81}\right\} = \sinh 9t \tag{14.34}$$

$$L^{-1}\left\{\frac{13}{s}\right\} = 13 \tag{14.35}$$

Exercises. Solve

1. $$L^{-1}\left\{\frac{s}{s^2-16}\right\}$$

2. $$L^{-1}\left\{\frac{6}{s^2-36}\right\}$$

3. $$L^{-1}\left\{\frac{7}{s}\right\}$$

4. $$L^{-1}\left\{\frac{4}{s+4}\right\}$$

5. $$L^{-1}\left\{\frac{7}{s^2+49}\right\}$$

6. $$L^{-1}\left\{\frac{27}{s^2+81}\right\}$$

List of inverse transforms

1. $\dfrac{a}{s} \to a$

2. $\dfrac{1}{s+a} \to e^{-at}$

3. $\dfrac{n!}{s^{n+1}} \to t^n$ (where n is a positive integer)

4. $\dfrac{1}{s^n} \to \dfrac{t^{n-1}}{(n-1)!}$ (where n is a positive integer)

5. $\dfrac{a}{s^2+a^2} \to \sin at$

6. $\dfrac{s}{s^2+a^2} \to \cos at$

7. $\dfrac{a}{s^2-a^2} \to \sinh at$

8. $\dfrac{s}{s^2-a^2} \to \cosh at$

14.1.3 Partial fractions

We have dealt with many of the standard expressions in the previous subsection. However, we still need to consider what happens when we get an expression like

$$L^{-1}\left\{\frac{2s+11}{s^2+11s+28}\right\} \tag{14.36}$$

We do not have any standard transform for this expression. But it can be shown that

$$\frac{2s+11}{s^2+11s+28}=\frac{1}{s+7}+\frac{1}{s+4} \tag{14.37}$$

We know how to deal with the right-hand side of Eq. (14.37), so now we can evaluate

$$\begin{aligned} L^{-1}\left\{\frac{2s+11}{s^2+11s+28}\right\} &= L^{-1}\left\{\frac{1}{s+7}+\frac{1}{s+4}\right\} \\ &= e^{7t}+e^{4t}, \end{aligned} \tag{14.38}$$

where $\frac{1}{s+7}$ and $\frac{1}{s+4}$ are called the partial fractions of the expression $\frac{2s+11}{s^2+11s+28}$. We will now consider how we go about finding the partial fractions of an expression.

Rules for finding partial fractions

The first step when trying to decompose an expression into partial fractions is to factorize the denominator into its prime factors. For example, $\frac{1}{s^2+9s+14}$ factorizes to $\frac{1}{(s+7)(s+2)}$. At this point, we need to consider what form the factors take.

1. A linear factor, e.g., $\frac{1}{(s+a)}$, gives a partial fraction of the form $\frac{A}{s+a}$, where A is a constant to be determined.

   Example 14.9.

   $$\frac{1}{(s+7)(s+2)}=\frac{A}{s+7}+\frac{B}{s+2}$$

2. A repeated factor of the form $(s+a)^2$ gives partial fractions $\frac{A}{s+a}+\frac{B}{(s+a)^2}$.

   Example 14.10.

   $$\frac{1}{s^2+8s+16}=\frac{1}{(s+4)(s+4)}=\frac{C}{s+4}+\frac{D}{(s+4)^2}.$$

3. If the repeated factor takes the form $\frac{1}{(s+a)^3}$, this gives partial fractions of $\frac{E}{s+a}+\frac{F}{(s+a)^2}+\frac{G}{(s+a)^3}$.

Example 14.11.

$$\begin{aligned}\frac{1}{(s+4)(s^2+8s+16)} &= \frac{1}{(s+4)(s+4)(s+4)}\\ &= \frac{E}{s+4}+\frac{F}{(s+4)^2}+\frac{G}{(s+4)^3}.\end{aligned}$$

4. A quadratic factor $\frac{1}{(s^2+as+b)}$ gives a partial fraction $\frac{Es+F}{s^2+as+b}$.

Example 14.12. We have

$$\frac{1}{(s^2+6s+7)} = \frac{Ps+Q}{s^2+6s+7}.$$

Example 14.13. Using rules 1 - 4 above, we can rewrite the following expressions in partial fraction format:

$$\frac{s-41}{(s-9)(s-1)} = \frac{A}{s-9}+\frac{B}{s-1} \tag{14.39}$$

$$\frac{s-41}{(s-9)(s-4)^2} = \frac{A}{s-9}+\frac{B}{s-4}+\frac{C}{(s-4)^2} \tag{14.40}$$

Exercises. Write the following expressions as partial fractions:

1.

$$\frac{s+4}{(s-9)(s-11)(s+5)}$$

2.

$$\frac{s+31}{(s-4)(s-9)(s+5)^2}$$

Once we have calculated the partial fractions, we then need to find the values of the constants. For example, suppose we have

$$\frac{s-29}{(s+3)(s-5)} = \frac{A}{s+3}+\frac{B}{s-5}. \tag{14.41}$$

To calculate the values of A and B, we begin by multiplying both sides by the denominator, to give us

$$(s-29) = A(s-5)+B(s+3) = (A+B)s+(3B-5A).$$

We now equate the coefficients of s and those of the constant to get

$$1 = A + B \Rightarrow 1 - A = B, \qquad \begin{aligned} -29 &= 3B - 5A \\ \Rightarrow 29 &= 5A - 3B \\ \Rightarrow 29 &= 5A - 3(1 - A) \\ \Rightarrow 29 &= 5A - 3 + 3A \\ \Rightarrow 32 &= 8A \\ \Rightarrow 4 &= A. \end{aligned}$$

This allows us to calculate a value for B, using $B = 1 - A$, to give us

$$A = 4, \qquad B = -3.$$

This gives

$$\frac{s - 29}{(s + 2)(s - 5)} = \frac{4}{s + 3} - \frac{3}{s - 5}.$$

To determine inverse Laplace transforms

Example 14.14. There are a number of steps to determine

$$L^{-1}\left\{\frac{3s - 5}{s^2 - 8s - 33}\right\}. \tag{14.42}$$

1. Factorize the denominator $\dfrac{3s - 5}{s^2 - 8s - 33} = \dfrac{3s - 5}{(s - 11)(s + 3)}$.
2. The partial fractions are of the form $\dfrac{A}{s + 3} + \dfrac{B}{s - 11}$, giving us the identity $\dfrac{3s - 5}{s^2 - 8s - 33} = \dfrac{A}{s + 3} + \dfrac{B}{s - 11}$.
3. Multiplying through both sides by the denominator gives us

$$\begin{aligned} &3s - 5 = A(s - 11) + B(s + 3) \\ &3s = As + Bs \text{ and } -5 = -11A + 3B \\ \Rightarrow &3 = A + B \\ \Rightarrow &A = 3 - B \text{ so } -5 = -11(3 - B) + 3B \\ &\text{which gives us } -5 = -33 + 11B + 3B = -33 + 14B \\ \Rightarrow &B = 2 \text{ and } A = 1. \end{aligned} \tag{14.43}$$

4. We have

$$L^{-1}\left\{\frac{3s - 5}{s^2 - 8s - 33}\right\} = \frac{1}{s + 3} + \frac{2}{s - 11}. \tag{14.44}$$

5. Finally, we calculate the inverse transform to get:

$$L^{-1}\left\{\frac{3s-5}{s^2-8s-33}\right\}=e^{-t}+2e^{11t} \tag{14.45}$$

14.1.4 The "cover up" rule

This method only works when the denominator has nonrepeated linear factors.

Example 14.15. The function

$$F(s)=\frac{5s+25}{(s+7)(s+2)} \tag{14.46}$$

has partial fractions of the form

$$\frac{A}{s+7}+\frac{B}{s+2}. \tag{14.47}$$

By the *cover up rule*, the constant A, the coefficient of $\frac{1}{s+7}$ is found by temporarily covering up the factor $s+7$ in the denominator of the function and finding the limiting value of what remains when $s+7$ (the factor covered up) tends to zero. So

$$A=\lim_{s\to -7}\frac{5s+25}{s+2}=\frac{-35+25}{-7+2}, \tag{14.48}$$

that is, $A=2$.
B is obtained by covering up the factor $(s-2)$ in the denominator of $F(s)$. This time $(s+2)\to 0$ so $s\to -2$, giving $B=3$.

14.1.5 The first shift theorem

If $F(s)$ is the Laplace transform of $f(t)$, then $F(s+a)$ is the Laplace transform of $e^{-at}f(t)$.

Example 14.16. Find the inverse transform of the following:

1. $L^{-1}\left\{\frac{1}{2s-3}\right\}=\frac{1}{2}e^{\frac{3t}{2}}$.
2. $L^{-1}\left\{\frac{5}{(s-4)^3}\right\}=5t^2e^{4t}$.
3. $L^{-1}\left\{\frac{3s+4}{s^2+9}\right\}=3\cos 3t+\frac{4}{3}\sin 3t$.

14.2 Solving first- and second-order differential equations

Laplace transforms provide a very useful method for solving differential equations. We can transform the differential equations (in the time (t) domain) into algebraic equations (in the s domain) via Laplace transforms. These algebraic equations are much easier to manipulate. We are then able to transform these algebraic equations in the s domain back into to the time domain via inverse Laplace transforms to get our final solutions.

14.2.1 Transforms of derivatives

$$L\{f'(t)\} = -f(0) + sL\{f(t)\} \tag{14.49}$$

$$L\{f''(t)\} = s^2F(s) - sf(0) - f'(0) \tag{14.50}$$

$$L\{f'''(t)\} = s^3F(s) - s^2f(0) - sf'(0) - f''(0). \tag{14.51}$$

14.2.2 Alternative notation

The present notation is quite cumbersome, so we shall introduce alternative notation at this point. Let $x = f(t)$ at $t = 0$. Then:

$$x = x_0 \text{ means } f(0) = x_0,$$

$$\frac{dx}{dt} = x_1 \text{ means } f'(0) = x_1,$$

$$\frac{d^2x}{dt^2} = x_2 \text{ means } f''(0) = x_2,$$

$$\frac{d^nx}{dt^n} = x_n \text{ means } f^n(0) = x_n.$$

We shall denote the Laplace transform of x by $\bar{x}$, i.e. $\bar{x} = L\{x\} = L\{f(t)\} = F(s)$. Now we can rewrite the list of transforms:

1. $L\{x\} = \bar{x}$
2. $L\{\dot{x}\} = s\bar{x} - x_0$
3. $L\{\ddot{x}\} = s^2\bar{x} - sx_0 - x_1$
4. $L\{\dddot{x}\} = s^3\bar{x} - s^2x_0 - sx_1 - x_2,$

where $\dot{x} = \dfrac{dx}{dt}$, etc.

14.2.3 Solving first-order differential equations

Example 14.17. Solve the equation

$$\frac{dx}{dt} - 4x = 7, \tag{14.52}$$

given that at $t = 0, x = 1$. We go through four stages.

1. Rewrite the equation in Laplace transforms. We have
   a. $L\{x\} = \bar{x}$
   b. $L\{\dot{x}\} = s\bar{x} - x_0$
   c. $L\{7\} = \frac{7}{s}$

   This gives us

$$(s\bar{x} - x_0) - 4\bar{x} = \frac{7}{s}. \tag{14.53}$$

2. Insert the initial conditions, that at $t = 0, x = 1$ $(x_0 = 1)$,

$$(s\bar{x} - 1) - 4\bar{x} = \frac{7}{s}. \tag{14.54}$$

3. Now we rearrange this to give an expression for $\bar{x}$,

$$\bar{x} = \frac{s+7}{s(s-4)}. \tag{14.55}$$

4. Now we get the inverse transform of this, but first we must take partial fractions,

$$\bar{x} = \frac{s+7}{s(s-4)} = \frac{A}{s} + \frac{B}{s-4}. \tag{14.56}$$

Using the cover up rule,

$$A = \lim_{s\to 0} \frac{s+7}{(s-4)} = -1.75 \text{ and } B = \lim_{s\to 4} \frac{s+7}{s} = 2.75.$$

$$\bar{x} = \frac{s+7}{s(s-4)} = \frac{-1.75}{s} + \frac{2.75}{s-4}$$

$$x = 2.75e^{4t} - 1.75 \tag{14.57}$$

Exercise. Solve the equation

$$\frac{dx}{dt} + 3x = 12e^{-2t}, \text{ given that at } t = 0, x = 6. \tag{14.58}$$

(**Solution:** $x = 12e^{-2t} - 6e^{-3t}$)

14.2.4 *Solving second-order differential equations using Laplace transforms*

Example 14.18. Solve the equation

$$\frac{d^2x}{dt^2} - 4\frac{dx}{dt} + 3x = e^{4t}, \text{ given that at } t = 0, x = 4 \text{ and } \frac{dx}{dt} = 6. \tag{14.59}$$

1. $L\{x\} = \bar{x}$.
2. $L\{\dot{x}\} = s\bar{x} - x_0$.
3. $L\{\ddot{x}\} = s^2\bar{x} - sx_0 - x_1$.
4. $L\{\dddot{x}\} = s^3\bar{x} - s^2x_0 - sx_1 - x_2$.

1. Get the Laplace transforms of both sides:

$$s^2\bar{x} - sx_0 - x_1 - 4(s\bar{x} - x_0) + 3\bar{x} = \frac{1}{s-4}. \tag{14.60}$$

2. Insert the initial conditions $x_0 = 4$ and $x_1 = 6$, i.e.,

$$\begin{aligned}
s^2\bar{x} - 4s - 6 - 4(s\bar{x} - 4) + 3\bar{x} &= \frac{1}{s-4}, \\
(s^2 - 4s + 3)\bar{x} - 4s + 10 &= \frac{1}{s-4}, \\
(s^2 - 4s + 3)\bar{x} &= \frac{1}{s-4} + 4s - 10, \\
(s-3)(s-1)\bar{x} &= \frac{4s^2 - 26s + 41}{s-4}, \\
\bar{x} &= \frac{4s^2 - 26s + 41}{(s-3)(s-1)(s-4)}.
\end{aligned} \tag{14.61}$$

3. We must solve this using partial fractions,

$$\bar{x} = \frac{A}{s-1} + \frac{B}{s-3} + \frac{C}{s-4}. \tag{14.62}$$

Cover up rule:

$$\begin{aligned}
A &= \lim_{s\to 1} \frac{4s^2 - 26s + 41}{(s-3)(s-4)} = \frac{4(1)^2 - 26(1) + 41}{(-2)(-3)} = \frac{19}{6} = 3.167 \\
B &= \lim_{s\to 3} \frac{4s^2 - 26s + 41}{(s-1)(s-4)} = \frac{4(3)^2 - 26(3) + 41}{(2)(-1)} = \frac{-1}{-2} = 0.5 \\
C &= \lim_{s\to 4} \frac{4s^2 - 26s + 41}{(s-1)(s-3)} = \frac{4(4)^2 - 26(4) + 41}{(3)(1)} = 0.33
\end{aligned}$$

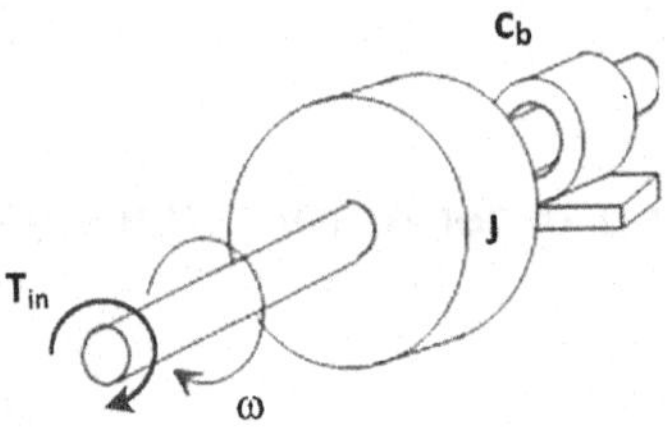

Figure 14.1 Flywheel-bearing arrangement.

$$\bar{x} = \frac{3.167}{s-1} + \frac{0.5}{s-3} + \frac{0.33}{s-4}$$
$$x = 3.167e^{t} + 0.5e^{3t} + 0.33e^{4t}. \quad (14.63)$$

14.3 Engineering applications of Laplace transforms: problems

14.3.1 Flywheel

Fig. 14.1 shows a disk flywheel of mass 6 kg and radius 0.4 m. This flywheel is driven by an electric motor that produces a constant torque $T_{in} = 8\text{Nm}$. The shaft bearing may be modeled as viscous rotary dampers with a damping coefficient of $c_b = 0.1\ \frac{\text{Nm s}}{\text{rad}}$. The differential equation relating the input torque, T_{in}, in Nm to the output angular velocity, ω, in rad/s for this arrangement is:

$$\frac{d\omega}{dt} + \frac{c_b}{J}\omega = \frac{1}{J}T_{in}, \quad (14.64)$$

where J is the moment of inertia of the flywheel in kg·m^2. The flywheel is initially at rest at $t = 0$ and the power is suddenly applied to the motor.

1. What is the transfer function of the system?
2. Starting with the transfer function from part 1, determine an expression for the angular velocity of the flywheel.
3. Determine the final maximum angular velocity of the flywheel.

Solution:

We know $m = 6$ kg, $r = 0.4$ m, $c_b = 0.1\ \frac{\text{Nm s}}{\text{rad}}$, and $T_{in} = 8\text{Nm}$. We also have:

$$\frac{d\omega}{dt} + \frac{c_b}{J}\omega = \frac{1}{J}T_{in}.$$

$$Gain = \frac{Output}{Input} \tag{14.65}$$

If the relationship is in differential form then simple division is not possible. To overcome this difficulty we can transform the relationship from the time domain to the s-domain, then we can define the relationship between the output and the input in terms of a transfer function.

$$Transfer\ Function = \frac{Laplace\ Transform\ of\ Output}{Laplace\ Transform\ of\ Input} \tag{14.66}$$

When the signal is in the time domain, it is written as $f(t)$ and when it is in the s-domain it is written as $F(s)$.

$$Y(s) \rightarrow G(s) \rightarrow X(s)$$

$$Transfer\ Function\ G(s) = \frac{X(s)}{Y(s)} \tag{14.67}$$

This can be rearranged to give

$$X(s) = G(s)Y(s) \tag{14.68}$$

Taking the Laplace transform of both sides, and inserting our value for T_{in}, we get:

$$s\bar{\omega} + \frac{c_b}{J}\bar{\omega} = \frac{8}{Js}$$

$$\bar{\omega}\left(s + \frac{c_b}{J}\right) = \frac{8}{Js}$$

$$\bar{\omega} = \frac{8}{s}\frac{\frac{1}{J}}{s + \frac{c_b}{J}}. \tag{14.69}$$

$$Transfer\ Function = \frac{Laplace\ Transform\ of\ Output}{Laplace\ Transform\ of\ Input} \tag{14.70}$$

$$Transfer\ Function\ G(s) = \frac{X(s)}{Y(s)} \tag{14.71}$$

Input= T_{in} and thus

$$Y(s) = \frac{8}{s} \tag{14.72}$$

$$X(s) = \bar{\omega} \tag{14.73}$$

$$G(s) = \bar{\omega}\frac{s}{8} = \frac{\frac{1}{J}}{s + \frac{c_b}{J}} \tag{14.74}$$

We need to calculate J. We have

$$\begin{aligned} J &= \frac{1}{2}mr^2 \\ &= \frac{1}{2} \times 6 \times \left(\frac{2}{5}\right)^2 = \frac{12}{25} \end{aligned} \tag{14.75}$$

This allows us to calculate

$$\begin{aligned} \bar{\omega} &= \frac{8}{s}\frac{\frac{1}{J}}{s + \frac{c_b}{J}} \\ &= \frac{8\left(\frac{25}{12}\right)}{s\left(s + \left(\frac{1}{10} \times \frac{25}{12}\right)\right)} \\ &= \frac{\frac{400}{24}}{s\left(s + \frac{5}{24}\right)} \\ &= \frac{80\left(\frac{5}{24}\right)}{s\left(s + \frac{5}{24}\right)}. \end{aligned} \tag{14.76}$$

Getting the inverse Laplace transform gives us The angular velocity

$$\omega(t) = 80\left(1 - e^{-\frac{5t}{24}}\right). \tag{14.77}$$

Finally the maximum angular velocity is when the exponential component decays to 0 thus

$$\omega_{max} = 80$$

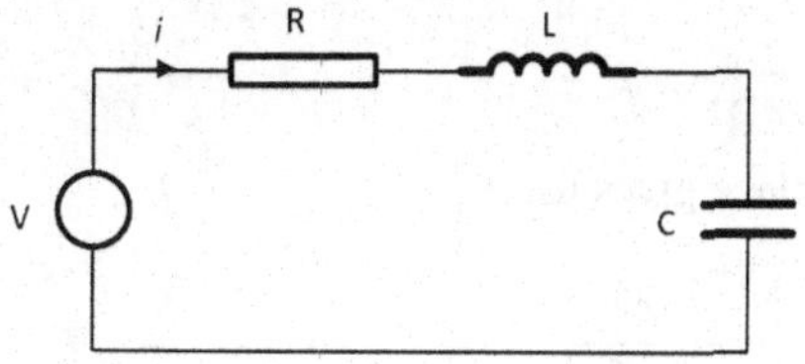

Figure 14.2 RLC series circuit.

14.3.2 RLC circuit

An RLC circuit is an electrical circuit in which there is a resistor (R), an inductor (L), and a capacitor (C). These may be connected in series or in parallel. The RLC circuit in Fig. 14.2 has a current i which varies with time t when subject to a step input of V and is described by

$$\frac{d^2i}{dt^2} + 10\frac{di}{dt} + 25i = 50V \tag{14.78}$$

Determine the solution to the equation if $i = 0$ when $t = 0$ and $\frac{di}{dt} = 0$ when $t = 0$.

Solution:
Get the Laplace transform of both sides:

$$s^2\bar{i} - si_1 - i_0 + 10(s\bar{i} - i_0) + 25\bar{i} = 50\frac{V(s)}{s}. \tag{14.79}$$

Insert the initial conditions to get:

$$s^2\bar{i} + 10(s\bar{i}) + 25\bar{i} = 50\frac{V(s)}{s}$$

$$(s^2 + 10s + 25)\bar{i} = 50\frac{V(s)}{s} = 50\frac{\bar{V}}{s}$$

$$(s+5)(s+5)\bar{i} = 50\frac{\bar{V}}{s}$$

$$\bar{i} = 50\frac{\bar{V}}{s(s+5)(s+5)}. \tag{14.80}$$

In the RLC circuit the input is the voltage V and the output is the current I.

$$\textit{Transfer Function} = \frac{\textit{Laplace Transform of Output}}{\textit{Laplace Transform of Input}} \tag{14.81}$$

$$\textit{Transfer Function } G(s) = \frac{X(s)}{Y(s)} = \tag{14.82}$$

$$\frac{\bar{i}}{\bar{V}} = \frac{50}{s(s+5)(s+5)} \tag{14.83}$$

Getting the inverse Laplace gives us:

$$\frac{\bar{i}}{\bar{V}} = \frac{X}{s} + \frac{Y}{s+5} + \frac{Z}{(s+5)^2}. \tag{14.84}$$

We then use the cover up rule to get

$$\begin{aligned} X &= \lim_{s \to 0} \frac{50}{(s+5)(s+5)} \\ &= \frac{50}{25} = 2. \end{aligned} \tag{14.85}$$

We cannot use the cover up rule to calculate Y and Z, as there are repeated linear factors. Therefore, we let $s = 1$ and then $s = 2$ to give

$$-22 = 6Y + Z \tag{14.86}$$

$$-48 = 14Y + 2Z \tag{14.87}$$

We can solve these simultaneous equations to give $Z = 10$ and $Y = -2$. Thus, our solution is given by

$$i(t) = -(10t + 2)Ve^{-5t} + 2V. \tag{14.88}$$

14.3.3 Problem: RC circuit

An RC circuit consists of a resistor with resistance R (ohms), a capacitor with capacitance C (in Farads), and an electromotive force (emf) with voltage E (volts).

1. Derive the following equation governing the amount of electrical charge q (in coulombs) on the capacitor:

$$\frac{dq}{dt} + \frac{1}{RC}q = \frac{E}{R}. \tag{14.89}$$

2. For the equation above $R = 1\Omega$, $C = 0.001$F, and $E = 20$V. If $q = 0.001$ when $t = 0$, find the particular solution of the equation.

Solution:
Kirchoff's loop law:
The algebraic sum of the voltage-drops in a simple closed electric circuit is zero. In other words, the voltage across the battery is equal to the sum of the voltages across the sum of each of the other components in the circuit.

1. The voltage across the battery is equal to E, the emf of the battery.
2. The voltage-drop across the resistor is $V = RI$ (Ohm's law).

3. The voltage-drop across the capacitor is given by $V = \frac{q}{c}$.
4. Finally, the current I is the rate of change of the current:

$$I = \frac{dq}{dt}. \tag{14.90}$$

5. Putting all this information together, we get

$$\begin{aligned} E &= RI + \frac{q}{C} \\ &= R\frac{dq}{dt} + \frac{q}{C} \end{aligned} \tag{14.91}$$

6. We can then rearrange this to obtain

$$\frac{dq}{dt} + \frac{1}{RC}q = \frac{E}{R} \tag{14.92}$$

for the equation.

$$\frac{dq}{dt} + 1000q = 20. \tag{14.93}$$

If $R = 1, C = 0.001$, and $E = 20$, and if $q = 0.001$ when $t = 0$, using Laplace transforms, we obtain

$$q = 0.02 - 0.019e^{-1000t}. \tag{14.94}$$

14.3.4 Problem: Newton's law of cooling

A mercury in glass thermometer is dipped into a liquid (which can be considered as a step input) at a temperature T_L of 70°C. From Newton's law of cooling, the differential equation relating the input temperature T_L in °C to the temperature reached by the thermometer, T_M, in °C can be given as

$$\frac{dT_M}{dt} + \frac{hA}{mc_p}T_M = \frac{hA}{mc_p}T_L, \tag{14.95}$$

where h is the heat transfer coefficient in $\frac{\text{W}}{\text{m}^2\text{K}}$, A is the surface area of the thermometer, m is the mass of mercury in the thermometer, and c_p is the specific heat capacity of the mercury.
The thermometer is initially at 0°C (i.e., $T_M = 0$°C and it can be assumed that $hA/mc_p = 0.1\ \text{s}^{-1}$).

1. What is the transfer function for this system?
2. Starting with the transfer function from above, determine an expression for the temperature reached by the thermometer $T_{M(t)}$.

Solution:
With the given conditions of $hA/mc_p = 0.1\ \mathrm{s}^{-1}$, we have

$$\frac{dT_M}{dt} + 0.1T_M = 0.1T_L \tag{14.96}$$

Taking the Laplace transform of both sides,

$$sT_{M(s)} + 0.1T_{M(s)} = 0.1T_{L(s)}. \tag{14.97}$$

The transfer function gives

$$\frac{T_{M(s)}}{T_L} = \frac{0.1}{s + 0.1} \tag{14.98}$$

For the given value of T_L, this gives

$$T_{M(s)} = \frac{70}{s} \times \frac{0.1}{s + 0.1}. \tag{14.99}$$

Using Laplace tables, we find that The equation for the temperature as a function of time is given by

$$T_{M(t)} = 70(1 - e^{-0.1t}). \tag{14.100}$$

14.3.5 Problem: servo positioning system

The relationship between the input, voltage V, and output, displacement x, for a servo positioning system is described by the following equation:

$$\frac{d^2x}{dt^2} + 5\frac{dx}{dt} + 6x = V. \tag{14.101}$$

For this system determine the transfer function and an expression or the position $x(t)$ for a step input of 12 V.

Solution:
We have

$$\frac{d^2x}{dt^2} + 5\frac{dx}{dt} + 6x = V. \tag{14.102}$$

We get the Laplace transform of both sides:

$$s^2X(s) + 5sX(s) + 6X(s) = V(s) \tag{14.103}$$

Rearranging this, we get

$$X(s)(s^2 + 5s + 6) = V(s)$$

$$G(s) = \frac{X(s)}{V(s)} = \frac{1}{(s+2)(s+3)} \tag{14.104}$$

$$V(s) = \frac{12}{s} = \frac{6}{s} \times 2, \tag{14.105}$$

$$X(s) = 2\frac{6}{s(s+2)(s+3)}, \tag{14.106}$$

$$x(t) = 2(1 - 3e^{-2t} + 2e^{-3t}). \tag{14.107}$$

14.3.6 Problem: robotic arm

A motorized robotic arm system has the following transfer function:

$$G(s) = \frac{6}{(s+4)^4} = \frac{X(s)}{Y(s)} \tag{14.108}$$

The robotic arm system is subjected to a step input of 8 V. What will be the output from the system?

Solution:
We have

$$G(s) = \frac{6}{(s+4)^4}. \tag{14.109}$$

If we have

$$Y(s) = \frac{8}{s} \tag{14.110}$$

then

$$\begin{aligned} X(s) &= \frac{48}{(s+4)^2 s} \\ &= \frac{A}{s} + \frac{B}{s+4} + \frac{C}{(s+4)^2} \end{aligned} \tag{14.111}$$

Thus,

$$\begin{aligned} 48 &= A(s^2 + 8s + 16) + B(s+4) + Cs \\ &= As^2 + 8AS + 16A + Bs^2 + 4Bs + Cs \\ &= s^2(A+B) + s(8A + 4B + C) + 16A \end{aligned} \tag{14.112}$$

This gives us

$$0 = A + B \text{ , } 8A + 4B + C = 0 \text{ and } 16A = 48 \tag{14.113}$$

which in turn implies

$$A = 3,\ B = -3 \text{ and } C = -12. \tag{14.114}$$

Finally, this gives:

$$X(s) = \frac{3}{s} - \frac{3}{s+4} - \frac{12}{(s+4)^2} \tag{14.115}$$

$$x(t) = 3(1 - e^{-4t} - 4te^{-4t}). \tag{14.116}$$

Specific mathematical software to solve some problems

15

Alberto Alonso Izquierdo[a], *Miguel Ángel González León*[a], *Jesús Martín-Vaquero*[a], *Deolinda M.L. Dias Rasteiro*[b], *Monika Kováčová*[c], *Daniela Richtáriková*[c], *Pablo Rodríguez-Gonzálvez*[d], *Manuel Rodríguez-Martín*[a], *Araceli Queiruga-Dios*[a]

[a]University of Salamanca, Salamanca, Spain, [b]Coimbra Polytechnic – ISEC, Coimbra, Portugal, [c]Slovak University of Technology in Bratislava, Bratislava, Slovakia, [d]University of León, León, Spain

15.1 Vibration and harmonic analysis

In Fig. 15.1, the determination of phase from a pure arctan function is presented. It is important in situations where extended inbuilt software statements with two arctan arguments (denominator, numerator) do not give suitable results, usually where phase determination is needed in the form of a function.

We consider the system excited by the sinusoidal force $F(t)$ with working hold-ups. Since it is a function which satisfies Dirichlet conditions, it can be decomposed into Fourier series, and the system's answer will be composed from harmonics responses using the superposition property. Here a reader can find Code 1 (using Mathematica, 2018), where excitation force $F(t)$ is decomposed into harmonics, and then harmonics responses $x_n(t)$ are computed. Although the computations are done in complex form, the final table shows the harmonics results also in real forms, using the transformation formulas. (See Figs. 15.2 and 15.3.)

```
Code 1:

 (*General computation for equation in complex form*)
 LS = m*x''[t] + b*x'[t] + k*x[t]; RS = c[n]*E^(I*n*Ω*t);rov = LS == RS;
 x[t_] = cx[n]*E^(I*n*Ω*t);
 r = Solve[rov, cx[n]];

 (*complex coeff of x(t)*)
 cx[n_] = r[[1, 1, 2]];

 (*complex harmonic of x(t)*)
 fxcw[n_,t_] = cx[n]*E^(I*n*Ω*t) + cx[-n] E^(-I*n*Ω*t);

 (*complex FS of x(t)*)
```

Calculus for Engineering Students. https://doi.org/10.1016/B978-0-12-817210-0.00022-9

Ranges	phase graph arctan function	phase α unit circle	cosine phase φ_n formula	sine phase ψ_n formula
phase $\in \left(-\frac{\pi}{2}, \frac{\pi}{2}\right)$ denominator > 0	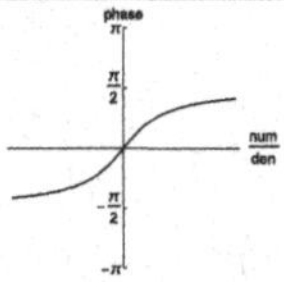	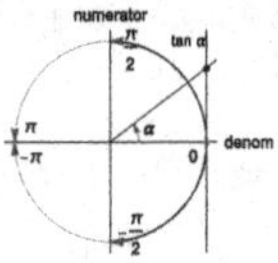	$\arctan \frac{-b_n}{a_n}$	$\arctan \frac{a_n}{b_n}$
phase $\in \left(\frac{\pi}{2}, \pi\right\rangle$ denominator < 0 numerator ≥ 0	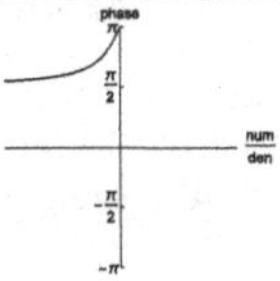	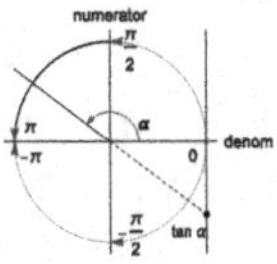	$\arctan \frac{-b_n}{a_n} + \pi$	$\arctan \frac{a_n}{b_n} + \pi$
phase $\in \left(-\pi, -\frac{\pi}{2}\right)$ denominator < 0 numerator < 0	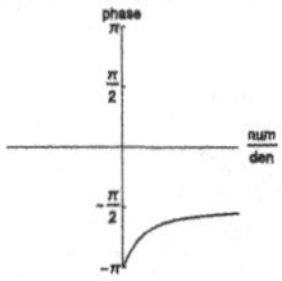	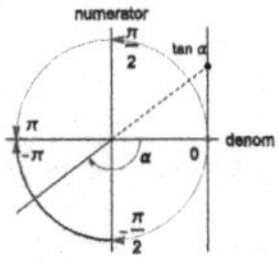	$\arctan \frac{-b_n}{a_n} - \pi$	$\arctan \frac{a_n}{b_n} - \pi$
phase $= -\frac{\pi}{2}$ denominator $= 0$ numerator < 0		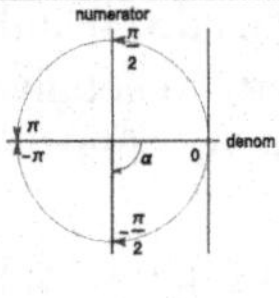	$-\frac{\pi}{2}$	$-\frac{\pi}{2}$
phase $= \frac{\pi}{2}$ denominator $= 0$ numerator > 0		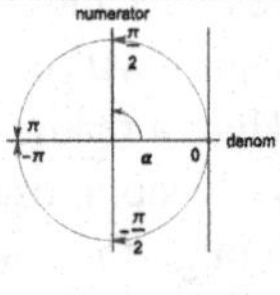	$\frac{\pi}{2}$	$\frac{\pi}{2}$
denominator $= 0$ numerator $= 0$			undefined or $\varphi_n = 0$	undefined or $\psi_n = 0$

Figure 15.1 Phase graphs, sine and cosine phase computations through pure arctan functions.

```
                    nn
FRcx[nn_, t_] :=    ∑    cx[n]*E^(-I*n*Ω*t) // Chop;
                  n=-nn

(*transform formulas for F(t) between complex and goniometric forms*)
aaF[n_] = c[n]+Conjugate[c[n]];
bbF[n_] = I(c[n]-Conjugate[c[n]]);
AmpFazF[n_,t_] = 2*Abs[c[n]]*Cos[n*Ω*t+Arg[c[n]]];
GonF[n_,t_]=aaF[n]*Cos[n*Ω*t]+bbF[n]*Sin[n*Ω*t];
```

```
(*transform formulas for x(t) between complex and goniometric forms*)
aax[n_] = (cx[n]+cx[-n]) // Chop;
bbx[n_] = I(cx[n]-cx[-n]) // Chop;
AmpFazx[n_,t_] := 2*Abs[cx[n]]*Cos[n*Ω *t+Arg[cx[n]]];
Gonx[n_,t_]:=aax[n]*Cos[n*Ω*t]+bbx[n]*Sin[n*Ω*t];


(*excitation force F(t) in complex form with values*)
FM=40;  Ω=10; T= 2π/Ω; t0=0.55*T;
F[t_]=FM*UnitStep[t]-2*FM*UnitStep[t-t0/2]+FM*UnitStep[t-t0];
c[n_]= 1/T* ∫_0^T F[t]*E^(-I*n*Ω*t) dt // Chop // Simplify;
c[0]=0;

(*frequency parameters*)
(*placement determines, whether Ω, T, t0 are fixed in F(t) or not*)
 Ω=10;
T= 2π/Ω;
t0=0.55*T;

(*excitation force F(t) in complex form with values*)
(*Ω, T, t0 are fixed *)
FM=40;
F[t_]=FM*UnitStep[t]-2*FM*UnitStep[t-t0/2]+FM*UnitStep[t-t0];
c[n_]= 1/T* ∫_0^T F[t]*E^(-I*n*Ω*t) dt // Chop // Simplify;

(*code for F(t) harmonics table*)
TabHF=
  Table[{i,Plot[(GonF[i,t]//Chop),{t,0,T},
     Ticks →{{{t0,"t0"},{T,"T"},{T+t0,"t0+T"},{2T,"2T"}},
       Automatic},PlotRange→{-30,30},AxexLabel→ {"t"},
     AxesStyle→{{Black,12},{Black,12}}],Column[{
       Row[{Subscript["F",i]//Chop,"(t)=",
         Style[(c[i]*E^(I*n*Ω*t)+c[-i]*E^(-I*n*Ω*t))//TraditionalForm,14]}],
       Row[{"    =",Style[GonF[i,t]//Chop//TraditionalForm,14]}],
       Row[{"    =",Style[AmpFazF[i,t]//TraditionalForm,14]}],
       Row[{Subscript["A",i],"= ",2*Abs[c[i]], "  ",
       Subscript["φ",i],  "= ",Arg[c[i]],"  ",
         Subscript["ω",i],"= ",i*Ω
       }]}]}],{i,0,7}];
TabHF1=
  Prepend[TabHF,{"n","harmonic wave",
    "harmonic expressions (exponential, goniometric,
      amplitude-phase)"}];
TabuHF1=Grid[TabHF1,Alignment→{{Center,Center,Left},Center},
  Background→{None,LightGray},Frame→ All,ItemSize→Full]
```

```
(*system values*)
mH=4;r=0.09;h=0.03;ro=7800;
d=0.008;l=0.5;G=0.8*10^11;
mD=π*r^2*h*ro;ID=1/2*mD*r^2;JP=π*d^4/32;
kt=G*JP/l;

(*system parameters*)
k=kt/r^2; m=mH+ID/r^2; ξ = 0.1; ω0 = Sqrt[k/m]; del=ξ*ω0;
b=2*m*del; FM=40;
ωd=Sqrt[ω0^2 - del^2];

(*code for x(t) harmonics table*)
Tabhx=
  Table[{i,Plot[(Ampfazx[i,t]),{t,0,T},
     Ticks →{{{t0,"t0"},{T,"T"},{T+t0,"t0+T"},{2T,"2T"}},
       Automatic},PlotRange→{-0.005,0.005},AxexLabel→{"t"},
     AxesStyle→{{Black,12},{Black,12}}],Column[{
       Row[{Subscript["x",i]//Chop,"(t)=",
         Style[(fxcw[i,t])//TraditionalForm,14]}],
       Row[{"    =",Style[Gonx[i,t]//Chop//TraditionalForm,14]}],
       Row[{"    =",Style[AmpFazx[i,t]//TraditionalForm,14]}],
       Row[{Subscript["A",i],"= ",2*Abs[cx[i]], "  ",
            Subscript["φ",i], "= ",Arg[c[i]],"  ",
         Subscript["ω",i],"= ",i*Ω
       }]}]},{i,0,7}];
TabHx1=
  Prepend[TabHx,{"n","harmonic wave",
    "harmonic expressions (exponential, goniometric,
      amplitude-phase)"}];
TabuHx1=Grid[TabHx1,Alignment→{{Center,Center,Left},Center},
  Background→{None,{LightGray}},Frame→ All,ItemSize→Full]
```

Code 2 demonstrates the creation of resulting vibration amplitude dependence on operational frequency. (See Fig. 15.4.) Since Code 1 works in exponential form, Code 2 presents computation of result harmonics $x_n(t)$ in goniometric form. In addition, we would like to point out that the operational frequency influences the length of the basic period T, and also the point t_0 in $F(t)$, so in the final loop, $F(t)$ has to change with respect to changing operational frequency ω_p.

```
Code 2:
 (*Equation in goniometrc form for one harmonic*)
 LS=m*x''[t]+b+x'[t]+k*x[t]; RS=aF[n]*Cos[n*Ω*t]+bF[n]*Sin[n*Ω*t];
 x[t_]=ax[n]*Cos[n*Ω*t]+bx[n]*Sin[n*Ω*t];
 q=Normal[coefficientArrays[LS,{Cos[n t Ω], Sin[n t Ω]}]];
```

n	harmonic wave	harmonic expressions (exponential, goniometric, amplitude-phase)
0	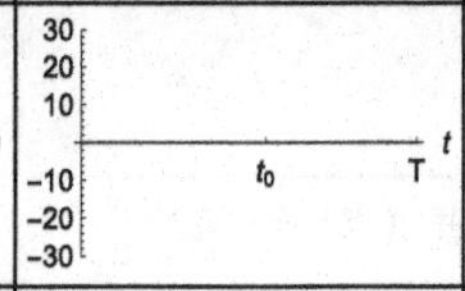	$F_0(t)=0$ $=0$ $=0$ $A_0=0 \quad \varphi_0=0 \quad \omega_0=0$
1	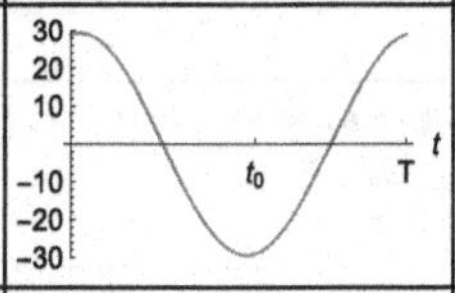	$F_1(t)=(14.5429+2.30337\,i)\,e^{-10\,i\,t}+(14.5429-2.30337\,i)\,e^{10\,i\,t}$ $=4.60674\sin(10\,t)+29.0858\cos(10\,t)$ $=29.4484\cos(0.15708-10\,t)$ $A_1=29.4484 \quad \varphi_1=-0.15708 \quad \omega_1=10$
2	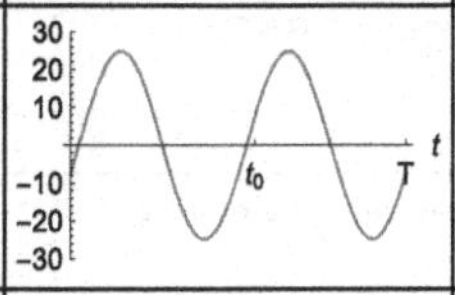	$F_2(t)=(-3.83824+11.8129\,i)\,e^{-20\,i\,t}-(3.83824+11.8129\,i)\,e^{20\,i\,t}$ $=23.6258\sin(20\,t)-7.67648\cos(20\,t)$ $=24.8416\cos(1.88496-20\,t)$ $A_2=24.8416 \quad \varphi_2=-1.88496 \quad \omega_2=20$
3	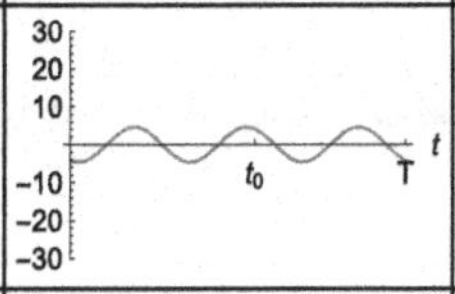	$F_3(t)=(-2.06476-1.05205\,i)\,e^{-30\,i\,t}-(2.06476-1.05205\,i)\,e^{30\,i\,t}$ $=-2.1041\sin(30\,t)-4.12952\cos(30\,t)$ $=4.63467\cos(30\,t+2.67035)$ $A_3=4.63467 \quad \varphi_3=2.67035 \quad \omega_3=30$
4	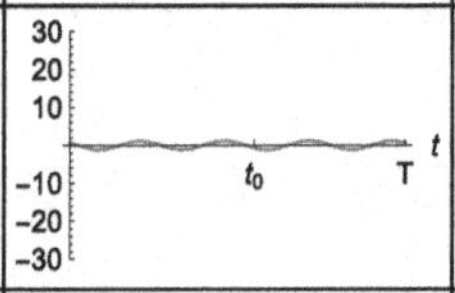	$F_4(t)=(0.357325-0.491816\,i)\,e^{-40\,i\,t}+(0.357325+0.491816\,i)\,e^{40\,i\,t}$ $=0.71465\cos(40\,t)-0.983632\sin(40\,t)$ $=1.21584\cos(40\,t+0.942478)$ $A_4=1.21584 \quad \varphi_4=0.942478 \quad \omega_4=40$
5	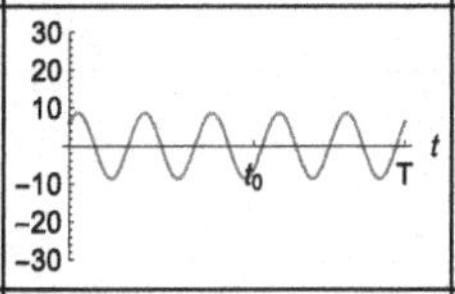	$F_5(t)=(3.07387+3.07387\,i)\,e^{-50\,i\,t}+(3.07387-3.07387\,i)\,e^{50\,i\,t}$ $=6.14774\sin(50\,t)+6.14774\cos(50\,t)$ $=8.69422\cos(0.785398-50\,t)$ $A_5=8.69422 \quad \varphi_5=-0.785398 \quad \omega_5=50$
6	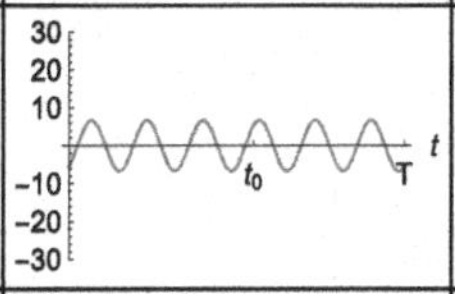	$F_6(t)=(-2.72589+1.98047\,i)\,e^{-60\,i\,t}-(2.72589+1.98047\,i)\,e^{60\,i\,t}$ $=3.96095\sin(60\,t)-5.45178\cos(60\,t)$ $=6.73877\cos(2.51327-60\,t)$ $A_6=6.73877 \quad \varphi_6=-2.51327 \quad \omega_6=60$
7	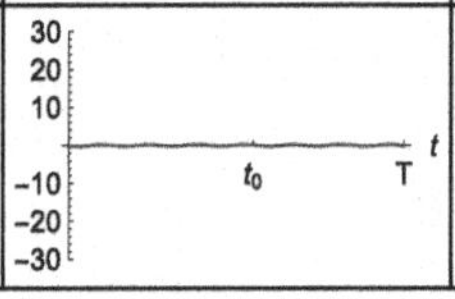	$F_7(t)=(-0.0900035-0.176642\,i)\,e^{-70\,i\,t}-(0.0900035-0.176642\,i)\,e^{70\,i\,t}$ $=-0.353284\sin(70\,t)-0.180007\cos(70\,t)$ $=0.396499\cos(70\,t+2.04204)$ $A_7=0.396499 \quad \varphi_7=2.04204 \quad \omega_7=70$

Figure 15.2 First seven real harmonics of $F(t)$, $\Omega=\omega_p=10$.

```
{{r1,r2}}=Solve[{q[[2,1]]==aF[n],q[[2,2]]==bF[n]},{ax[n],bx[n]}]//
   Simplify;

(*Fourier coefficients of solution*)
ax[n_]=r1[[2]];
```

n	harmonic wave	harmonic expressions (exponential, goniometric, amplitude-phase)
0	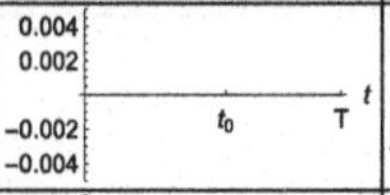	$x_0(t) = 0. + 0.\,i$ $= 0$ $= 0$ $A_0 = 0 \quad \varphi_0 = 0 \quad \omega_0 = 0$
1	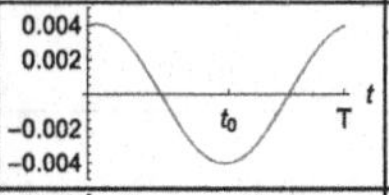	$x_1(t) = (0.00197816 + 0.000446454\,i)\,e^{-10\,i\,t} + (0.00197816 - 0.000446454\,i)\,e^{10\,i\,t}$ $= 0.000892907 \sin(10\,t) + 0.00395633 \cos(10\,t)$ $= 0.00405583 \cos(0.221972 - 10\,t)$ $A_1 = 0.00405583 \quad \varphi_1 = -0.15708 \quad \omega_1 = 10$
2	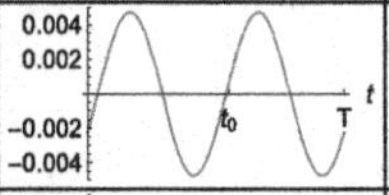	$x_2(t) = (-0.00112638 + 0.00208689\,i)\,e^{-20\,i\,t} - (0.00112638 + 0.00208689\,i)\,e^{20\,i\,t}$ $= 0.00417377 \sin(20\,t) - 0.00225276 \cos(20\,t)$ $= 0.00474292 \cos(2.06573 - 20\,t)$ $A_2 = 0.00474292 \quad \varphi_2 = -1.88496 \quad \omega_2 = 20$
3	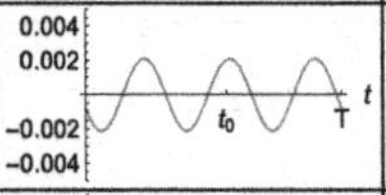	$x_3(t) = (-0.000409202 - 0.000979816\,i)\,e^{-30\,i\,t} - (0.000409202 - 0.000979816\,i)\,e^{30\,i\,t}$ $= -0.00195963 \sin(30\,t) - 0.000818404 \cos(30\,t)$ $= 0.00212366 \cos(30\,t + 1.96641)$ $A_3 = 0.00212366 \quad \varphi_3 = 2.67035 \quad \omega_3 = 30$
4	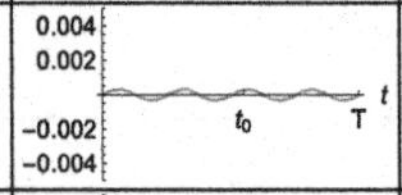	$x_4(t) = (-0.0000161293 + 0.000162148\,i)\,e^{-40\,i\,t} - (0.0000161293 + 0.000162148\,i)\,e^{40\,i\,t}$ $= 0.000324296 \sin(40\,t) - 0.0000322586 \cos(40\,t)$ $= 0.000325896 \cos(1.66994 - 40\,t)$ $A_4 = 0.000325896 \quad \varphi_4 = 0.942478 \quad \omega_4 = 40$
5	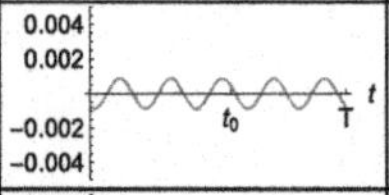	$x_5(t) = (-0.000380387 - 0.000229301\,i)\,e^{-50\,i\,t} - (0.000380387 - 0.000229301\,i)\,e^{50\,i\,t}$ $= -0.000458603 \sin(50\,t) - 0.000760774 \cos(50\,t)$ $= 0.00088831 \cos(50\,t + 2.59911)$ $A_5 = 0.00088831 \quad \varphi_5 = -0.785398 \quad \omega_5 = 50$
6	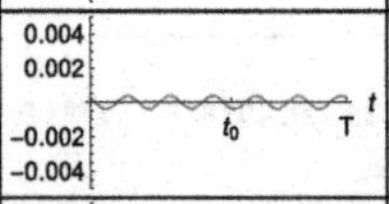	$x_6(t) = (0.000136064 - 0.000137692\,i)\,e^{-60\,i\,t} + (0.000136064 + 0.000137692\,i)\,e^{60\,i\,t}$ $= 0.000272128 \cos(60\,t) - 0.000275383 \sin(60\,t)$ $= 0.000387156 \cos(60\,t + 0.791343)$ $A_6 = 0.000387156 \quad \varphi_6 = -2.51327 \quad \omega_6 = 60$
7	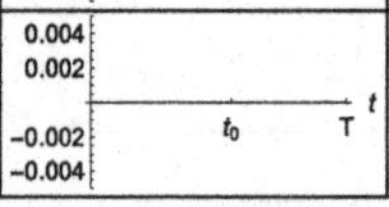	$x_7(t) = (4.20811 \times 10^{-6} + 6.20191 \times 10^{-6}\,i)\,e^{-70\,i\,t} + (4.20811 \times 10^{-6} - 6.20191 \times 10^{-6}\,i)\,e^{70\,i\,t}$ $= 0.0000124038 \sin(70\,t) + 8.41623 \times 10^{-6} \cos(70\,t)$ $= 0.0000149896 \cos(0.974633 - 70\,t)$ $A_7 = 0.0000149896 \quad \varphi_7 = 2.04204 \quad \omega_7 = 70$

Figure 15.3 First seven harmonics of $x(t)$, $\Omega = \omega_p = 10$.

```
bx[n_]=r2[[2]];

(*solution nth harmonic*)
fxgw[n_,t_]=ax[n]*Cos[n*Ω*t]+bx[n]*Sin[n*Ω*t]

(*partial series*)
FRgx[nn_,t_]:= Sum_{n=1}^{nn} (ax[n]*Cos[n*Ω*t]+bx[n]*Sin[n*Ω*t])

(*amplitude*)
amp[n_]=Sqrt[ax[n]^2 + bx[n]^2]

(*phase*)
fis[n_]=ArcTan[bx[n],ax[n]];
fic[n_]=ArcTan[ax[n],-bx[n]];

(*excitation force: sin with hold ups*)
```

```
f[t_]=FM*Sin[2π/t0 * t]*HeavisideTheta[-t+10]//Chop;

(*force fourier coefficients*)
a0=2/T ∫_0^T f[t]dt
aF[n_]=2/T ∫_0^T f[t]*Cos[n*Ω *t] dt//Chop
bF[n_]=2/T ∫_0^T f[t]*Sin[n*Ω*t] dt//Chop

(*partial series*)
FRg[N_,t_]:=2/T+ Sum_{n=1}^{N} (aF[n]*Cos[n*Ω*t]+bf[n]*Sin[n*Ω*t])

(*system values*)
mH=4; r=0.09; h=0.03; ro=7800;
d=0.008; l=0.5; G=0.8*10^11
mD=π*r^2*h*ro; ID=1/2*mD*r^2; JP=π*d^4/32;
kt=G*JP/l;

(*system parameters*)
k=kt/r^2; m=mH+ID/r^2;ξ = 0.1;ω0 = Sqrt[k/m]; del=ξ*ω0;
b=2*m*del; FM=40;
ωd=Sqrt[ω0^2 - del^2]; T=2π/Ω; t0=0.55*T;

(*sets of ordered pairs {[ω,max]},{[ω,sum]},{[ω,ef]}*)
max=Table[{Ω=coeff*ω0,
    Max[Table[amp[n],{n,1,5}]]},{coeff,0.01,1.3,0.001}],
sum=Table[{Ω=coeff*ω0,
    Sum[amp[n],{n,1,5}]},{coeff,0.01,1.3,0.001}],
ef=Table[{Ω=coeff*ω0,
    Sqrt[Sum[amp[n]^2, {n,1,5}]]},{coeff,0.01,1.3,0.001}];

(*graph of ordered pairs {[ω,max]},{[ω,sum]},{[ω,ef]}*)
freqChain=ListPlot[{sum,max,ef},PlotStyle→ {Red,Green,Blue},
  Ticks→{{{Ω,"",{.02,0}},{ω0,"Ω0",{.6,0}},{ω0/2,"Ω0/2",{.6,0}},
     {ω0/3,"Ω0/3",{.6,0}},{ω0/5,"Ω0/5",{.6,0}}},Automatic},
  AxesLabel→{"ω ",None},LabelStyle→ Black,
  AxesStyle→ Directive[Black,12]]
```

Fig. 15.4 shows the system forced by harmonic function with working hold-ups, goniometric form. The impact of the force operational frequency on the amplitude of the resulting vibration $x(t)$ is exhibited by three representations of final wave amplitude: sum, max, and ef.

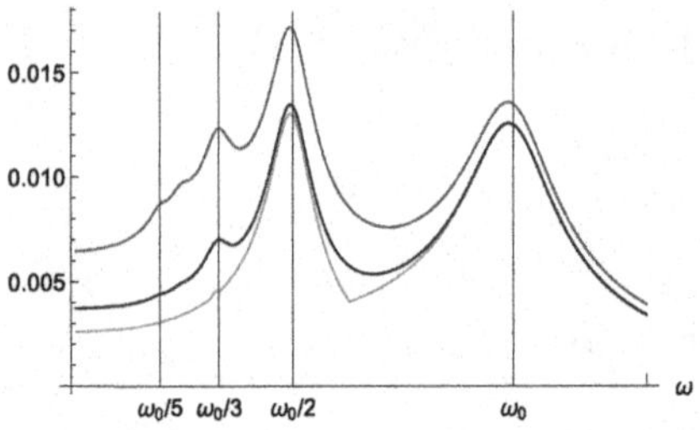

Figure 15.4 System forced by harmonic function with working hold-ups, goniometric form.

15.2 Critical forces – how to solve nonlinear equations and their systems

We provide an example of code for computation of the Newton's method for systems of nonlinear equations: collision avoidance for unmanned vehicles with fixed trajectories.

```
(*Matrix F(x)*)
F = {f1[x, y], f2[x, y]};

(*trajectories*)
f1[x_, y_] = x^2 - 4 x - y + 2;
f2[x_, y_] = 2 x^2 + y^2 - 3;

(*Jacobi Matrix J(x)*)
J[x_, y_] = D[F, {{x, y}}];

(*Inverse of Jacobi Matrix J(x)^-1*)
INV[x_, y_] = Inverse[J[x, y]];
B = 2.54;

(*coordinates of initial vector*)
x[0] = 0.2; y[0] = 0.4;
numberofloops = 20;
tolerance = 10^-15;
numberofiterations = Catch[
(*computation of x^i and y^i in a loop*)
  Do[
    {x[i+1], y[i+1]} =
    {x[i], y[i]}-INV[x[i], y[i]].{f1[x[i], y[i]], f2[x[i], y[i]]} // N;
    If[Max[Abs[x[i+1] - x[i]] , Abs[y[i+1]-y[i]]] < tolerance,
    Throw[i+1]],
   {i, 0, numberofloops}]
 ]
```

```
(*Table 1 print out*)
TableForm[Table[
 {i, NumberForm[x[i], 16], NumberForm[y[i], 16],
 Max[Abs[x[i]-x[i-1]], Abs[y[i]-y[i - 1]]]]},{i, 0,
       numberofiterations}],
TableHeadings → {None, {"i", "x_i","y_i","x̄_{i+1} − x̄_i"}},
TableSpacing → {1, 5}]
```

15.3 Shortest path problem and computer algorithms

The following functions for solving shortest path problems have been developed using MATLAB® software (Gdeisat and Lilley, 2012).

```
function [path, minDist, label] = shortestPath(startNode, endNode,
                                               point, suc, dist)
% Shortest path from a startNode to a endNode
numberOfNodes=length(point)-1;
for node_i=1:numberOfNodes
  label(node_i) = inf;
  list(node_i) = 0;
end
label(startNode) = 0;
list(startNode) = inf;
start  = startNode;
ending = startNode;
node_i = start
path(startNode) = 1

while ~(node_i == inf)

  for arc_ij = point(node_i):point(node_i+1)-1
    node_j=suc(arc_ij);

    if label(node_j) > label(node_i) + dist(arc_ij)
       label(node_j) = label(node_i) + dist(arc_ij)
       path(node_j) = node_i

          if (list(node_j)==0)
              list(ending) = node_j
              ending       = node_j
              list(node_j) = inf
          end
```

```
    end

  end

  aux       = start
  start     = list(start)
  node_i    = list(aux)
  list(aux) = 0
end

path(1)= [];
path(endNode) = endNode;
minDist=label(endNode);
disp(['label: ' num2str(label)]);
disp(['path : ' num2str(path)]);
disp(['minDist: ' minDist]);

end

function [path,maxdist,label] = longestPath(startNode, endNode, point,
                                              suc, dist)

% Longest Path from a startNode to an endNode
 numberOfNodes=length(point)-1;
   for i=1:numberOfNodes
     label(i) = 0;
     list(i)  = 0;
   end
   label(startNode)= 0;
   list(startNode)= 0;
   start  = startNode;
   ending = startNode;
   node_i = startNode;

   while ~(node_i == inf)
       node_i = start;
       for arc_ij = point(node_i):point(node_i+1)-1
         node_j=suc(arc_ij);

         if label(node_j) < label(node_i) + dist(arc_ij)
            label(node_j) = label(node_i) + dist(arc_ij);
            path(node_j) = node_i;

            if (list(node_j)==0)
               list(ending) = node_j;
```

```
                ending        = node_j;
                list(ending) = inf;
            end

        end
      end
      aux        = start;
      node_i     = list(start);
      start      = list(start);
      list(aux) = 0;
   end
   disp(['label: ' num2str(label)]);
   disp(['path : ' num2str(path)]);

   path(1)=[];
   path(2)=[];
   path(endNode-1)=endNode;
   maxdist=label(endNode);

   disp(['label: ' num2str(label)]);
   disp(['path : ' num2str(path)]);
   disp(['maxDist: ' maxdist]);

end

function [path,minDist,maxCap,label] = maximumCapacityInShortestPath
Set(startNode,endNode,point,suc,dist,cap)

% Maximum Capacity Path in the set of Shortest Paths from a
     startNode to a endNode

   numberOfNodes=length(point)-1;
   for i=1:numberOfNodes
       label(i) = 0;
       labelCap(i)=inf;
       list(i)  = 0;
   end
   label(startNode)= 0;
   labelCap(startNode)=inf;
   list(startNode)= inf;

   start  = startNode;
   ending = startNode;
   node_i = startNode;
```

```
  while ~(node_i == inf)
      node_i = start;
      for arc_ij = point(node_i):point(node_i+1)-1
        node_j=suc(arc_ij);

        if label(node_j) > label(node_i) + dist(arc_ij)
           label(node_j) = label(node_i) + dist(arc_ij);
           path(node_j) = node_i;

           if (list(node_j)==0)
              list(ending) = node_j;
              ending       = node_j;
              list(ending) = inf;
           end

        end
        if label(node_j) == label(node_i) + dist(arc_ij)

          if labelCap(node_j) < min(labelCap(node_i),cap(arc_ij))
          labelCap(node_j) = min(labelCap(node_i),cap(arc_ij));

           if (list(node_j)==0)
              list(ending) = node_j;
              ending       = node_j;
              list(ending) = 0;
           end

          end
      end
      aux        = start;
      node_i     = list(start);
      start      = list(start);
      list(aux) = 0;
      end

  path(1)=[];
  path(endNode)=endNode;
  minDist=label(endNode);
  maxCap=labelCap(endNode);
  disp(['label: ' num2str(label)]);
  disp(['path : ' num2str(path)]);
  disp(['minDist: ' minDist]);
  disp(['maxCap: ' maxCap]);
end
```

```
function [path,maxCap,label] = maximumCapacityPath(startNode, endNode,
                                                    point, suc, cap)
% Maximum Capacity Path from a startNode to an endNode

   path=[];
   numberOfNodes=length(point)-1;
   for i=1:numberOfNodes
       label(i) = -inf;
       list(i)  = 0;
   end
   label(startNode)= inf;
   list(startNode)= inf;

   start  = startNode;
   ending = startNode;
   node_i = startNode;

   while ~(node_i == inf)
       node_i = start;
       for arc_ij = point(node_i):point(node_i+1)-1
         node_j=suc(arc_ij);

        if label(node_j) < min(label(node_i),cap(arc_ij))
           label(node_j) = min(label(node_i),cap(arc_ij));
           path(node_j)  = node_i;

            if (list(node_j)==0)
               list(ending) = node_j;
               ending       = node_j;
               list(ending) = inf;
            end

         end
            maxCap=label(length(label));
      end
      aux       = start;
      node_i    = list(start);
      start     = list(start);
      list(aux) = 0;
  end

  path(1)=[];
  path(2)=[];
  path(endNode-1)=endNode;
```

```
    disp(['label: ' num2str(label)]);
    disp(['path : ' num2str(path)]);
    disp(['maxCap: ' maxCap]);
end

 function [ path,minDist,maxCap,label ]
     shortestPathInMaxCapacitySet(startNode,endNode,point,suc,dist,cap)

 % Shortest Path in the set o maximum capacity paths from startNode to
    endNode

    path=[];
    numberOfNodes=length(point)-1;
    for i=1:numberOfNodes
        label(i) = 0;
        list(i)  = 0;
    end
    label(startNode)= inf;
    list(startNode)= inf;
    start  = startNode;
    ending = startNode;
    node_i = startNode;

    while ~(node_i == inf)
        node_i = start;
        for arc_ij = point(node_i):point(node_i+1)-1
          node_j=suc(arc_ij);

          if label(node_j) < min(label(node_i),cap(arc_ij))
             label(node_j) = min(label(node_i),cap(arc_ij));
             path(node_j)  = node_i;

              if (list(node_j)==0)
                 list(ending) = node_j;
                 ending       = node_j;
                 list(ending) = inf;
              end

          end

        end
        aux       = start;
        node_i    = list(start);
        start     = list(start);
        list(aux) = 0;
```

```
    end

    maxCap=label(endNode);
    disp(['label: ' num2str(label)]);
    disp(['path : ' num2str(path)]);

    path(1)=[];
    path(2)=[];
    path(endNode-1)=endNode;

    path=[];
     numberOfNodes=length(point)-1;
    for i=1:numberOfNodes
        label(i) = inf;
        list(i)  = 0;
    end
    label(startNode)= 0;
    list(startNode)= inf;
    start  = startNode;
    ending = startNode;
    node_i = startNode;

    while ~(node_i == inf)
        node_i = start;
        for arc_ij = point(node_i):point(node_i+1)-1
          node_j=suc(arc_ij);

          if label(node_j) > label(node_i) + dist(arc_ij) &&
                              cap(arc_ij)>=maxCap
             label(node_j) = label(node_i) + dist(arc_ij);
             path(node_j) = node_i;

             if (list(node_j)==0)
                list(ending) = node_j;
                ending       = node_j;
                list(ending) = inf;
             end

          end
        end
        aux        = start;
        node_i     = list(start);
        start      = list(start);
        list(aux) = 0;
    end
```

```
     path(1)=[];
    path(endNode)=endNode;
    minDist=label(endNode);

    disp(['label: ' num2str(label)]);
    disp(['path : ' num2str(path)]);
    disp(['minDist: ' minDist]);
    disp(['maxCap: ' maxCap]);

end
```

15.4 Snails, snakes, and first-order ordinary differential equations

15.4.1 The epidemiological, and also malware propagation, model of Kermack and McKendrick

In this section, we would like to complete some of the explanations we provided in Section 11.3.1. SIR models are frequently numerically solved using numerical methods for ODEs. Recently, traditional Runge–Kutta methods have been found that might not preserve some conditions that SIR, SEIR, or MSEIR models normally require. For example, naturally, these compartmental models require that all populations obtain positive values at every time. However, several different works (Mickens, 2005; Anguelov et al., 2014; Martín-Vaquero et al., 2017) showed that traditional methods, for large step lengths, can provide negative values for any of these populations, which makes no sense from a biological point of view, but it is also very problematic mathematically, since these negative values may produce overflows (or very large and meaningless values) after several iterations. For this reason there is a large scientific literature around nonstandard finite difference (NSFD) methods, created to preserve some qualitative properties of the solutions (see Mickens, 1994; Mickens, 2000 and related references).

However, for educational purposes, we decided to describe here only the well-known Runge–Kutta methods and its solutions. Obviously, we had to consider smaller step lengths to guarantee positive solutions. Additionally, we use NSFD algorithms for security reasons to check our numerical solutions obtained with the RK4 described in Chapter 11. This algorithm written in Mathematica is presented below.

```
(*  Constants and initial data  *)
a = 0.00218;
b = 0.4404;

t0 = 0;
s0 = 762;
i0 = 1;
```

```
r0 = 0;
y0 = {s0, i0, r0};

f[t_, su_, in_, re_] = {-a*su*in, a*su*in - b*in, b*in};
h = 0.1;
tend = 40;

(*  RK4 algorithm  *)
For[i = 1, i <= Floor[tend/h] , i++,
  k1 = f[ t0, y0[[1]], y0[[2]], y0[[3]] ];
  prk2 = y0 + h/2*k1 ;
  k2 = f[ t0 + h/2, prk2[[1]], prk2[[2]], prk2[[3]] ];
  prk3 = y0 + h/2*k2 ;
  k3 = f[ t0 + h/2, prk3[[1]], prk3[[2]], prk3[[3]] ];
  prk4 = y0 + h*k3;
  k4 = f[ t0 + h, prk4[[1]], prk4[[2]], prk4[[3]] ];

  y1 = y0 + h/6*(k1 + 2*k2 + 2*k3 + k4);
  y0 = y1;
  t0 = t0 + h;
  ];

(*  Numerical solution  *)
Print[y0];
(*  Error compared with the numerical solution obtained by NDSolve *)
Print[Norm[ y0 - {19.182964543489646`, 0.0008218033840308274`,
    743.8162136531262`}]]
```

Mathematica also has its own created commands to obtain these solutions:

```
a = 0.00218;
b = 0.4404;
sol = NDSolve[{ s'[t] == -a*s[t]*i[t], i'[t] == a*s[t]*i[t] - b*i[t],
 r'[t] == b*i[t], s[0] == 762, i[0] == 1, r[0] == 0}, {s, i, r}, {t,0,1000}]
```

And numerical solutions can be represented with the command Plot:

```
Plot[{Evaluate[s[t] /. sol], Evaluate[i[t] /. sol],
  Evaluate[r[t] /. sol]}, {t, 0, 40}, PlotRange → All,
 PlotLegends → {"Susceptibles", "Infectious" , "Recovered"}]
```

15.4.2 *Coevolution and chirality: a story of snails and snakes*

The Mathematica code employed to generate Fig. 11.7 is given as follows:

```
(* Simulation time interval *)
tmin = 0; tmax = 6000;
(* Ecological parameters *)
k = 850; (* snail carrying capacity *)
a1 = 0.02 (* per capita predation coefficient on dextral snails *);
a2 = 0.125*a1; (* per capita predation coefficient on sinistral snails *)
b1 = 0.0001; (* snake consumption efficiency per time unit on dextral snails *)
b2 = 0.125*b1; (* snake consumption efficiency per time unit on sinistral snails *)
r = 0.025; (* intrinsic snail growth rate *)
s = 0.01; (* intrinsic snake mortality rate *)
(* Initial populations *)
x10 = 450; (* Initial population of dextral snails *)
y0 = 0.9; (* Initial population of snakes *)
x20 = 0.0001; (* Initial population of sinistral snails *)
(* Stationary populations *)
p1 = {s/b1,0,r*(b1*k-s)/(a1*b1*k)}; (* Dextral snail/snake population stationary point *)
p2 = {0,s/b2,r*(b2*k-s)/(a2*b2*k)}; (* Sinistral snail/snake population stationary point *)
(* Numerical resolution with Mathematica *)
sol = NDSolve[{x1'[t] == r*x1[t]*(1 - (x1[t] + x2[t])/k) - a1*x1[t]*y[t],
               x2'[t] == r*x2[t]*(1 - (x1[t] + x2[t])/k) - a2*x2[t]*y[t],
               y'[t] == -s*y[t] + b1*x1[t]*y[t] + b2*x2[t]*y[t],
               x1[0.] == x10, x2[0.] == x20, y[0.] == y0},
               {x1[t], x2[t], y[t]}, {t, tmin, tmax}];
(* Graphical representation of the stationary points and the initial population *)
pto0 = Graphics3D[{PointSize[0.03], Point[0, 0, 0] }];
pto1 = Graphics3D[{PointSize[0.03], Blue, Point[{x10, x20, y0}], Point[p1]}];
pto2 = Graphics3D[{PointSize[0.03], Red, Point[p2]}];
pto3 = Graphics3D[{PointSize[0.03], Line[{{k, 0, 0}, {0, k, 0}}]}];
(* Labels in the Figure *)
lp0 = Graphics3D[Text[StyleForm["P0"], {0.2, 0.2, 0.2}]];
lp1 = Graphics3D[Text[StyleForm["P1", Blue], {150, 0.1, 1.4}]];
lp2 = Graphics3D[Text[StyleForm["P2", Red], {0.1, 750, 0.4}]];
lxy = Graphics3D[Text[StyleForm["P(μ)"], {600, 600, 0.1}]];
l0 = Graphics3D[Text[StyleForm["(x10, x20, y)", Blue], {500, 0.1, 0.7}]];
(* Graphical representation of the simulation *)
p = ParametricPlot3D[{Evaluate[{x1[t], x2[t], y[t]} /. sol]}, {t, tmin, tmax},
    ColorFunction -> Function[{x, y, z, u}, RGBColor[0.9*{y,0,x}/Sqrt[x^2 + y^2]]],
    BoxRatios -> {1, 1, 1}, ViewPoint -> {4, 3,2}];
Show[p, pto0, pto1, pto2, pto3, lp0, lp1, lp2, lxy, l0, Ticks -> None,
   AxesStyle -> {Blue, Red, Green}, PlotRange -> {{-20, k}, {-20, k}, {0, 3}},
    AxesEdge -> {{-1, -1}, {-1, -1}, {-1, -1}}, AxesLabel -> {"x1", "x2", "y" }]
```

15.5 Oscillations in higher-order differential equations and systems of differential equations

In code, the differential equation (5.20) is solved numerically as a system of first-order linear differential equations using the RK4 method. The equation describes the problem from Chapter 5 on a simplified manipulator forced by a polygonal periodic chain (Fig. 5.14C in Chapter 5).

```
(*system values*)
mH = 4; r = 0.09; h = 0.03; ro = 7800;
d = 0.008; l = 0.5; G = 0.8*10^11;
mD = π*r^2*h*ro; ID = 1/2*mD*r^2; JP = π*d^4/32;
kt = (G*JP)/l;

(*system parameters*)
k = kt/r^2; m = mH + ID/r^2; ξ = 0.1; ω0 = Sqrt[k/m]; del = ξ*ω0;
b = 2*m*del;
FM = 40; ωd = Sqrt[ω0^2 - del^2];
T = 2π / Ω; t0 = 0.55*T;
Ω = 10;

(*piecewise determined force function within two periods*)
F[t_] := FM/m /; 0 <= t < t0 / 2
F[t_] := -FM/m /; t0/2 <= t < t0
F[t_] := 0 /; t0 <= t < T
F[t_] := FM/m /; T <= t < t0 / 2 + T
F[t_] := -FM/m /; t0 / 2 + T <= t < t0 + T
F[t_] := 0 /; t0 + T <= t < T + T
Plot[F[t], {t, 0, 2 T}]

(*differential equation system matrix*)
A = {{0, 1}, {-k/m, -b/m}}
g[t_] = {0, F[t]};

(*interval division parameters*)
step = 0.001;
aa = 0;
bb = Floor[2 T, step];
pin = (bb - aa)/step;

(*initial values*)
t[0] = 0;
z[0] = {0, 0};

(*Runge-Kutta4 calculation loop*)
```

```
Do[   k1 = A.z[i] + g[t[i]];
      k2 = A.(z[i] + step/2 *k1 ) + g[t[i] + step/2];
      k3 = A. (z[i] + step/2*k2) + g[t[i] + step/2] ;
      k4 = A.(z[i] + step*k3) + g[t[i] + step];
      kk = (k1 + 2 k2 + 2 k3 + k4)/6;
    z[i + 1] = z[i] + step*kk;
    t[i + 1] = t[i] + step // N,
         {i, 0, pin}]

(*results table print out*)
Ta = Table[{i, t[i], NumberForm[ z[i][[1]], 10],
   NumberForm[z[i][[2]], 10]}, {i, 0, pin, 200}];
   TableForm[Ta,
   TableHeadings → {None, {"i", "t", "z1(t)", "z2(t)"}},
   TableSpacing → {1, 3}]

(*result wave graph*)
points1 = Table[{t[i], z[i][[1]]}, {i, 0, pin}];
gE1 = ListPlot[points1, PlotStyle → {PointSize[0.03], Orange},
               AxesOrigin → {0, 0},
               AxesLabel → {"t","x(t)"},
               LabelStyle → Directive[Bold, Medium],
               AxesStyle → {{Black, 12}, {Black, 12}}]

(*result velocity graph*)
points2 = Table[{t[i], z[i][[2]]}, {i, 0, pin}];
gv1 = ListPlot[points2, PlotStyle → {PointSize[0.015], Black},
 AxesOrigin → {0, 0}, AxesLabel → {"t","x'(t)"},
 LabelStyle → Directive[Bold, Medium],
 AxesStyle → {{Black, 12}, {Black, 12}}]
```

The solution of $y(t) = -17.732\cos(1.936t) + 7.622\sin(1.936t) + 72.6$ in Problem 3 on bungee cords can be obtained in a symbolic way directly from second-order differential equations using MATLAB with the following code:

```
syms y(t)
Dy = diff(y);

% defining the second degree differential equation
ode = diff(y,t,2) == (-3.75*y)+272.31

% applying the initial conditions
cond1 = y(3.778) == 70;
cond2 = Dy(3.778) == 37.06;
conds = [cond1 cond2];
```

```
ySol(t) = dsolve(ode,conds);
ySol = simplify(ySol1)
```

Please note that the obtained expression can be expressed in another algebraical form, e.g.,

```
9077/125 -
(872*69^(1/2)*sin(arctan((3*15^(1/2))/85) - (15^(1/2)*t)/2 +
(1889*15^(1/2))/1000))/375
```

but it really represents the same function.

The solution of y_{max} can be obtained in a symbolic way directly from second-order differential equations in functions of the parameters previously established using MATLAB with the following easy code:

```
syms y(t)
Dy = diff(t);
%Previously defined the variables in the workspace or in this script
ode = diff(y,x,2) == g+((k*L)/m)-((k*y)/m)
cond1 = y((2*L/g)^0.5) == L;
cond2 = Dy((2*L/g)^0.5) == (2*L*g)^0.5;
conds = [cond1 cond2];
ySol(t) = dsolve(ode,conds);
ySol = simplify(ySol)
```

References

Anguelov, R., Dumont, Y., Lubuma, J.-S., Shillor, M., 2014. Dynamically consistent nonstandard finite difference schemes for epidemiological models. Journal of Computational and Applied Mathematics 255, 161–182.

Gdeisat, M., Lilley, F., 2012. MATLAB® by Example: Programming Basics. Newnes.

Martín-Vaquero, J., del Rey, Á.M., Encinas, A.H., Guillén, J.D.H., Dios, A.Q., Sánchez, G.R., 2017. Higher-order nonstandard finite difference schemes for a MSEIR model for a malware propagation. Journal of Computational and Applied Mathematics 317, 146–156.

Mathematica, 2018. V. 11.3. Wolfram Research Inc., Champaign, IL, USA.

Mickens, R., 1994. Nonstandard Finite Difference Models of Differential Equations. World Scientific.

Mickens, R., 2000. Applications of nonstandard finite difference schemes. https://books.google.nl/books?id=0GTVCgAAQBAJ.

Mickens, R.E., 2005. Advances in the Applications of Nonstandard Finite Difference Schemes. World Scientific, Singapore.

Index

Printed in the United States
By Bookmasters